Proceedings
of the
STEKLOV INSTITUTE
OF MATHEMATICS

1983, ISSUE 4

Analytic Number Theory, Mathematical Analysis and their Applications

Collection of Papers
Dedicated to Academician
IVAN MATVEEVICH VINOGRADOV
on his Ninetieth Birthday

Edited by
N. N. Bogolyubov and K. K. Mardzhanishvili
with the assistance of
A. A. Karatsuba, K. I. Oskolkov and V. I. Blagodatskikh

Translation of
ТРУДЫ
ордена Ленина
МАТЕМАТИЧЕСКОГО ИНСТИТУТА
имени В. А. СТЕКЛОВА

Том 158 (1981)

АКАДЕМИЯ НАУК

СОЮЗА СОВЕТСКИХ СОЦИАЛИСТИЧЕСКИХ РЕСПУБЛИК

ТРУДЫ

ордена Ленина

МАТЕМАТИЧЕСКОГО ИНСТИТУТА

имени В. А. СТЕКЛОВА

CLVIII

АНАЛИТИЧЕСКАЯ ТЕОРИЯ ЧИСЕЛ, МАТЕМАТИЧЕСКИЙ АНАЛИЗ И ИХ ПРИЛОЖЕНИЯ

сборник статей
Посвящается академику
Ивану Матвеевичу Виноградову
к его девяностолетию

Издательство "Наука"
Москва 1981

Library of Congress Cataloging in Publication Data

Analiticheskaia͡ teorii͡a chisel, mathematicheskiĭ analiz i ikh prilozhenii͡a. English.
Analytic number theory, mathematical analysis and their applications.
(Proceedings of the Steklov Institute of Mathematics, ISSN 0081-5438; 1983, issue 4)
Translation of: Analiticheskaia͡ teorii͡a chisel, matematicheskiĭ analiz i ikh prilozhenii͡a.
Number in Russian series statement: t. 158.
Includes bibliography.
1. Numbers, Theory of—Addresses, essays, lectures. 2. Mathematical analysis—Addresses, essays, lectures. 3. Vinogradov, Ivan Matveevich. 1891– . I. Bogoli͡ubov, N. N. (Nikolaĭ Nikolaevich), 1909– . II. Mardzhanishvili, K. K. III. Title. IV. Series: Trudy Ordena Lenina Matematicheskogo instituta imeni V. A. Steklova. English; 1983, issue 4.
QA1.A413 no. 158 [QA241] 510s [512′.7] 84-265
ISBN 0-8218-3077-5

December 1983

Translation authorized by the All-Union Agency for Author's Rights, Moscow

ABSTRACT. This collection consists of original papers on various branches of mathematics: analytic number theory, algebra, partial differential equations, probability theory, and differential games.

1980 *Mathematics Subject Classification.* Primary 10B15, 10B40, 10C07, 10C15, 10E10, 10E15, 10E25, 10G10, 10H05, 10H25, 10J25, 12A20, 12E05, 12F20, 17B20, 17B40, 30B50, 30F35, 35-02, 35B40, 35J10, 35K05, 35K55, 35K60, 35P99, 41A65, 47A50, 49A45, 60F05, 60G40, 60G42, 60G44, 60J50, 60J80, 90D26; Secondary 03C10, 10B35, 10C05, 10E05, 10E30, 12A25, 12A55, 12A70, 12A75, 12B05, 12D05, 12F10, 12L05, 32N15, 33A40, 33A70, 41A60, 45E10, 49A45, 60E07, 60J30, 60K05, 65B15, 80A20.

PROCEEDINGS OF THE STEKLOV INSTITUTE OF MATHEMATICS IN THE ACADEMY OF SCIENCES OF THE USSR

TABLE OF CONTENTS

RUSSIAN TABLE OF CONTENTS*

* The American Mathematical Society scheme for transliteration of Cyrillic may be found at the end of index issues of *Mathematical Reviews.*

Correspondence between *Trudy Mat. Inst. Steklov.* and *Proc. Steklov Inst. Math.*

Russian imprint	Vol.	English imprint
1966	74	1967
1965	75	1967
1965	76	1967
1965	77	1967
1965	78	1967
1965	79	1966
1965	80	1968
1966	81	1968
1966	82	1967
1965	83	1967
1965	84	1968
1966	85	1967
1965	86	1967
1966	87	1967
1967	88	1969
1967	89	1968
1967	90	1969
1967	91	1969
1966	92	1968
1967	93	1970
1968	94	1969
1968	95	1971
1968	96	1970
1968	97	1969
1968	98	1971
1967	99	1968
1971	100	1974
1972	101	1975
1967	102	1970
1968	103	1970
1968	104	1971
1969	105	1971
1969	106	1972

Russian imprint	Vol.	English imprint
1971	107	1976
1968	108	1971
1971	109	1974
1970	110	1972
1970	111	1972
1971	112	1973
1970	113	1972
1970	114	1974
1971	115	1974
1971	116	1973
1972	117	1974
1972	118	1976
1973	119	1976
1974	120	1976
1972	121	1974
1973	122	1975
1973	123	1975
1976	124	1978, Issue 3
1973	125	1975
1973	126	1975
1975	127	1977
1972	128	1974
1973	129	1976
1978	130	1979, Issue 4
1974	131	1975
1973	132	1975
1973	133	1977
1975	134	1977
1975	135	1978, Issue 1
1975	136	1978, Issue 2
1976	137	1978, Issue 4
1975	138	1977

Russian imprint	Vol.	English imprint Year	Issue
1980	139	1982	1
1976	140	1979	1
1976	141	1979	2
1976	142	1979	3
1977	143	1980	1
1979	144	1980	2
1980	145	1981	1
1978	146	1980	3
1980	147	1981	2
1978	148	1980	4
1978	149	1981	3
1979	150	1981	4
1980	151	1982	2
1980	152	1982	3
1981	153	1982	4
1981	155	1983	1
1980	156	1983	2
1981	157	1983	3
1981	158	1983	4

Труды Матем. Инст. Стеклов.
158(1981)

Proceedings of the Steklov
Institute of Mathematics
1983, Issue 4

A GEOMETRIC ESTIMATE OF THE NUMBER OF REPRESENTATIONS OF A REAL NUMBER BY A POSITIVE QUADRATIC FORM AND AN ESTIMATE OF THE REMAINDER OF THE MULTIDIMENSIONAL ZETA-FUNCTION

UDC 513.82 + 511

E. P. BARANOVSKIĬ, S. S. RYSHKOV AND S. SH. SHUSHBAEV

ABSTRACT. The following theorem is proved using simple geometric considerations:

THEOREM. *Let $f = f(x_1, \dots, x_n)$ be a positive definite quadratic form in n variables with arithmetic minimum $m(f)$. Then the number $r(f; q)$ of integral solutions of the equation $f = q$ $(q \geqslant m(f))$ does not exceed the number $N(\alpha)$ of spherical segments of angular radius $\alpha = \arcsin(\sqrt{m(f)}/2\sqrt{q})$ which can be arranged without overlap on the $(n-1)$-dimensional sphere.*

It follows from this theorem together with Rankin's inequality that for all $q \geqslant m = m(f)$

$$r(f; q) < \sqrt{\pi}\,\frac{(n+1)(n+3)}{\sqrt{n}}\sqrt{1 - \frac{m}{2q}}\,2^{n/2-1}\left(\frac{q}{m}\right)^{(n-1)/2}$$

This inequality is applied to the theory of the multidimensional zeta-function.

Bibliography: 11 titles.

1. THEOREM 1. *Let $f = f(x_1, \dots, x_n)$ be a positive definite quadratic form (PQF) in n variables with arithmetic minimum $m(f)$. Then the number $r(f; q)$ of integral solutions of the equation $f = q$ $(q \geqslant m(f))$ does not exceed the maximal number $N(\alpha)$ of spherical segments with angular radius $\alpha = \arcsin(\sqrt{m(f)}/2\sqrt{q})$, which can be arranged without overlap on the $(n-1)$-dimensional sphere.*

PROOF. Consider a basis $\mathfrak{S}_f = \{\bar{e}_1, \dots, \bar{e}_n\}$ for the form f and the associated lattice Γ_f. The equation $f(x_1, \dots, x_n) = q$, where q is a positive integer, represents (in the metric corresponding to $\mathfrak{S}_f$) the sphere $S(n; q)$ of radius $\sqrt{q}$ with center at the origin. Thus the task of finding the number of integral points which satisfy the equation $f(x_1, \dots, x_n) = q$ reduces to finding the number of points of the lattice Γ_f which lie on $S(n; q)$. Since the distance between two arbitrary points of Γ_f is not less than $\sqrt{m(f)}$, the spheres of radius $\sqrt{m(f)}/2$ with centers at the lattice points on the sphere do not intersect. Therefore the number of lattice points in Γ_f which lie on $S(n; q)$ can be estimated by means of the number of spherical segments $C(n; q) = C(n; \alpha)$ of angular radius $\alpha = \arcsin(\sqrt{m(f)}/2\sqrt{q})$ which can be arranged without overlap on $S(n; q)$. The theorem is proved.

1980 *Mathematics Subject Classification.* Primary 10C15; Secondary 10E30.

THEOREM 2. *For every* $q \geqslant m = m(f)$

$$r(f;q) < \sqrt{\pi}\,\frac{(n+1)(n+3)}{\sqrt{n}}\sqrt{1-\frac{m}{2q}}\cdot 2^{n/2-1}\left(\frac{q}{m}\right)^{(n-1)/2}. \tag{1}$$

PROOF. By Rankin's theorem [1] on the maximal number $N(\alpha)$ of spherical segments of angular radius α we have

$$N(\alpha) \leqslant \sqrt{\pi}\,\Gamma\left(\frac{n-1}{2}\right)\sin\beta\tan\beta/2\Gamma\left(\frac{n}{2}\right)\int_0^\beta (\sin\theta)^{n-2}(\cos\theta-\cos\beta)\,d\theta, \tag{2}$$

where $\beta = \arcsin(\sqrt{2}\,\sin\alpha)$, $0 < \alpha < \pi/4$. By [1],

$$\int_0^\beta (\sin\theta)^{n-2}(\cos\theta-\cos\beta)\,d\theta = \int_0^\beta (\sin\theta)^{n-2}\cos\theta\,d\theta - \cos\beta\int_0^\beta (\sin\theta)^{n-2}\,d\theta$$

$$= \frac{(\sin\beta)^{n-1}}{n-1} - \frac{(\sin\beta)^{n-1}}{n-1}\left\{1-\frac{\tan^2\beta}{n+1}+\frac{3\tilde{\theta}\tan^4\beta}{(n+1)(n+3)}\right\}$$

$$= \frac{(\sin\beta)^{n+1}}{n^2-1}\,\frac{1}{\cos^2\beta}\left(1-\frac{3\tilde{\theta}}{n+3}\tan^2\beta\right), \tag{3}$$

where $0 \leqslant \tilde{\theta} \leqslant 1$. It is easy to see that for $0 < \sin\beta \leqslant \sqrt{2}/2(\sin\alpha \leqslant 1/2)$ one has $0 \leqslant \beta \leqslant \pi/4$ and $0 < \tan\beta \leqslant 1$. Therefore

$$1-(3\tilde{\theta}/(n+3))\tan^2\beta \geqslant 1-3/(n+3).$$

Hence if $0 < \sin\alpha < 1/2$ it follows from (3) that

$$\int_0^\beta (\sin\theta)^{n-2}(\cos\theta-\cos\beta)\,d\theta \geqslant \frac{(\sin\beta)^{n+1}n}{\cos^2\beta\cdot(n^2-1)(n+3)}. \tag{4}$$

By Stirling's formula we obtain

$$\frac{\Gamma((n-1)/2)}{\Gamma(n/2)} = \frac{((n-1)/2)^{(n-2)/2}\exp(-(n-1)/2+\theta^*/6(n-1))}{(n/2)^{(n-1)/2}e^{-n/2+\theta/6n}}$$

$$= \frac{\sqrt{2en}\,e^{\theta^*/6(n-1)}}{(n-1)(1+1/(n-1))^{n/2}e^{\theta/6n}}$$

$$< \frac{\sqrt{2en}^{\,\theta^*/6(n-1)}}{(n-1)(1+1/(n-1))^{n/2}}, \tag{5}$$

where $0 < \theta^* < 1$. Next,

$$\frac{e^{\theta^*/6(n-1)}}{(1+1/(n-1))^{n/2}} = \frac{e^{\theta^*/6(n-1)}}{\exp((n/2)\ln(1+1/(n-1)))}$$

$$= \exp\left(\frac{\theta^*}{6(n-1)}-\frac{n}{2}\left(\frac{1}{n-1}-\frac{1}{2(n-1)^2}+\frac{\theta^*}{(n-1)^3}\right)\right),$$

$$\frac{\theta^*}{6(n-1)}-\frac{n}{2(n-1)}+\frac{n}{4(n-1)^z}-\frac{\theta^* n}{2(n-1)^3}$$

$$< \frac{1}{6(n-1)}-\frac{n}{2(n-1)}+\frac{n}{4(n-1)^2}$$

$$= -\frac{1}{2}-\frac{n-4}{12(n-1)^2} \leqslant -\frac{1}{2} \quad \text{for } n \geqslant 4.$$

For $n = 2, 3$ we obtain respectively

$$e^{1/6}\sqrt{e}/2 = 0.973\ldots < 1, \qquad \left[e^{1/12}/1.5^{1.5}\right]\sqrt{e} = 0.975\ldots < 1.$$

Consequently for all $n \geqslant 2$

$$\frac{e^{\theta^*/6(n-1)}}{(1+1/(n-1))^{n/2}} < \frac{1}{\sqrt{e}}. \tag{6}$$

Using (6), we can write (5) in the form

$$\frac{\Gamma((n-1)/2)}{\Gamma(n/2)} < \frac{\sqrt{2n}}{n-1}. \tag{7}$$

From (2) we obtain by means of (4) and (7)

$$N(\alpha) < \sqrt{\pi}\,(n+1)(n+3)\cos\beta \Big/ \left[\sqrt{2n}\,(\sin\beta)^{n-1}\right]. \tag{8}$$

Since $\sin\beta = \sqrt{2}\,\sin\alpha$, $\sin\alpha = \sqrt{m}/(2\sqrt{q})$ and $\cos\beta = \sqrt{1-m/2q}$ it follows from (8) that

$$N(\alpha) < \sqrt{\pi}\,\frac{(n+1)(n+3)}{\sqrt{n}}\sqrt{1-\frac{m}{2q}}\cdot 2^{n/2-1}\left(\frac{q}{m}\right)^{(n-1)/2}. \tag{9}$$

Inequality (1) of Theorem 2 follows from Theorem 1 and (9).

COROLLARY. *Let f be an arbitrary PQF with integral coefficients (integral PQF). Then*

$$r(f;q) < \sqrt{\pi}\,\frac{(n+1)(n+3)}{\sqrt{2n}}(2q)^{(n-1)/2}. \tag{10}$$

Indeed, an integral PQF takes only integral values for integral arguments, and therefore the arithmetic minimum of every integral PQF is not less than 1. Now apply Theorem 2.

REMARK 1. Inequality (1) shows that for n large, $r(f;q)$ is asymptotically equal to $2^{n(0.5+o(1))}(q/m)^{(n-1)/2}$.

If one applies the inequality for $N(\alpha)$ from [2] and [3], then Theorem 1 shows that $r(f;q)$ is asymptotically equal to $2^{n(0.401+o(1))}(q/m)^{(n-1)/2}$.

In the special case $f = x_1^2 + \cdots + x_n^2$ one finds in [4]–[6] estimates for $r(f;q)$ with better asymptotic behavior.

2. Using inequality (10), one can estimate the remainder of the multidimensional zeta function $\zeta(f;s) = \Sigma_{x\neq 0}\{f(x)\}^{-ns/2}$ (see [7] and [8]).

THEOREM 3. *Let $f = f(x_1,\ldots,x_n)$ be an arbitrary integral PQF. Then the following inequality holds for the remainder $Q(f \geqslant k) = \Sigma_{q\geqslant k} r(f;q)q^{-ns/2}$ of the zeta function $\zeta(f;s)$:*

$$Q(f\geqslant k) < \sqrt{\pi}\,\frac{(n+1)(n+3)}{\sqrt{n}\,(ns-n-1)}2^{n/2}(k-1)^{-(ns-n-1)/2}. \tag{11}$$

PROOF. We represent the zeta function in the form

$$\zeta(f; s) = \sum_{q=1}^{\infty} r(f; q) q^{-ns/2},$$

where $r(f; q)$ is the number of integral solutions of the equations $f(x_1, \ldots, x_n) = q$. Inequality (10) yields

$$Q(f \geqslant k) = \sum_{q=k}^{\infty} \frac{r(f; q)}{q^{ns/2}} < \sqrt{\pi}\, \frac{(n+1)(n+3)}{\sqrt{n}} 2^{n/2-1} \sum_{q=k}^{\infty} q^{-(ns-n+1)/2}$$

$$< \sqrt{\pi}\, \frac{(n+1)(n+3)}{\sqrt{n}} 2^{n/2-1} \int_{(k-1)}^{\infty} x^{-(ns-n+1)/2}\, dx$$

$$= \sqrt{\pi}\, \frac{(n+1)(n+3)}{\sqrt{n}\,(ns-n-1)} 2^{n/2} (k-1)^{-(ns-n-1)/2}.$$

Theorem 3 is established.

REMARK 2. Bearing in mind that

$$Q(f \geqslant k) = r(f; k) k^{-ns/2} + Q(f \geqslant k+1),$$

one can obtain from inequality (11) a slight but occasionally useful improvement:

$$Q(f \geqslant k) < \sqrt{\pi}\, \frac{(n+1)(n+3)}{\sqrt{n}} 2^{n/2-1} \frac{(ns-n-1+2k)}{ns-n-1} k^{-(ns-n+1)/2}. \quad (12)$$

We note that the very same ideas make it easy to construct estimates similar to (11) and (12) for different lattice functions which are related in form to zeta functions.

We note also that the known estimates (see [9] and [10]) of the remainder of the multidimensional zeta function $\zeta(f; s)$ are suitable for any, i.e. not necessarily integral, PQF in n variables. But as a rule they are either much cruder or they require more detailed information about the form, such as knowledge of the Voronoĭ-Dirichlet region of the associated lattice. Note that in the theory of the multidimensional zeta functions one mostly has to study the zeta function of integral forms.

3. We give an example of using (11) and (12) in the problem of extrema of a multidimensional zeta function (see [7], [8], [10] and [11]).

In [11] the following theorem is proved:

THEOREM. *Let $\varphi_0^{(n)} = \sum_{1 \leqslant i \leqslant j \leqslant n} x_i x_j$ be the first perfect form of Voronoĭ and ξ_n the real root of the equation*

$$\zeta(\varphi_0^{(n)}; \xi_n) = (n+1)^2/2n + ((n+1)^2/4)\xi_n, \qquad \xi_n > 1. \quad (13)$$

Then $S(\varphi_0^{(n)}) \leqslant \xi_n$.

In the same paper [11] it is shown that

$$\frac{(4n^2 - 2n - 2)}{n^2 + n} \leqslant \xi_n < 4 \quad (14)$$

(up to $n = 12$ the right-hand side of the inequality was proved with essential use of a computer for the solution of the transcendental equation (13)).

We show that the right-hand side of (14) can be obtained by means of inequality (11) for $Q(\varphi_0^{(n)} \geqslant 4)$ without use of computers for $n \geqslant 3$; moreover the extensive theoretical computations mentioned in [11] are not required.

THEOREM 4. *For all* $n \geqslant 10$

$$\xi_n < 4 - 6/(n+1) \tag{15}$$

and for $n = 3, 4, \ldots, 9$

$$\begin{array}{ccccccc} n=3 & 4 & 5 & 6 & 7 & 8 & 9 \\ \xi < 3.857 & 3.2728 & 3.1336 & 3.1589 & 3.2485 & 3.3329 & 3.3898 \end{array} \tag{16}$$

PROOF. From (13) and the left-hand side of (14) one obtains without difficulty the following inequality for ξ_n:

$$\xi_n \leqslant \left(4/(n+1)^2\right)\zeta\left(\varphi_0^{(n)}; s_0\right) - 2/n, \tag{17}$$

where $s_0 \geqslant (4n^2 - 2n - 2)/(n^2 + n)$.

From (17) we find by applying (12) to $Q(\varphi_0^{(u)} \geqslant 3)$ that

$$\begin{aligned} \xi_n \leqslant 4 &- \frac{4}{n+1} - \frac{2}{n} + \frac{n(n-1)(n-2)}{n+1} 2^{-ns_0/2} \\ &+ \sqrt{\pi}\, \frac{(n+1)(n+3)}{\sqrt{\pi}} \frac{ns_0 - n + 5}{ns_0 - n - 1} 2^{n/2-1} \cdot 3^{-(ns_0 - n + 1)/2}. \end{aligned} \tag{18}$$

Since for $n \geqslant 10$ we have $(4n^2 - 2n - 2)/(n^2 + n) \geqslant 3,4$ we obtain from (18)

$$\begin{aligned} \xi_n < 4 - \frac{1}{n+1} \Bigg\{ & 4 + \frac{2(n+1)}{n} - \frac{n(n-1)(n-2)(n+1)}{2^{1.7n}} \\ & - \frac{4\sqrt{\pi}\,(n+3)}{\sqrt{n}} \left(1 + \frac{6}{2.4n - 1}\right) 2^{n/2-1} \cdot 3^{-(2.4n+1)/2} \Bigg\} \\ = 4 - \frac{1}{n+1} \Bigg\{ & 6 + \left[\frac{1}{n} - \frac{n(n-1)(n-2)(n+1)}{2^{1.7n}}\right] \\ & + \left[\frac{1}{n} - \frac{4\sqrt{\pi}\,(n+3)}{\sqrt{n}} \left(1 + \frac{6}{2.4n - 1}\right) 2^{n/2-1} \cdot 3^{-(2.4n+1)/2}\right] \Bigg\}. \end{aligned} \tag{19}$$

By (19), in order to prove (15) it suffices to verify that

$$\delta_1(n) = 1/n - n(n-1)(n-2)(n+1)/2^{1.7n} > 0$$

and

$$\delta_2(n) = 1/n - 4\sqrt{\pi}\,(n+3)[1 + 6/(2.4n - 1)] \cdot 2^{n/2-1} \cdot 3^{-(2.4n+1)/2}/\sqrt{n} > 0.$$

To this end we consider the functions $\delta_1^*(n) = n \cdot 2^{1.7n}\delta_1(n)$ *and* $\delta_2^*(n) = n \cdot 3^{(2.4n+1)/2}\delta_2(n)$, and prove that they are positive.

Indeed, the inequality $2^{1.7n} > n^2(n^2-1)(n-2)$ is a consequence of $2^{1.7n} > n^5$. The latter inequality is equivalent to $1.7n\log 2 > 5\log n$, which is evident, because the function $1.7n - (5/\log 2)\log n$ is increasing; its smallest value 0.039 is attained for $n = 10$ and therefore it is possible. Consequently, $\delta_1^*(n) > 0$ and hence $\delta_1(n) > 0$.

Similarly, the inequality

$$3^{(2.4n+1)/2} > 2\sqrt{\pi n}\,(n+3)\left(1+\frac{6}{2.4n-1}\right)2^{n/2}$$

follows from $(3^{2.4}/2^{0.5})^n > 2.6\sqrt{\pi/3}\,\sqrt{\pi}\,(n+3)$ since $1 + 6/(2.4n-1) < 1.3$ for $n > 10$. The latter inequality is equivalent to

$$n(2.4\log 3 - 0.5\log 2) > \log 2.6 + (3/2)\log n + 0.435\log(1+3/n)$$

and

$$0.99454n > 0.415 + 3(0.5\log n + 0.44/n)$$

which is evident, because the function $0.99454n - 3(0.5\log n + 0.44n) - 0.415$ is increasing and its smallest value 7.8984 is attained for $n = 10$; therefore it is positive. Consequently, $\delta_2^*(n) > 0$; hence $\delta_2(n) > 0$. Therefore $\delta_1(n) > 0$ and $\delta_2(n) > 0$ for all $n \geqslant 10$, and thus (15) is proved.

Putting $f = \varphi_0^{(n)}$ and $k = 4$ in Theorem 3 and estimating $Q(\varphi_0^{(n)} \geqslant 4)$ by (11), from (17) we obtain

$$\begin{aligned}\xi_n \leqslant 4 - \frac{4}{n+1} - \frac{2}{n} + \frac{n(n-1)(n-2)}{n+1}2^{-ns_0/2} \\ + \left[\frac{n(n-1)\cdots(n-5)}{9(n+1)^2} + \frac{2n(n-1)\cdots(n-4)}{3(n+1)^2}\right. \\ \left. + \frac{4n(n-1)(n-2)}{(n+1)^2} + \frac{12n(n-1)}{(n+1)^2}\right]3^{-ns_0/2} \\ + \frac{4\sqrt{\pi}\,(n+3)}{\sqrt{n}\,(n+1)(ns_0-n-1)}2^{n/2}\cdot 3^{-(ns_0-n-1)/2}.\end{aligned} \tag{20}$$

If in (20) we put $s_0 = 7/3$ and $n = 3, 4, \ldots, 9$, we obtain (16). Theorem 4 is proved.

To obtain the inequality $\xi_2 < 3.7945$ it is sufficient to use the fact that the Dirichlet-Vornoĭ region for the form $\varphi_0^{(2)}$ is a regular hexagon, together with the estimate of $Q(\varphi_0^{(2)} \geqslant k)$ from [9]. The computations are similar to the case $n \geqslant 3$.

We note, finally, that by making the left-hand side of (14) somewhat cruder and using (15) we obtain for $n \geqslant 10$ the following obvious bounds for ξ_n:

$$4 - 6.2/(n+1) < \xi_n < 4 - 6/(n+1).$$

BIBLIOGRAPHY

1. R. A. Rankin, *The closest packing of spherical caps in n dimensions*, Proc. Glasgow Math. Assoc. **2** (1955), 139–144.

2. V. M. Sidel'nikov, *New bounds for densest packing of spheres in n-dimensional Euclidean space*, Mat. Sb. **95 (137)** (1974), 148–158; English transl. in Math. USSR Sb. **24** (1974).

3. G. A. Kabatyanskiĭ and V. I. Levenshteĭn, *Bounds for packings on the sphere and in space*, Problemy Peredachi Informatsii **14** (1978), no. 1, 3–25; English transl. in Problems of Information Transmission **14** (1978).

4. A. G. Postnikov, *Additive problems with a growing number of terms*, Izv. Akad. Nauk SSSR Ser. Mat. **20** (1956), 751–765. (Russian)

5. R. A. Rankin, *A minimum problem for the Epstein zeta-function*, Proc. Glasgow Math. Assoc. **1** (1953), 149–158.

6. S. Kh. Sirazhdinov, T. A. Azlarov and T. M. Zuparov, *Additive problems with an increasing number of terms*, "Fan", Tashkent, 1975. (Russian)

7. B. N. Delone and S. S. Ryshkov, *A contribution to the theory of the extrema of a multidimensional* ζ*-function*, Dokl. Akad. Nauk SSSR **173** (1967), 991–994; English transl. in Soviet Math. Dokl. **8** (1967).

8. ______, *Extremal problems in the theory of positive quadratic forms*, Trudy Mat. Inst. Steklov. **112** (1971), 203–223; English transl. in Proc. Steklov Inst. Math. **112** (1971).

9. L. V. Voĭtishek [Vojtíšek], *Computation of Epstein's zeta function*, Voprosy Vychisl. i Priskl. Mat. (Tashkent) Vyp. 38 (1970), 16–21. (Russian)

10. S. Sh. Shushbaev, *Extrema of five-dimensional Epstein zeta functions*, Izv. Vyssh. Uchebn. Zaved. Matematika **1980**, no. 12 (223), 60–66; English transl. in Soviet Math. (Iz. VUZ) **24** (1980).

11. ______, *On the Rankin-Sobolev problem on the multidimensional zeta function* (*estimating the origin of the ray of extremality of the principal perfect form*), Trudy Mat. Inst. Stkelov. **152** (1980), 232–235; English transl. in Proc. Steklov Inst. Math. **1982**, no. 3 (152).

Translated by P. HAFNER

Труды Матем. Инст. Стеклов.
158(1981)

Proceedings of the Steklov
Institute of Mathematics
1983, Issue 4

A LOCAL LIMIT THEOREM FOR CRITICAL BELLMAN-HARRIS BRANCHING PROCESSES

UDC 519.2

V. A. VATUTIN

ABSTRACT. Let $z(t)$ be the number of particles in a Bellman-Harris branching process with lifetime distribution function $G(t)$ and offspring generating function $f(s)$. The asymptotic behavior of the quantities $\mathbf{P}\{z(t)=n\}$ $(n=1,2,\dots)$ as $t\to\infty$ in dependence on the behavior of the "tail" of the function $G(t)$ at infinity is determined under the assumption that $f'(1)$ and that $f(s)$ has the form

$$f(s)=s+(1-s)^{1+\alpha}L(1-s),$$

where $\alpha\in(0,1]$ and $L(x)$ is slowly varying as $x\to+0$.

Bibliography: 14 titles.

§1. Introduction

Let us consider the following model for a branching process. At time $t=0$ there is a single particle of age zero. The lifetime τ of this particle is a random variable with distribution function $G(t)=\mathbf{P}\{\tau\leqslant t\}$. At the end of its life the particle produces a random number ξ of offspring of age zero with generating function $f(s)=\mathbf{E}s^{\xi}=\sum_0^\infty p_k s^k$. All subsequent particles evolve just as the original particle did. It is assumed that the lifetime and the number of offspring of a particle born at time t are mutually independent and do not depend on the previous history of the process nor on the number and composition of the particles existing in the process at the time t. The process described above is called a *Bellman-Harris process*.

Let $z(t)$ be the number of particles in a branching process at time t. In recent years several authors (see [2], [3], [11] and [14]) have directed their efforts towards investigating the asymptotic behavior of the probability $\mathbf{P}\{z(t)=n\}$ for fixed n as $t\to\infty$ in critical Bellman-Harris branching processes, i.e., in processes whose offspring generating function satisfies the relation $f'(1)=1$. We recall some of the results.

Let $f_n(s)$ be the nth iterate of $f(s)$. It was shown in [13] that if

$$f'(1)=1,\qquad f(s)=s+(1-s)^{1+\alpha}L(1-s), \tag{1.1}$$

where $\alpha\in(0,1]$ and $L(x)$ is slowly varying (SV) as $x\to+0$, then there exists a function $L^*(j)$ that is SV as $j\to\infty$ such that

$$1-f_j(0)=j^{-1/\alpha}L^*(j). \tag{1.2}$$

1980 *Mathematics Subject Classification*. Primary 60F05, 60J80; Secondary 45E10, 60K05.

This equality turned out to be very useful in investigating the quantities $\mathbf{P}\{z(t)>0\}$ and $\mathbf{P}\{z(t)=n\}$, for it was later established that the asymptotic behavior of these probabilities depends in an essential way on the ratio of the functions $1-f_j(0)$ and $1-G(j)$ as $j\to\infty$.

THEOREM A [2], [3]. *Suppose that* (1.1) *holds and that* $1-G(t)=t^{-\beta}L_2(t)$, *where* $\beta\in[0,1+\alpha^{-1}]$ *and the function* $L_2(t)$ *is SV as* $t\to\infty$. *If*

$$\lim_{j\to\infty}\frac{f(1-G(j))}{1-f_j(0)}=c,\qquad 0<c\leqslant\infty, \tag{1.3}$$

then the following assertions are true:

1) *There exists a function* $L_3(t)$ *that is SV as* $t\to\infty$ *and such that*

$$\mathbf{P}\{z(t)>0\}\sim t^{-\beta/1+\alpha}L_3(t),\qquad t\to\infty.$$

2) *There exists a sequence of positive constants* c_n, $n=1,2,\dots$, *such that*

$$\mathbf{P}\{z(t)=n\}\sim c_n\mathbf{P}\{z(t)>0\},\qquad t\to\infty.$$

The case where condition (1.3) is violated was considered in [11] and [14]. The best result obtained is as follows.

THEOREM B. *Suppose that* $f'(1)=1$, $\mathrm{GCD}\{k\colon p_k>0\}=1$, *and* $\sum_{k=0}^{\infty}k^{2+\delta}p_k<\infty$ *for some* $\delta>0$. *Suppose, moreover, that the following conditions hold*:

1) $\mu=\mathbf{E}\tau<\infty$.

2) $G(t)$ *has a differentiable density* $g(t)$, *and* $g'(t)$ *is continuous, integrable, bounded at* 0, *and satisfies* $g'(t)=O(t^{-2})$ *as* $t\to\infty$.

3) $\int_0^\infty t^3g(t)\ln t\,dt<\infty$.

Then there exists a sequence of positive constants A_n $(n=1,2,\dots)$ *such that as* $t\to\infty$

$$\mathbf{P}\{z(t)=n\}\sim t^{-1}A_n\mathbf{P}\{z(t)>0\}\sim(2A_n\mu/f''(1))t^{-2}. \tag{1.4}$$

REMARK. The last relation in (1.4) follows from Theorem 1 in [9].

The absence of any natural consistency between Theorem A 2) and Theorem B leads to the supposition that the asymptotic behavior of $\mathbf{P}\{z(t)=n\}$ as $t\to\infty$ will differ from that described in Theorems A and B under conditions that are in some sense intermediate between the conditions of these theorems. This supposition is confirmed by Theorem 1, proved below.

In the arguments to follow $c,c_1,c_2,\dots$ will denote positive constants, and $L_1(t)$, $L_2(t),\dots$ or $L_{j_1\cdots j_k}(t)$ will denote functions that are SV as $t\to\infty$, different in different formulas.

We call $G(t)$ a distribution function of *absolutely continuous type* and write $G(t)\in V$ if there exists a $k\geqslant1$ such that the kth convolution of $G(t)$ with itself contains an absolutely continuous component.

Let $P_n(t)=\mathbf{P}\{z(t)=n\}$.

THEOREM 1. *If* $G(t)\in V$,

$$\lim_{j\to\infty}\frac{j(1-G(j))}{1-f_j(0)}=0, \tag{1.5}$$

$$1-G(t)=(t+1)^{-1-\rho}L_1(t),\qquad \rho\in[\alpha^{-1},1+\alpha^{-1}], \tag{1.6}$$

and either

$$\sum_{k=n}^{\infty} p_k \sim \frac{L(n)}{n^{1+\alpha}}, \qquad 0 < \alpha \leqslant 1, n \to \infty, \tag{1.7}$$

or for some $\delta > 0$(¹)

$$\sum_{k=n}^{\infty} p_k = o(n^{-2-\delta}), \qquad n \to \infty, \tag{1.8}$$

then

$$P_n(t) = \begin{cases} t^{-\rho-(n-1)(\rho-\alpha^{-1})} L_n(t) & \text{if } n(\rho - \alpha^{-1}) \leqslant 1, \\ t^{-1-\alpha^{-1}} L_n(t) & \text{if } n(\rho - \alpha^{-1}) > 1. \end{cases} \tag{1.9}$$

REMARK 1. For $0 < \alpha < 1$ condition (1.7) is equivalent to (1.1); for $\alpha = 1$ both (1.7) and (1.8) are stronger than (1.1).

REMARK 2. It follows from the main theorem in [1] that under conditions (1.1) and (1.5)

$$Q(t) = \mathbf{P}\{z(t) > 0\} \sim t^{-1/\alpha} L_1(t), \qquad t \to \infty, \tag{1.10}$$

and, consequently, for any $\gamma \in [0, 1]$ and any N there exist a branching process $z(t)$ and an SV function $L_N(t)$ such that

$$\mathbf{P}\{z(t) = N\} \sim t^{-\gamma} L_N(t) \mathbf{P}\{z(t) > 0\} \quad \text{as } t \to \infty.$$

The following theorem is a natural complement to Theorem 1 and generalizes the analogous theorems in [5] and [13].

THEOREM 2. *Suppose that* (1.1) *holds and* $G(t) \in V$. *If*

$$1 - G(t) = o(t^{-1-\alpha^{-1}-\delta}), \qquad \delta > 0, t \to \infty, \tag{1.11}$$

then

$$P_n(t) \sim t^{-1-\alpha^{-1}} L_n(t), \qquad t \to \infty. \tag{1.12}$$

This article is arranged as follows. It is shown in §2 that the study of the asymptotic behavior of the quantities $P_n(t)$ as $t \to \infty$ can be reduced to the investigation of the asymptotic behavior of the solution of the equation of renewal type

$$\Theta(t) = \xi(t) + \gamma(t) \Theta * G(t),$$

where $\xi(t) \geqslant 0$, $0 \leqslant \gamma(t) \leqslant 1$ and $\gamma(t) \to 1$ as $t \to \infty$. In §3 we find coarse upper and lower estimates for the function $\Theta(t)$, and then use them in §4 to obtain the asymptotic behavior of integrals of the form $\int_0^t u^k \Theta(u)\, du$. In §5 we refine the upper estimate for $\Theta(t)$, and this enables us in §6 to determine the asymptotic behavior of $\Theta(t)$ as $t \to \infty$ under various restrictions on $\xi(t)$. Theorems 1 and 2 are proved in §7, and it is shown in §8 how to follow the scheme in the proof of Theorem 2 and get an improvement of Theorem B.

(¹) In this case we assume that $\alpha = 1$.

§2. Derivation of the main integral equation

LEMMA 1. *Suppose that* (1.1) *holds. Then the following assertions are true*:

1) $1 - f'(s) \sim (\alpha + 1)(1 - s)^{\alpha} L(1 - s)$, $s \uparrow 1$.

2) $f''(s) \sim (\alpha + 1)\alpha(1 - s)^{\alpha - 1} L(1 - s)$, $s \uparrow 1$.

3) *For any* $\delta > 0$ *there exists a* $c_r = c_r(\delta)$ *such that*

$$f^{(r)}(s) \leqslant c_r(1 - s)^{1+\alpha-\delta-r}, \qquad r \geqslant 3.$$

PROOF. It is clear that $f'(s) = 1 - \varphi(1 - s)$, where $\varphi(u) \searrow 0$ as $u \searrow 0$. Consequently, $1 - f(s) = 1 - s - \int_0^{1-s} \varphi(u)\, du$, which, together with (1.1), gives

$$\int_0^x \varphi(u)\, du \sim x^{1+\alpha} L(x), \qquad x \to +0.$$

From this and the monotonicity of $\varphi(x)$ it follows (cf. [12], p. 60, Theorem 2.4) that $\varphi(x) \sim (1 + \alpha)x^{\alpha} L(x)$ as $x \to +0$. Assertion 2) is proved similarly.

Let us prove 3). Using (1.1), we see that $B_\delta = \sum_{k=1}^{\infty} k^{1+\alpha-\delta} p_k < \infty$ if $\delta > 0$. Thus,

$$f^{(r)}(s) \leqslant \sum_{k=r}^{\infty} k^r p_k s^{k-r} \leqslant B_\delta \max_{k \geqslant 0} (k + r)^{r-1-\alpha+\delta} s^k.$$

Obviously, the function $h(k) = (k + r)^{r-1-\alpha+\delta} s^k$ has a maximum at the point $k = -r - (r - 1 - \alpha + \delta)/\ln s$, and

$$\max_k h(k) \leqslant c(-\ln s)^{1+\alpha-\delta-r} \leqslant c_1(1 - s)^{1+\alpha-\delta-r}$$

for $s \in (1/2, 1)$, which proves 3), since $f^{(r)}(1/2) < \infty$.

LEMMA 2. *If* (1.8) *holds, then*

$$f^{(r)}(s) \leqslant c_r(1 - s)^{2+\nu-r}, \qquad r \geqslant 3,\ \nu < \delta. \tag{2.1}$$

If (1.7) *holds, then as* $s \uparrow 1$

$$f^{(r)}(s) \sim \frac{L_r((1 - s)^{-1})}{(1 - s)^{r-1-\alpha}}, \qquad r \geqslant 3. \tag{2.2}$$

The proof of (2.1) is analogous to that of Lemma 1.3), and (2.2) is a consequence of Theorem 2 in [7], Chapter VIII, §9.

Let $F(t, s) = \mathbf{E}s^{z(t)}$. It is known (see [8], p. 134) that $F(t, s)$ is the unique bounded solution of the integral equation

$$F(t, s) = s(1 - G(t)) + \int_0^t f(F(t - u, s))\, dG(u). \tag{2.3}$$

Differentiating this equation with respect to s at the point $s = 0$, we find that

$$P_1(t) = 1 - G(t) + \int_0^t f'(F(t - u, 0))P_1(t - u)\, dG(u).$$

From this, setting $M_1(t) = f'(F(t, 0))P_1(t)$, we get

$$M_1(t) = f'(F(t, 0))(1 - G(t)) + f'(F(t, 0)) \int_0^t M_1(t - u)\, dG(u). \tag{2.4}$$

In exactly the same way, by setting $M_n(t) = n!f'(F(t, 0))P_n(t)$ and using (2.3) and the formula of Faà di Bruno ([4], 0.430) for the derivative of a composite function, it can be shown that

$$M_n(t) = \psi_n(t) + f'(F(t, 0))M_n * G(t), \qquad n \geqslant 2, \tag{2.5}$$

where

$$\psi_n(t) = f'(F(t,0)) \sum_{r=2}^{n} \int_0^t Y_{n,r}(t-u) f^{(r)}(F(t-u,0))\, dG(u), \tag{2.6}$$

and

$$Y_{n,r}(t) = \sum_{\substack{j_1+\cdots+j_{n-1}=r \\ j_1+2j_2+\cdots+(n-1)j_{n-1}=n}} \frac{n!}{f_1!\cdots j_{n-1}!} P_1^{j_1}(t)\cdots P_{n-1}^{j_{n-1}}(t). \tag{2.7}$$

Since it was established in [1] that under conditions (1.1) and (1.5)

$$(1-F(t,0))^{\alpha} L(1-F(t,0)) \sim \mu(\alpha t)^{-1}, \tag{2.8}$$

it follows from 1) of Lemma 1 that (2.5) can be rewritten in the form

$$M_n(t) = \psi_n(t) + \left(1 - \frac{b\mu - \sigma(t)}{t + b\mu}\right) M_n * G(t), \qquad n \geq 2, \tag{2.9}$$

where $\sigma(t) \to 0$ as $t \to \infty$, and $b = 1 + \alpha^{-1}$.

Instead of (2.9) it will be more convenient for us to investigate an equation of more general type

$$\Theta(t) = \xi(t) + \gamma(t)\Theta * G(t), \tag{2.10}$$

where $\xi(t)$ is a bounded nonnegative measurable function, $0 \leq \gamma(t) \leq 1$, and, finally,

$$\gamma(t) = 1 - \frac{b\mu - \sigma(t)}{t + b\mu}, \qquad b > 0, \qquad \sigma(t) \to 0 \quad \text{as } t \to \infty. \tag{2.11}$$

We remark that for $b = 2$ equation (2.10) was studied in [10].

§3. Determination of coarse upper and lower estimates for $\Theta(t)$

LEMMA 3. *If $G(t) \in V$, $G(+0) = 0$, $\xi(t)$ is Lebesgue-integrable on $[0,\infty)$, and $\xi(t) \to 0$ as $t \to \infty$, then (2.10) has a unique solution in the class of measurable functions that are bounded on finite intervals.*

The proof is analogous to that of Proposition 1 in [10].

LEMMA 4. *Suppose that there exists a T such that for any $T_1 \geq T$*

$$\inf_{t\in[T,T_1]} \Theta(t) = p(T_1) > 0. \tag{3.1}$$

Then for any $\varepsilon > 0$ there exists a constant $c = c(\varepsilon) > 0$ such that

$$\Theta(t) \geq c(t+1)^{-b-\varepsilon}, \qquad t \geq T_2 = T_2(\varepsilon). \tag{3.2}$$

REMARK. Obviously, (3.1) is satisfied if there exists a T such that $\inf_{t\in[T,T_1]} \xi(t) > 0$ for any $T_1 \geq T$.

PROOF. We represent $\Theta(t)$ in the form

$$\Theta(t) = K(t)/(t+1)^{b+\varepsilon}. \tag{3.3}$$

Let $K_n = \inf_{n-1\leq t<n} K(t)$, $n = 1,2,\ldots$. By the conditions of the lemma, there exists an $N < \infty$ such that $K_n > 0$ for any $n \geq N-1$. Assume that (3.2) does not hold, i.e., there exists an $\varepsilon > 0$ such that

$$\liminf_{n\to\infty} K_n = 0. \tag{3.4}$$

Suppose that the sequence $\{n_i\}$ $(i = 1, 2, \dots)$ is such that

$$\lim_{i\to\infty} K_{n_i} = 0 \quad \text{and} \quad K_{n_i} = \min_{N\leqslant n\leqslant n_i} K_n.$$

Using the representation (3.3), we get that for any i there exists a $\tau_i \in [n_i - 1, n_i)$ such that

$$\frac{K_{n_i}}{(\tau_i+1)^{b+\varepsilon}} + \frac{K_{n_i}}{(n_i+1)^{b+2+\varepsilon}} \geqslant \Theta(\tau_i) \geqslant \xi(\tau_i) + \gamma(\tau_i)K_{n_i}\int_0^{\tau_i - N} \frac{dG(u)}{(\tau_i - u + 1)^{b+\varepsilon}}$$
$$\geqslant \frac{\gamma(\tau_i)K_{n_i}}{(\tau_i+1)^{b+\varepsilon}} \int_0^{\tau_i - N}\left(1 + \frac{u(b+\varepsilon)}{\tau_i + 1}\right) dG(u),$$

since $(1-x)^{-\alpha} \geqslant 1 + \alpha x$, $\alpha \geqslant 0$. From this and the conditions of the lemma it follows that for $n_i \geqslant T_3 = T_3(\varepsilon)$

$$\frac{K_{n_i}}{(\tau_i+1)^{b+\varepsilon}}\left(1 + \frac{1}{(n_i+1)^2}\right) \geqslant \frac{\gamma(\tau_i)K_{n_i}}{(\tau_i+1)^{b+\varepsilon}}\left(1 + \frac{(b+\varepsilon/2)\mu}{\tau_i+1}\right)$$
$$\geqslant \left(1 - \frac{(b+\varepsilon/4)\mu}{\tau_i+1}\right)\frac{K_{n_i}}{(\tau_i+1)^{b+\varepsilon}}\left(1 + \frac{(b+\varepsilon/2)\mu}{\tau_i+1}\right)$$

or

$$\frac{1}{(n_i+1)^2} \geqslant \frac{\varepsilon\mu}{4(\tau_i+1)} - \frac{(b+\varepsilon/4)(b+\varepsilon/2)\mu^2}{(\tau_i+1)^2},$$

which is not true for sufficiently large values of n_i. Thus, the assumption (3.4) is false, and the lemma is proved.

LEMMA 5. *If $\rho \leqslant b$ and, for any $\varepsilon > 0$,*

$$\xi(t) = o(t^{-\rho-1+\varepsilon}), \qquad t \to \infty, \tag{3.5}$$

and

$$1 - G(t) = o(t^{-\rho-1+\varepsilon}), \qquad t \to \infty, \tag{3.6}$$

then for any $\varepsilon > 0$ there is a constant $c = c(\varepsilon) > 0$ such that

$$\Theta(t) \leqslant c(t+1)^{-\rho+\varepsilon}. \tag{3.7}$$

PROOF. We define the function $K(t)$ by the equality

$$\Theta(t) = K(t)/(t+1)^{\rho-\varepsilon}.$$

If $\limsup_{t\to\infty} K(t) < \infty$, then the lemma is proved. Suppose that

$$\limsup_{t\to\infty} K(t) = \infty. \tag{3.8}$$

Let

$$K_n = \sup_{n-1\leqslant t<n} K(t), \tag{3.9}$$

and let the sequence $\{n_i\}$ $(i = 1, 2, \ldots)$ be such that

$$K_{n_i} = \max_{1 \leqslant n \leqslant n_i} K_n. \tag{3.10}$$

Obviously, $K_{n_i} \to \infty$ as $i \to \infty$. From (3.9) and (3.10) it follows that for any n_i there is a $\tau_i \in [n_i - 1, n_i)$ such that

$$\frac{1}{(\tau_i + 1)^{\rho-\varepsilon}}\left(K_{n_i} - \frac{1}{(n_i + 1)^2}\right) \leqslant \Theta(\tau_i) \leqslant \xi(\tau_i) + \gamma(\tau_i) K_{n_i} \int_0^{\tau_i} \frac{dG(u)}{(\tau_i - u + 1)^{\rho-\varepsilon}}. \tag{3.11}$$

According to (3.6),

$$\int_{t/2}^{t} \frac{dG(u)}{(t - u + 1)^{\rho-\varepsilon}} \leqslant \int_{t/2}^{t} dG(u) = o(t^{-\rho-1+\varepsilon/2}), \qquad t \to \infty,$$

while

$$\int_0^{t/2} \frac{dG(u)}{(t - u + 1)^{\rho-\varepsilon}} = \frac{1}{(t + 1)^{\rho-\varepsilon}}\left(1 + \frac{\mu(\rho - \varepsilon)}{t + 1}\right) + O\left(\frac{1}{(t + 1)^2}\int_0^{t/2} u^2\, dG(u)\right), \qquad t \to \infty,$$

and for any $\varepsilon > 0$

$$\int_0^t u^2\, dG(u) = O(1) + o(t^{1-\rho+\varepsilon}), \qquad t \to \infty.$$

The estimates obtained show that for sufficiently large values of i

$$\int_0^{\tau_i} \frac{dG(u)}{(\tau_i - u + 1)^{\rho-\varepsilon}} \leqslant \frac{1}{(\tau_i + 1)^{\rho-\varepsilon}}\left(1 + \frac{\mu(\rho - \varepsilon/2)}{\tau_i + 1}\right),$$

which, in view of (3.11) and (2.11), gives

$$\frac{1}{(\tau_i + 1)^{\rho-\varepsilon}}\left(K_{n_i} - \frac{1}{(n_i + 1)^2}\right) \leqslant \xi(\tau_i) + \left(1 - \frac{(b - \varepsilon/4)\mu}{\tau_i + 1}\right)\frac{K_{n_i}}{(\tau_i + 1)^{\rho-\varepsilon}}\left(1 + \frac{(\rho - \varepsilon/2)\mu}{\tau_i + 1}\right).$$

From this we obtain

$$K_{n_i}\mu(b - \rho + \varepsilon/4) \leqslant (\tau_i + 1)^{\rho+1-\varepsilon}\xi(\tau_i) + (\tau_i + 1)^{-1}.$$

The latter, together with (3.5), contradicts (3.8).

§4. Integral characteristics of $\Theta(t)$

We need some new notation for the presentation to follow. Let

$$\Theta_k(t) = t^k\Theta(t), \quad G_k(t) = \int_0^t u^k\, dG(u), \qquad k = 1, 2, \ldots.$$

With a view to uniformity we sometimes write $\Theta_0(t)$ instead of $\Theta(t)$, and $G_0(t)$ instead of $G(t)$. Let $\xi_{k+1}(t) = t^k(t + b\mu)\xi(t)$, $\Delta_k(t) = t^k\sigma(t)\Theta * G(t)$ (cf. (2.11)),

and

$$\varphi_k(t) = \sum_{j=2}^{k+1} C_{k+1}^j \Theta_{k+1-j} * G_j(t)$$

(here and below, $\Sigma_{j=m}^n = 0$ if $m > n$).

The Laplace transform of a function $X(t)$ will be denoted by $\hat{X}(\lambda)$, with two exceptions:

$$\hat{G}_k(\lambda) = \int_0^\infty e^{-\lambda t} t^k \, dG(t)$$

and

$$\hat{\varphi}_k(\lambda) = \sum_{j=2}^{k+1} c_{k+1}^j \hat{\Theta}_{k+1-j}(\lambda)\hat{G}_j(\lambda).$$

LEMMA 6. *If condition* (3.1) *holds, if*

$$1 - G(t) = o(t^{-b-1}), \qquad t \to \infty, \tag{4.1}$$

and if for some $\varepsilon > 0$

$$\xi(t) = o(t^{-b-1-\varepsilon}), \qquad t \to \infty, \tag{4.2}$$

then for $\beta = [b]$

$$\int_0^t u^\beta \Theta(u)\, du \sim t^{\beta+1-b} L_1(t), \qquad t \to \infty. \tag{4.3}$$

PROOF. We rewrite (2.10) in the form

$$t\Theta(t) = (t + b\mu)\xi(t) - b\mu\Theta(t) + t\Theta * G(t) + \sigma(t)\Theta * G(t). \tag{4.4}$$

Multiplying the left-hand and right-hand sides of this equality by t^β and applying the Laplace transformation to the relation thus obtained, we find that

$$\begin{aligned} -\hat{\Theta}_\beta'(\lambda) = \hat{\xi}_{\beta+1}(\lambda) - b\mu\hat{\Theta}_\beta(\lambda) - \hat{\Theta}_\beta'(\lambda)\hat{G}(\lambda) \\ + (\beta+1)\hat{\Theta}_\beta(\lambda)\hat{G}_1(\lambda) + \hat{\Delta}_\beta(\lambda)\hat{\varphi}_\beta(\lambda) \end{aligned} \tag{4.5}$$

or

$$\frac{\hat{\Theta}_\beta'(\lambda)}{\hat{\Theta}_\beta(\lambda)} = -\frac{(\beta+1)\hat{G}_1(\lambda) - b\mu}{1 - \hat{G}(\lambda)} - \frac{\hat{\Delta}_\beta(\lambda) + \hat{\xi}_{\beta+1}(\lambda) + \hat{\varphi}_\beta(\lambda)}{(1 - \hat{G}(\lambda))\hat{\Theta}_\beta(\lambda)}. \tag{4.6}$$

We show that (4.6) can be represented in the form

$$\frac{\hat{\Theta}_\beta'(\lambda)}{\hat{\Theta}_\beta(\lambda)} = -\frac{\beta+1-b}{\lambda} + \frac{\nu(\lambda)}{\lambda}, \tag{4.7}$$

where $\nu(\lambda) \to 0$ as $\lambda \downarrow 0$. Obviously,

$$1 - \hat{G}(\lambda) \sim \lambda\mu, \qquad \lambda \downarrow 0, \tag{4.8}$$

$$\mu - \hat{G}_1(\lambda) \to 0, \qquad \lambda \downarrow 0. \tag{4.9}$$

By Lemma 1, for any $\varepsilon > 0$ there exists a λ_0 such that

$$\hat{\Theta}_\beta(\lambda) \geqslant \lambda^{-(\beta+1-b-\varepsilon)}, \qquad \lambda \in (0, \lambda_0), \tag{4.10}$$

and, consequently,

$$\hat{\Theta}_\beta(\lambda) \to \infty, \qquad \lambda \downarrow 0. \tag{4.11}$$

It now follows from (3.2) and (4.2) that

$$\hat{\xi}_{\beta+1}(\lambda) = o\big(\hat{\Theta}_\beta(\lambda)\big), \qquad \lambda \downarrow 0, \tag{4.12}$$

and from (4.11) and (4.1) that as $\lambda \downarrow 0$

$$\hat{\Delta}_\beta(\lambda) = o\big(\hat{\Theta}_\beta(\lambda)\big) \tag{4.13}$$

and

$$\sum_{j=2}^{\beta} C^j_{\beta+1}\hat{\Theta}_{\beta+1-j}(\lambda)\hat{G}_j(\lambda) = o\big(\hat{\Theta}_\beta(\lambda)\big).$$

Finally, let us consider the expression $\hat{\Theta}_0(\lambda)\hat{G}_{\beta+1}(\lambda)$ for $b \geqslant \beta \geqslant 1$. According to Lemma 5, $\hat{\Theta}_0(\lambda) = o(\lambda^{-\varepsilon})$ for any $\varepsilon > 0$ as $\lambda \downarrow 0$. On the other hand,

$$\begin{aligned}\hat{G}_{\beta+1}(\lambda) &= (-1)^{\beta+2}\Big(\lambda\int_0^\infty e^{-\lambda t}t^{\beta+1}(1-G(t))\,dt \\ &\qquad - (\beta+1)\int_0^\infty e^{-\lambda t}t^{\beta}(1-G(t))\,dt\Big) \\ &= o\Big(\log\frac{1}{\lambda}\Big), \qquad \lambda \downarrow 0. \end{aligned} \tag{4.14}$$

Thus,

$$\hat{\varphi}_\beta(\lambda) = o\big(\hat{\Theta}_\beta(\lambda)\big), \qquad \lambda \downarrow 0. \tag{4.15}$$

Combining (4.8), (4.9) and (4.11)–(4.13) with (4.15), we get (4.7). From (4.7) we find that

$$\hat{\Theta}_\beta(\lambda) = \frac{c(\lambda_0)}{\lambda^{\beta+1-b}}\exp\left\{\int_\lambda^{\lambda_0}\frac{\nu(z)}{z}\,dz\right\}. \tag{4.16}$$

But the function

$$L_2\frac{1}{\lambda} = \exp\left\{\int_\lambda^{\lambda_0}\frac{\nu(z)}{z}\,dz\right\}$$

is SV as $\lambda \downarrow 0$, and this, together with (4.16) and a Tauberian theorem ([7], Chapter XIII, §5), gives us (4.3).

LEMMA 7. *Suppose that either* $\max\{0, b-1\} < \rho \leqslant \gamma < b$ *or (if* $b > 1$*)* $\rho = \gamma = b - 1$. *If for any* $\varepsilon > 0$

$$\xi(t) = o(t^{-\rho-1+\varepsilon}), \qquad t \to \infty, \tag{4.17}$$

$$\hat{\xi}_{\kappa+1}(\lambda) \sim \Gamma(\kappa+2-\gamma)L_1(\lambda^{-1})/\lambda^{\kappa+1-\gamma}, \qquad \lambda \downarrow 0,\ \kappa = [\gamma], \tag{4.18}$$

and

$$1 - G(t) = (t+1)^{-1-\rho}L_2(t), \tag{4.19}$$

then

$$\Theta_\kappa(\lambda) \to \infty, \qquad \lambda \downarrow 0.$$

PROOF. Setting $\kappa = \beta$, we have from (4.5) that

$$\hat{\Theta}'_\kappa(\lambda) + \frac{(\kappa+1)\hat{G}_1(\lambda) - b\mu}{1-\hat{G}(\lambda)}\hat{\Theta}_\kappa(\lambda) + \frac{\hat{\Delta}_\kappa(\lambda)}{1-\hat{G}(\lambda)} = -\frac{\hat{\xi}_{\kappa+1}(\lambda)}{1-\hat{G}(\lambda)} - \frac{\hat{\varphi}_\kappa(\lambda)}{1-\hat{G}(\lambda)}. \tag{4.20}$$

Assume that

$$\lim_{\lambda \downarrow 0} \hat{\Theta}_\kappa(\lambda) = c < \infty. \tag{4.21}$$

In this case

$$\lim_{\lambda \downarrow 0} \hat{\Theta}_j(\lambda) < \infty, \qquad j = 0, 1, \ldots, \kappa - 1. \tag{4.22}$$

By using (4.19) it is not hard to check that

$$\lim_{\lambda \downarrow 0} \hat{G}_j(\lambda) < \infty, \qquad j = 0, 1, \ldots, \kappa, \tag{4.23}$$

while

$$\hat{G}_{\kappa+1}(\lambda) = O(1) + O(\lambda^{\rho-\kappa} L_2(\lambda^{-1})), \qquad \lambda \downarrow 0, \tag{4.24}$$

and, since $1 - \gamma > -\rho$, it follows that $\hat{G}_{\kappa+1}(\lambda) = o(\hat{\xi}_{\kappa+1}(\lambda))$ as $\lambda \downarrow 0$. Thus,

$$\hat{\varphi}_\kappa(\lambda) = o(\hat{\xi}_{\kappa+1}(\lambda)), \qquad \lambda \downarrow 0. \tag{4.25}$$

By the same arguments,

$$\limsup_{\lambda \downarrow 0} |\hat{\Delta}_\kappa(\lambda)| < \infty. \tag{4.26}$$

It follows from (4.22)–(4.26) that (4.20) can be rewritten as

$$\hat{\Theta}'_\kappa(\lambda) = -\frac{\Gamma(\kappa + 2 - \gamma)}{\lambda^{\kappa+2-\gamma}} L_1(\lambda^{-1})(1 + o(1)), \qquad \lambda \downarrow 0,$$

whence

$$\hat{\Theta}_\kappa(\lambda) = \frac{\Gamma(\kappa + 1 - \gamma)}{\lambda^{\kappa+1-\gamma}} L_1(\lambda^{-1})(1 + o(1)), \qquad \lambda \downarrow 0,$$

which contradicts (4.21). This proves the lemma.

LEMMA 8. *If the conditions of Lemma 7 hold and* $\gamma < b$, *then*

$$\int_0^t u^\kappa \Theta(u)\, du \sim \frac{1}{\mu(b - \gamma)(\kappa + 1 - \gamma)} t^{\kappa+1-\gamma} L_1(t), \qquad t \to \infty, \tag{4.27}$$

where the function $L_1(t)$ *is the same as in* (4.18).

PROOF. From Lemma 5 and (4.23) it follows that

$$\hat{\Delta}_\kappa(\lambda) = o(\hat{\Theta}_\kappa(\lambda)), \qquad \lambda \downarrow 0. \tag{4.28}$$

For the same reason (we assume that $\kappa \geqslant 1$)

$$\sum_{j=2}^{\kappa} C^j_{\kappa+1} \hat{\Theta}_{\kappa+1-j}(\lambda) \hat{G}_j(\lambda) = o(\hat{\Theta}_\kappa(\lambda)), \qquad \lambda \downarrow 0. \tag{4.29}$$

Let us now consider separately the two cases $\rho \geqslant 1$ and $\rho < 1$. If $\rho \geqslant 1$, then, by Lemma 5, $\hat{\Theta}_0(\lambda) = o(\lambda^{-\varepsilon})$ as $\lambda \downarrow 0$ for any $\varepsilon > 0$, and this, together with (4.24) and the inequality $1 - \gamma > -\rho$, gives us that

$$\hat{\Theta}_0(\lambda) \hat{G}_{\kappa+1}(\lambda) = o(\hat{\xi}_{\kappa+1}(\lambda)), \qquad \lambda \downarrow 0. \tag{4.30}$$

Let $\rho < 1$. In view of the conditions in the lemma and the inequality $\kappa \geqslant 1$, this means that $b - 1 \leqslant \rho < 1 = \kappa \leqslant \gamma < b < 2$. We show that

$$\hat{\Theta}_0(\lambda) = o\big(L_3(\lambda^{-1})\big), \qquad \lambda \downarrow 0. \tag{4.31}$$

Indeed,

$$\hat{\Theta}_0'(\lambda) + \frac{\hat{G}_1(\lambda) - b\mu}{1 - \hat{G}(\lambda)}\hat{\Theta}_0(\lambda) + \frac{\hat{\Delta}_0(\lambda)}{1 - \hat{G}(\lambda)} = -\frac{\hat{\xi}_1(\lambda)}{1 - \hat{G}(\lambda)}. \tag{4.32}$$

If now $\lim_{\lambda \downarrow 0} \hat{\Theta}_0(\lambda) < \infty$, then (4.31) is proved. Suppose that $\hat{\Theta}_0(\lambda) \to \infty$ as $\lambda \downarrow 0$. In this case (4.32) can be written as

$$\hat{\Theta}_0'(\lambda) + \frac{1 - b + \nu(\lambda)}{\lambda}\hat{\Theta}_0(\lambda) = -\frac{\hat{\xi}_1(\lambda)}{1 - \hat{G}_0(\lambda)}, \tag{4.33}$$

where $\nu(\lambda) \to 0$ as $\lambda \downarrow 0$. Let $X(\lambda)$ be the solution of the equation

$$X'(\lambda) + \frac{1 - b + \nu(\lambda)}{\lambda}X(\lambda) = 0, \qquad X(1) = 1. \tag{4.34}$$

Then

$$X(\lambda) = \lambda^{b-1}\int_\lambda^1 \frac{\nu(u)}{u}\,du = \lambda^{b-1}L_2\Big(\frac{1}{\lambda}\Big).$$

From (4.18) with $\kappa = 1$ we get that $\hat{\xi}_1(\lambda) \sim L_4(\lambda^{-1})$ as $\lambda \downarrow 0$, and $\lim_{\lambda \downarrow 0} \hat{\xi}_1(\lambda) < \infty$ if $\gamma > 1$. From this, representing $\hat{\Theta}_0(\lambda)$ in the form $\hat{\Theta}_0(\lambda) = C(\lambda)X(\lambda)$, we find that as $\lambda \downarrow 0$

$$C'(\lambda) = -\frac{\hat{\xi}_1(\lambda)}{(1 - \hat{G}(\lambda))X(\lambda)} \sim -\frac{L_4(\lambda^{-1})}{\mu\lambda^b L_2(\lambda^{-1})} \tag{4.35}$$

and, consequently,

$$C(\lambda) \sim \frac{L_4(\lambda^{-1})}{\mu(b-1)\lambda^{b-1}L_2(\lambda^{-1})}, \qquad \lambda \downarrow 0.$$

Thus, $\hat{\Theta}_0(\lambda) \sim L_3(\lambda^{-1})$, $\lambda \downarrow 0$, and, hence, (4.30) is also valid for $\kappa = 1$.

From (4.28)–(4.30) and (4.18) it follows that (4.20) can be transformed to the form

$$\hat{\Theta}_\kappa'(\lambda) + \frac{\kappa + 1 - b + \nu(\lambda)}{\lambda}\hat{\Theta}_\kappa(\lambda) = -\frac{\Gamma(\kappa + 2 - \gamma)L_1(\lambda^{-1})(1 + \delta(\lambda))}{\mu\lambda^{\kappa+2-\gamma}},$$

where $|\nu(\lambda)| + |\delta(\lambda)| \to 0$ as $\lambda \downarrow 0$. Arguing as in the investigation of (4.33), we get that

$$\hat{\Theta}_\kappa(\lambda) = Y(\lambda)R(\lambda),$$

where

$$Y(\lambda) = \lambda^{-(\kappa+1-b)}\exp\Big\{\int_\lambda^1 \frac{\nu(u)}{u}\,du\Big\} = \lambda^{-(\kappa+1-b)}L_2(\lambda^{-1}),$$

and

$$R'(\lambda) = -\frac{\Gamma(\kappa + 1 - \gamma)L_1(\lambda^{-1})(1 + \delta(\lambda))}{\mu\lambda^{b+1-\gamma}L_2(\lambda^{-1})}. \tag{4.36}$$

This and the inequality $b > \gamma$ imply that

$$R(\lambda) \sim \frac{\Gamma(\kappa + 1 - \gamma)L_1(\lambda^{-1})}{\mu(b - \gamma)\lambda^{b-\gamma}L_2(\lambda^{-1})}, \qquad \lambda \downarrow 0.$$

Consequently,

$$\hat{\Theta}_\kappa(\lambda) \sim \frac{\Gamma(\kappa + 1 - \gamma)L_1(\lambda^{-1})}{\mu(\beta - \gamma)\lambda^{\kappa+1-\gamma}}, \qquad \lambda \downarrow 0, \tag{4.37}$$

which (see [7], Chapter XIII, §5) is equivalent to (4.27).

LEMMA 9. *If the conditions of Lemma 7 hold and* $\gamma = b$, *then*

$$\int_0^t u^\kappa \Theta(u)\, du \sim t^{\kappa+1-b}L_2(t), \qquad t \to \infty. \tag{4.38}$$

The proof is analogous to that of Lemma 8. The only difference is that, first, for $b = \gamma = 1$ formula (4.35) implies that $C(\lambda) \sim L_3(\lambda^{-1})$ as $\lambda \downarrow 0$, and, second, for $b = \gamma$ it follows from (4.36) that $R(\lambda) \sim L_4(\lambda^{-1})$ as $\lambda \downarrow 0$.

LEMMA 10. *Suppose that the conditions of Lemma 7 hold. If the number* γ *is not an integer, then*

$$\lim_{\lambda \downarrow 0} \hat{\Theta}_i(\lambda) < \infty, \qquad i = 0, 1, \ldots, \kappa - 1. \tag{4.39}$$

But if γ *is an integer, then* (4.39) *holds for all* $i < \kappa - 1$, *and*

$$\hat{\Theta}_{\kappa-1} \sim L_4(\lambda^{-1}), \qquad \lambda \downarrow 0.$$

It is not hard to derive this lemma from the equality $\hat{\Theta}_i'(\lambda) = -\hat{\Theta}_{i+1}(\lambda)$, the relation (4.37), and its analogue for the case $b = \gamma$.

§5. Refinement of the upper estimate for $\Theta(t)$

It is not possible to draw a conclusion about the asymptotic behavior of $\Theta(t)$ as $t \to \infty$ directly from Lemmas 8–10. However, these lemmas enable us to refine the upper estimate for $\Theta(t)$ obtained in §3.

Let $G^0(t)$ be the distribution function having a unit atom at 0, and let $H(t) = \sum_{j=0}^\infty G^{*j}(t)$ be the renewal function.

LEMMA 11. *Suppose that the conditions of Lemma 7 hold,* $G(t) \in V$, *and for any* $\varepsilon > 0$

$$\xi_{\kappa+1} * H(t) = o(t^{\kappa+1-\gamma+\varepsilon}), \qquad t \to \infty. \tag{5.1}$$

For any $\varepsilon > 0$ *there is a constant* $c = c(\varepsilon) > 0$ *such that*

$$\Theta(t) \leqslant c(t + 1)^{-\gamma+\varepsilon}. \tag{5.2}$$

REMARK. For $\rho = \gamma$ this lemma is a particular case of Lemma 6; therefore, we assume in what follows that $\rho < \gamma$.

PROOF. By (4.4), $\Theta_{\kappa+1}(t) = \xi_{\kappa+1} * H(t) - b\mu\Theta * H(t) + (\kappa+1)\Theta_\kappa * G * H(t) + \varphi_\kappa * H(t) + \Delta_\kappa * H(t)$. Obviously,

$$\left|\Theta_{\kappa+1}(t) - \xi_{\kappa+1} * H(t) - \frac{\kappa+1}{\mu}\int_0^t \Theta_\kappa * G_1(u)\,du + b\int_0^t \Theta_\kappa(u)\,du\right|$$
$$\leqslant (\kappa+1)\int_0^t \Theta_\kappa * G_1(t-u)\left|dH(u) - \frac{du}{\mu}\right|$$
$$+\int_0^t \Theta_x(t-u)\left|dH(u) - \frac{du}{\mu}\right| + \varphi_\kappa * H(t) + |\Delta_\kappa| * H(t). \tag{5.3}$$

According to Lemma 4, $\Theta(t) \leqslant c(t+1)^{-\rho+\varepsilon}$ for any $\varepsilon > 0$. We now assume that for some $i \geqslant 1$ such that $i\rho <$ and any $\varepsilon > 0$ we have proved that

$$\Theta(t) \leqslant c(t+1)^{-i\rho+\varepsilon}. \tag{5.4}$$

Let us show that then for any $\varepsilon > 0$

$$\Theta(t) \leqslant \begin{cases} c_1(t+1)^{-(i+1)\rho+\varepsilon} & \text{if } (i+1)\rho < \gamma, \\ c_2(t+1)^{-\gamma+\varepsilon} & \text{if } (i+1)\rho \geqslant \gamma. \end{cases} \tag{5.5}$$

First of all we remark that, by (4.19), $\int_0^\infty u^{\rho+1-\varepsilon}\,dG(u) < \infty$ for any $\varepsilon > 0$. From this and the condition $G(t) \in V$ it follows (see [6], Corollary 3) that

$$\int_t^{t+1}\left|dH(u) - \frac{du}{\mu}\right| = o(t^{-\rho+\varepsilon}), \qquad t \to \infty. \tag{5.6}$$

Thus, if (5.4) is valid, then, as a consequence of inequalities (5.5),

$$\int_0^t \Theta_\kappa(t-u)\left|dH(u) - \frac{du}{\mu}\right| \leqslant c\int_0^t (t-u+1)^{\kappa-i\rho+\varepsilon}\left|dH(u) - \frac{du}{\mu}\right|$$
$$\leqslant c_1\sum_{k=0}^{[t]}(t-k+1)^{\kappa-i\rho+\varepsilon}(k+1)^{-\rho+\varepsilon}$$
$$\leqslant c_2(t+1)^{\kappa-i\rho+\varepsilon}\sum_{k=0}^{[t]/2}(k+1)^{-\rho+\varepsilon} + c_3(t+1)^{-\rho+\varepsilon}\sum_{k=0}^{[t]/2}(k+1)^{\kappa-i\rho+\varepsilon}$$
$$\leqslant c_4\left((t+1)^{\kappa-i\rho+\varepsilon} + (t+1)^{\kappa+1-(i+1)\rho+2\varepsilon}\right)$$
$$\leqslant c_5\left((t+1)^{\kappa+1-\gamma+\varepsilon} + (t+1)^{\kappa+1-(i+1)\rho+2\varepsilon}\right) \tag{5.7}$$

(to get the last two inequalities we must consider separately the cases $\rho \leqslant 1$ and $\rho > 1$). Using (5.7), we find that

$$\int_0^t dG_1(u)\int_0^{t-u}\Theta_\kappa(t-u-v)\left|dH(v) - \frac{dv}{\mu}\right|$$
$$\leqslant c_5\int_0^t\left((t-u+1)^{\kappa+1-\gamma+\varepsilon} + (t-u+1)^{\kappa+1-(i+1)\rho+2\varepsilon}\right)dG_1(u)$$
$$\leqslant c_6(t+1)^{\kappa+1}\left((t+1)^{-\gamma+\varepsilon} + (t+1)^{-(i+1)\rho+2\varepsilon}\right). \tag{5.8}$$

Consider now the function $\varphi_\kappa * H(t)$. We assume that $\kappa \geqslant 1$, since otherwise $\varphi_\kappa \equiv 0$. Obviously, for $2 \leqslant j \leqslant \kappa + 1$

$$\Theta_{\kappa+1-j} * G_j(t) \leqslant \int_0^t (t-u+1)^{\kappa-1}\Theta(t-u)(u+1)^{\kappa+1}\,dG(u).$$

Therefore,

$$\varphi_\kappa * H(t) \leqslant cS * R * H(t), \tag{5.9}$$

where $S(t) = (t+1)^{\kappa-1}\Theta(t)$ and $R(t) = \int_0^t (u+1)^{\kappa+1}dG(u)$. Clearly,

$$S * R * H(t) \leqslant \frac{1}{\mu}\int_0^t S * R(u)\,du + \int_0^t dR(u)\int_0^{t-u} S(t-u-v)\left|dH(v) - \frac{dv}{\mu}\right|. \tag{5.10}$$

From Lemma 10 and a Tauberian theorem ([7], Chapter XIII, §5) it follows that for any $\varepsilon > 0$ there exists a constant $c = c(\varepsilon) > 0$ such that

$$\int_0^t S(u)\,du \leqslant c(t+1)^\varepsilon, \qquad t \geqslant 0, \tag{5.11}$$

and from (4.19) and the Tauberian theorem it follows that

$$R(t) \leqslant c_1\left(1 + (t+1)^{\kappa-\rho+\varepsilon}\right), \qquad t \geqslant 0. \tag{5.12}$$

Combining (5.11) and (5.12), we find that

$$\int_0^t S * R(u)\,du \leqslant c_2(t+1)^\varepsilon\left(1 + (t+1)^{\kappa-\rho+\varepsilon}\right). \tag{5.13}$$

By the induction hypothesis, $S(t) \leqslant c_1(t+1)^{\kappa-1-i\rho+\varepsilon}$. From this, just as in (5.7), we get that

$$\int_0^t S(t-u)\left|dH(u) - \frac{du}{\mu}\right| \leqslant c_3\left((t+1)^{\kappa-\gamma+\varepsilon} + (t+1)^{\kappa-(i+1)\rho+2^\varepsilon}\right),$$

and this, together with (5.12), leads to the estimate

$$\begin{aligned} D &= \int_0^t dR(u)\int_0^{t-u} S(t-u-v)\left|dH(v) - \frac{dv}{\mu}\right| \\ &\leqslant c_4\left(1 + (t+1)^{\kappa-\gamma+\varepsilon} + (t+1)^{\kappa-(i+1)\rho+2\varepsilon}\right)\left(1 + (t+1)^{\kappa-\rho+\varepsilon}\right). \end{aligned}$$

It is very simple to verify that if $\varepsilon > 0$ is fixed and sufficiently small, then as $t \to \infty$

$$D = \begin{cases} O(t^\varepsilon) & \text{if } \kappa < \rho, \\ O(t^{\kappa-\rho+2\varepsilon}) & \text{if } \rho \leqslant \kappa < (i+1)\rho, \\ O(t^{2\kappa-(i+2)\rho+3\varepsilon}) & \text{if } \kappa \geqslant (i+1)\rho. \end{cases}$$

Using these estimates and recall that $\kappa - \rho < 1$ and $1 - \gamma > -\rho$, we find that

$$D \leqslant c\max\left\{(t+1)^{\kappa+1-\gamma+\varepsilon}, (t+1)^{\kappa+1-(i+1)\rho+\varepsilon}\right\}. \tag{5.14}$$

Substituting (5.13) and (5.14) into (5.10) and taking (5.9) into account, we get

$$\varphi_\kappa * H(t) \leqslant c(t+1)^{\kappa+1}\left((t+1)^{-\gamma+\varepsilon} + (t+1)^{-(i+1)\rho+\varepsilon}\right). \tag{5.15}$$

Analogous estimates show that

$$|\Delta_\kappa| * H(t) \leqslant c\varphi_\kappa * H(t) + \int_0^t |\sigma(t-u)| \Theta_\kappa * G(t-u)\, dH(u)$$

$$= \nu(t) + o\left(\int_0^t \Theta_\kappa(u)\, du\right), \qquad t \to \infty, \tag{5.16}$$

where $\nu(t) \leqslant c(t+1)^{\kappa+1}((t+1)^{-\gamma+\varepsilon} + (t+1)^{-(i+1)\rho+\varepsilon})$.

Combining (5.7), (5.8), (5.15) and (5.16), we find that

$$\left|\Theta_{\kappa+1}(t) - \xi_{\kappa+1} * H(t) + b\int_0^t \Theta_x(u)\, du - \frac{\kappa+1}{\mu}\int_0^t \Theta_\kappa * G_1(u)\, du\right|$$

$$= O(t^{\kappa+1}(t^{-(i+1)\rho+\varepsilon} + t^{-\gamma+\varepsilon})) + o\left(\int_0^t \Theta_\kappa(u)\, du\right).$$

Finally, by Lemmas 8 and 9,

$$\int_0^t \Theta_\kappa * G_1(u)\, du = \mu(1+o(1))\int_0^t \Theta_x(u)\, du, \qquad t \to \infty. \tag{5.17}$$

Therefore (cf. (4.27), (4.38) and (5.1)),

$$\Theta_{\kappa+1}(t) = O(t^{\kappa+1}(t^{-(i+1)\rho+\varepsilon} + t^{-\gamma+\varepsilon})), \qquad t \to \infty,$$

which proves (5.5) and, hence, also (5.2).

§6. Asymptotic behavior of $\Theta(t)$ as $t \to \infty$

In this section we use the results obtained in §§4 and 5 to study the behavior of $\Theta(t)$ as $t \to \infty$ under various restrictions on the functions $\xi(t)$ and $G(t)$.

THEOREM 3. *Suppose that either* $\max\{0, b-1\} < \rho \leqslant \gamma < b$ *or (if* $b > 1$*)* $\rho = \gamma = b - 1$. *If conditions* (4.17) *and* (4.19) *hold,* $G(t) \in V$, *and, moreover,*

$$\xi_{\kappa+1} * H(t) \sim \frac{1}{\mu(\kappa+1-\gamma)} t^{\kappa+1-\gamma} L_1(t), \qquad t \to \infty, \tag{6.1}$$

then

$$\Theta(t) \sim \frac{1}{\mu(b-\gamma)} t^{-\gamma} L_1(t), \qquad t \to \infty,$$

where $L_1(t)$ *is the same as in* (6.1).

PROOF. Let us turn again to (5.3). Using (5.2), as in (5.7), we find that for sufficiently small $\varepsilon > 0$

$$\int_0^t \Theta_\kappa(t-u) \left| dH(u) - \frac{du}{\mu} \right| \leqslant c_1 \sum_{k=0}^{[t]} (t-k+1)^{\kappa-\gamma+\varepsilon}(k+1)^{-\rho+\varepsilon}$$

$$\leqslant c_2((t+1)^{\kappa-\gamma+\varepsilon} + (t+1)^{\kappa+1-\gamma-\rho+2^\varepsilon})$$

$$< c_3(t+1)^{\kappa+1-\gamma-\varepsilon}.$$

Similarly,

$$\int_0^t dG_1(u) \int_0^{t-u} \Theta_\kappa(t-u-v) \left| dH(v) - \frac{dv}{\mu} \right| \leqslant c_1(t+1)^{\kappa+1-\gamma-\varepsilon}$$

and

$$\varphi_\kappa * H(t) + |\Delta_\kappa| * H(t) = o(t^{\kappa+1-\gamma-\varepsilon}) + o\left(\int_0^t \theta_\kappa(u)\,du\right), \qquad t \to \infty.$$

From these relations, Lemma 8, (5.17) and (6.1), it follows that

$$\Theta_{\kappa+1}(t) = \left[\frac{1}{\mu(\kappa+1-\gamma)} t^{\kappa+1-\gamma} L_1(t) + \frac{\kappa+1}{\mu(b-\gamma)(\kappa+1-\gamma)} t^{\kappa+1-\gamma} L_1(t) - \frac{b}{\mu(\kappa+1-\gamma)} t^{\kappa+1-\gamma} L_1(t)\right](1+o(1))$$

$$\sim \frac{1}{\mu(b-\gamma)} t^{\kappa+1-\gamma} L_1(t), \qquad t \to \infty.$$

The theorem is proved.

THEOREM 4. *Suppose that* $\max\{0, b-1\} < \rho \leqslant \gamma = b$, *and that all the remaining conditions of Theorem* 3 *hold. Then*

$$\Theta(t) \sim (\kappa + 1 - b) t^{-b} L_2(t), \qquad t \to \infty,$$

where $L_2(t)$ *is the same as in* (4.38).

PROOF. Just as in Theorem 3 (in this case $\kappa = [\gamma] = [b]$) we have

$$\Theta_{\kappa+1}(t) = \xi_{\kappa+1} * H(t) + (1+o(1))(\kappa+1-b)\int_0^t \Theta_\kappa(u)\,du, \qquad t \to \infty.$$

Thus, as $t \to \infty$

$$\Theta(t) \sim \frac{1}{\mu(\kappa+1-b)} t^{-b} L_1(t) + (\kappa+1-b) t^{-b} L_2(t). \tag{6.2}$$

Comparing (6.2) with (4.38), we see that $L_1(t) = o(L_2(t))$ as $t \to \infty$, which concludes the proof.

THEOREM 5. *If conditions* (3.1), (4.1) *and* (4.2) *hold and* $G(t) \in V$, *then*

$$\Theta(t) \sim (\beta + 1 - b) t^{-b} L_1(t), \qquad t \to \infty,$$

where $L_1(t)$ *is the same as in* (4.3).

The proof is similar to that of Theorem 3.

THEOREM 6. *Suppose that* (4.19) *holds for some* $\rho \in (\max\{0, b-1\}, b)$ *and that* $G(t) \in V$. *If* $\xi(t) = o(t^{-\rho-1+\varepsilon})$ *as* $t \to \infty$ *for any* $\varepsilon > 0$,

$$\xi_{\beta+1} * H(t) = o(t^{\beta+1-b-\delta}), \qquad \delta > 0, \tag{6.3}$$

and there exists a T *such that* $\inf_{t\in[T,T_1]} \xi(t) > 0$ *for any* $T_1 \geqslant T$, *then*

$$\Theta(t) \sim t^{-b} L_1(t), \qquad t \to \infty. \tag{6.4}$$

In large part the proof of this theorem coincides with that of Theorem 3. It is first necessary to show that for any $\varepsilon > 0$ there exist positive constants c_1 and c_2 such that

$$c_1(t+1)^{-b-\varepsilon} \leqslant \Theta(t) \leqslant c_2(t+1)^{-b+\varepsilon}, \qquad t \geqslant T_2(\varepsilon). \tag{6.5}$$

From (6.3), (6.5), the equality

$$\int_0^\infty e^{-\lambda t} \xi_{\beta+1} * H(t)\,dt = \hat{\xi}_{\beta+1}(\lambda) \hat{H}(\lambda),$$

and the relation $\hat{H}(\lambda) \sim \mu^{-1}\lambda^{-1}$ as $\lambda \downarrow 0$, it can be concluded that $\hat{\xi}_{\beta+1}(\lambda) = o(\hat{\Theta}_{\beta+1}(\lambda))$ as $\lambda \downarrow 0$. From this, as in Lemma 9, it follows that

$$\int_0^t u^{\beta+1}\Theta(u)\,du \sim t^{\beta+1-b}L_2(t), \qquad t \to \infty. \tag{6.6}$$

Similarly to the way it was done in Lemma 11, (6.6) enables us to get the estimate

$$\Theta(t) \leqslant c(t+1)^{-b+\varepsilon}, \qquad \varepsilon > 0. \tag{6.7}$$

From (6.6) and (6.7) we can get (6.4) by repeating almost word-for-word the arguments given in the proof of Theorem 3.

§7. Proofs of Theorems 1 and 2

We prove Theorem 1 by induction. Let us first study equation (2.4), which, by (2.8) and Lemma 1, can be written in the form

$$M_1(t) = f'(F(t,0))(1-G(t)) + \left(1 - \frac{b\mu - \sigma(t)}{t + b\mu}\right) M_1 * G(t), \tag{7.1}$$

where $b = 1 + \alpha^{-1}$, and $\sigma(t) \to 0$ as $t \to \infty$. We show that if the conditions of Theorem 1 hold, then the conditions of Theorem 3 hold for (7.1) with $\rho < 1 + \alpha^{-1}$, and those of Theorem 4 hold for (7.1) with $\rho = 1 + \alpha^{-1}$.

Obviously, we need only verify the relation

$$\int_0^t (t-u+b\mu)(t-u)^\kappa(1-G(t-u))\,dH(u) \sim ct^{\kappa+1-\rho}L_1(t), \tag{7.2}$$

where $\kappa = [\rho]$, and $L_1(t)$ is the same as in (1.6). Turning to (5.6), we get

$$\begin{aligned}\int_0^t (t-u+b\mu)(t-u)^\kappa(1-G(t-u))&\left|dH(u) - \frac{du}{\mu}\right| \\ &\leqslant c\sum_{k=0}^{[t]} (t-k+b\mu)(t-k+1)^{\kappa-\rho-1}L_1(t-k)(k+1)^{-\rho+\varepsilon} \\ &\leqslant c_1(t+1)^{\kappa+1-\rho-\varepsilon}, \qquad 0 < 3\varepsilon < \min\{1,\rho\}.\end{aligned}$$

From this and the equality

$$\int_0^t (u+b\mu)u^\kappa(1-G(u))\,du = \frac{1}{(\kappa+1-\rho)}t^{\kappa+1-\rho}L_1(t)(1+o(1)), \quad t \to \infty,$$

we get (7.2). Thus, (1.9) is proved for $n = 1$.

Let $n \geqslant 2$, and consider (2.9). We again want to use the theorems proved in §6, but to do this we need to know the asymptotic behavior as $t \to \infty$ of the function

$$\psi_\kappa(t) = f'(F(t,0))\sum_{r=2}^{n}\int_0^t Y_{n,r}(t-u)f^{(r)}(F(t-u,0))\,dG(u). \tag{7.3}$$

First of all we remark that if (1.7) holds, then, by (1.10) and Lemmas 1 and 2,

$$f^{(r)}(F(t,0)) \sim t^{\alpha^{-1}(r-1)-1}L_r^*(t), \qquad r \geqslant 2,\ t \to \infty, \tag{7.4}$$

and if (1.8) holds, then

$$f^{(r)}(F(t,0)) \leqslant c_r(t+1)^{r-2-\nu}, \qquad r \geqslant 3,\ t \geqslant T,\ \nu < \delta, \tag{7.5}$$

and

$$\lim_{t\to\infty} f''(F(t,0)) = f''(1) < \infty. \tag{7.6}$$

We now consider the function $Y_{n,r}(t)$ for fixed r. Let $\Delta = \rho - \alpha^{-1}$, and let n be such that $n\Delta \leqslant 1$. Suppose we have proved that $P_i(t) \sim t^{-\rho-(i-1)\Delta}L_i(t)$ as $t \to \infty$ for all $i < n$. We show that this relation is also valid for $i = n$. Indeed, if $j_1 + j_2 + \cdots + j_{n-1} = r$ and $j_1 + 2j_2 + \cdots + (n-1)j_{n-1} = n$, then, by the induction hypothesis,

$$P_1^{j_1}(t)P_2^{j_2}(t)\cdots P_{n-1}^{j_{n-1}}(t) \sim t^{-r\rho-(n-r)\Delta}L_{j_1j_2\cdots j_{n-1}}(t), \qquad t\to\infty.$$

Substituting this in (2.7) and using the fact that a sum of SV functions is again an SV function ([12], p. 18), we find that

$$Y_{n,r}(t) \sim t^{-r\rho-(n-r)\Delta}L_r(t), \qquad t\to\infty. \tag{7.7}$$

The combination (7.4)–(7.7) gives us that

$$X_n(t) = \sum_{r=2}^{n} Y_{n,r}(t) f^{(r)}(F(t,0)) \sim t^{-\rho-1-(n-1)\Delta}L_n(t), \qquad t\to\infty. \tag{7.8}$$

Let $\gamma = \rho + (n-1)\Delta$ and $\kappa = [\gamma]$. Since

$$\psi_n(t) = f'(F(t,0))\int_0^t X_n(t-u)\,dG(u),$$

we can show by using (7.8) and recalling the conditions of Theorem 1 that for any $\varepsilon > 0$

$$\lim_{t\to\infty} t^{\rho+1-\varepsilon}\psi_n(t) = 0, \tag{7.9}$$

and

$$\int_0^t (t-u)^\kappa (t-u+b\mu)\psi_n(t-u)\,dH(u) \sim c_n t^{\kappa+1-\gamma}L_n(t), \quad t\to\infty. \tag{7.10}$$

Turning to Theorem 3 if $n\Delta < 1$ and to Theorem 4 if $n\Delta = 1$, we can conclude from (7.9), (7.10) and (2.9) that

$$M_n(t) \sim t^{-\rho-(n-1)\Delta}L_n(t), \qquad t\to\infty.$$

This proves Theorem 1 for the case $n\Delta \leqslant 1$ and, hence, for the case $\rho = \alpha^{-1}$.

Let $N = \max\{n\colon n\Delta \leqslant 1\} < \infty$. We show that

$$P_{N+1}(t) \sim t^{-1-\alpha^{-1}}L_{N+1}(t), \qquad t\to\infty. \tag{7.11}$$

Just as in the case $n < N$,

$$\psi_{N+1}(t) = f'(E(t,0))\int_0^t X_{N+1}(t-u)\,dG(u),$$

where

$$X_{N+1}(t) \sim t^{-\rho-1-N\Delta}L_{N+1}(t), \qquad t\to\infty. \tag{7.12}$$

Obviously, $\psi_{N+1}(t) = o(t^{-\rho-1+\varepsilon})$ as $t\to\infty$ for $\varepsilon > 0$, and $\psi_{N+1}(t) \geqslant ct^{-1-\rho-\varepsilon-N\Delta} > 0$ for sufficiently large t.

Setting $\beta = 1 + [\alpha^{-1}]$ and $\delta = \max\{1, b\mu\}$ and using the inequality $(x+y)^a \leqslant 2^a(x^a + y^a)$, $a > 0$, we get

$$\begin{aligned}
&\int_0^t (t-u+1)^\beta (t-u+b\mu)\psi_{N+1}(t-u)\,dH(u)\\
&\qquad \leqslant \int_0^t (t-u+\delta)^{\beta+1}\psi_{N+1}(t-u)\,dH(u)\\
&\qquad \leqslant \int_0^t dH(u)\int_0^{t-u}(t-u+\delta)^{\beta+1}X_{N+1}(t-u-v)\,dG(v)\\
&\qquad \leqslant 2^{\beta+1}\int_0^t dG(v)\int_0^{t-v}(t-u-v+\delta)^{\beta+1}X_{N+1}(t-u-v)\,dH(u)\\
&\qquad\quad +2^{\beta+1}\int_0^t v^{\beta+1}\,dG(v)\int_0^{t-v}X_{N+1}(t-u-v)\,dH(u). \qquad (7.13)
\end{aligned}$$

Since $\beta - \rho - N\Delta = \beta - \alpha^{-1} - (N+1)\Delta < \beta - \alpha^{-1} - 1$, the use of (7.12) and a repetition of the arguments in the derivation of (5.7) show that

$$\begin{aligned}
&\int_0^t dG(v)\int_0^{t-v}(t-u-v+\delta)^{\beta+1}X_{N+1}(t-u-v)\,dH(u)\\
&\qquad \leqslant c\int_0^t (t-v+1)^{\beta-\alpha^{-1}-\varepsilon}\,dG(v) \leqslant c_1\left((t+1)^{\beta-\alpha^{-1}-\varepsilon} + (t+1)^{-\rho-1+\varepsilon}\right). \qquad (7.14)
\end{aligned}$$

On the other hand, by (7.12) and the fact that $G(t) \in V$,

$$\begin{aligned}
&\int_0^t u^{\beta+1}\,dG(u)\int_0^{t-u}X_{N+1}(t-u-v)\,dH(v)\\
&\qquad \leqslant c\int_0^\infty X_{N+1}(v)\,dv\int_0^t u^{\beta+1}\,dG(u). \qquad (7.15)
\end{aligned}$$

Finally, from (1.6) it follows (see [7], Chapter VIII, §9) that either

$$\int_0^\infty u^{\beta+1}\,dG(u) < \infty, \qquad (7.16)$$

or as $t \to \infty$

$$\int_0^t u^{\beta+1}\,dG(u) \sim ct^{\beta+1}(1 - G(t)) = ct^{\beta-\rho}L_1(t). \qquad (7.17)$$

Combining (7.13)–(7.17), we find that if $\rho > \alpha^{-1}$, then

$$\int_0^t (t-u+1)^\beta (t-u+b\mu)\psi_{N+1}(t-u)\,dH(u) = o\left(t^{\beta-\alpha^{-1}-\varepsilon}\right), \qquad t \to \infty.$$

Further, we note that $\lim_{t\to\infty} t^{\rho+1-\varepsilon}\psi_{N+1}(t) = 0$ for $\varepsilon > 0$ and that $\psi_{N+1}(t) \geqslant ct^{-\rho-1-\varepsilon-N\Delta}$ for sufficiently large t. This give us that all the conditions of Theorem 6 hold for equation (2.9) with $n = N+1$ and $b = 1 + \alpha^{-1}$, and that proves (7.11).

Suppose that Theorem 1 has been proved for all $n < R$, $R > N+1$. We show that it is true for $n = R$.

Indeed, if $\sum_{i=1}^{R-1} j_i = r$, $\sum_{i=1}^{R-1} ij_i = R$, $S_N = \sum_{i=1}^{N} j_i$, and $S_N^* = r - S_N$, then, by the induction hypothesis,

$$P_1^{j_1}(t)P_2^{j_2}(t)\cdots P_{R-1}^{j_{R-1}}(t) \sim t^{-\lambda}L_{j_1 j_2\cdots j_{R-1}}(t), \qquad t \to \infty,$$

where

$$\lambda = \rho S_N + \Delta \sum_{i=1}^{N} (i-1) j_i + (1+\alpha^{-1}) S_N^*$$

$$= \rho r + (1-\Delta) S_N^* + \Delta \sum_{i=1}^{N} i j_i - \Delta S_N = \rho r - r\Delta + S_N^* + \Delta \sum_{i=1}^{N} i j_i.$$

If now $S_N^* = 0$, then $\lambda = r\rho + \Delta(R-r) \geqslant r\rho + \Delta(N-r)$, but if $S_N^* > 0$, then $S_N^* \geqslant 1 \geqslant \Delta N$, and again $\lambda \geqslant \rho r + \Delta(N-r)$. Thus,

$$Y_{R,r}(t) \leqslant t^{-\rho r - \Delta(N-r)} L_r(t), \qquad t \geqslant T.$$

From this, just as in the case $n = N+1$, it can be shown that $M_R(t) \sim t^{-1-\alpha^{-1}} L_R(t)$ as $t \to \infty$. This proves Theorem 1.

PROOF OF THEOREM 2. Let $n = 1$. We consider equation (2.4) and show that all the conditions of Theorem 5 hold for it. By the conditions of Theorem 2, it suffices to verify that there exists a T such that for any $T_1 \geqslant T$

$$\inf_{t \in [T, T_1]} M_1(t) > 0. \tag{7.18}$$

If $G(t) < 1$ for any $t < \infty$, this is obvious. Now let $S_2 = \min\{t: G(t) = 1\} < \infty$, and $S_1 = \inf\{t: G(t) > 0\}$ (we assume that $G(t)$ is right-continuous). Since $G(t) \in V$, it follows that $S_1 < S_2$, and there exist points τ_1 and τ_2, $S_1 < \tau_1 < \tau_2 < S_2$, such that $G(\tau_2) - G(\tau_1) > 0$. Let ε be in the interval $(0, \min\{S_2 - \tau_2, \tau_1 - S_1\})$. Since $M_1(t) \geqslant f'(F(t,0))(1 - G(t))$ and $F(t,0) > 0$ if $G(t) > 0$, we have

$$c(\varepsilon) = \inf_{t \in [S_1+\varepsilon, S_2-\varepsilon]} M_1(t) > 0. \tag{7.19}$$

From (7.19) and (2.4) it follows that for $\varepsilon \in (0, \min\{S_2 - \tau_2, \tau_1 - S_1\})$

$$M_1(t) \geqslant f'(F(t,0)) \int_{t-S_2+\varepsilon}^{t-S_1-\varepsilon} M_1(t-u)\, dG(u)$$

$$\geqslant f'(F(t,0)) c(\varepsilon) (G(t - S_1 - \varepsilon) - G(t - S_2 + \varepsilon)).$$

Let

$$g_1 = G(S_1 + \varepsilon/2) - G(S_1 - \varepsilon/2) = G(S_1 + \varepsilon/2) > 0,$$

$$g_2 = G(\tau_2) - G(\tau_1) > 0,$$

$$g_3 = G(S_2) - G(S_2 - \varepsilon/2) = 1 - G(S_2 - \varepsilon/2) > 0,$$

and $g = \min_{1 \leqslant i \leqslant 3} g_i$. It is not hard to verify that if $t \in (2S_1 + (3/2)\varepsilon, 2S_2 - (3/2)\varepsilon)$, then

$$G(t - S_1 - \varepsilon) - G(t - S_2 + \varepsilon) \geqslant g > 0.$$

In view of this,

$$\inf_{t \in [2S_1 + (3/2)\varepsilon, 2S_2 - (3/2)\varepsilon]} M_1(t) > 0. \tag{7.20}$$

If now $2S_1 < S_2$, then it follows from (7.19) and (7.20) that

$$\inf_{t \in [S_1+\varepsilon, 2S_2 - (3/2)\varepsilon]} M_1(t) > 0, \qquad \varepsilon \in (0, (2/5)(S_2 - 2S_1)).$$

By choosing $\varepsilon > 0$ sufficiently small and repeating the arguments given above the necessary number of times, it can easily be concluded from this that (7.18) holds for $T = S_1 + \varepsilon$.

Suppose now that $2S_1 \geqslant S_2$. Then

$$\int_{t-2S_2+(3/2)s}^{t-2S_1-(3/2)\varepsilon} M_1(t-u)\,dG(u) \geqslant c\left(G\left(t-2S_1-\frac{3}{2}\varepsilon\right)-G\left(t-2S_2+\frac{3}{2}\varepsilon\right)\right) \geqslant c_1 > 0$$

if $t - 2S_1 - (3/2)\varepsilon > S_1 + \varepsilon$ and $t - 2S_2 + (3/2)\varepsilon < S_2 - \varepsilon$, i.e., if

$$t \in (3S_1 + (5/2)\varepsilon, 3S_2 - (5/2)\varepsilon).$$

Since $S_1 < S_2$, the quantity $m = \min\{k\colon kS_1 < (k-1)S_2\}$ is finite. Continuing the process described above, we can show that for sufficiently small $\varepsilon > 0$

$$\inf_{t\in[(m-1)S_1+\varepsilon,\, mS_2-\varepsilon]} M_1(t) > 0,$$

and this leads to (7.18) for $T = (m-1)S_1 + \varepsilon$, as in the case $m = 2$.

Thus, all the conditions of Theorem 5 hold for equation (2.4), and Theorem 2 is proved for the case $n = 1$.

Let us now use induction. Assuming that (1.12) has been proved for all $n < N$ and using (2.7), we get

$$Y_{N,r}(t) \leqslant t^{-(1+\alpha^{-1})r} L_r(t), \qquad r = 2, 3, \ldots, N,\ t \geqslant T.$$

Turning to Lemma 1.3) and (2.8), we find that

$$f^{(r)}(F(t,0)) \leqslant ct^{\alpha^{-1}(r-1+\delta)-1}, \qquad \delta > 0,\ r \geqslant 2,\ t \geqslant T,$$

which, together with the preceding inequality, gives us

$$Y_{N,r}(t) f^{(r)}(F(t,0)) \leqslant ct^{-\alpha^{-1}(1-\delta)-r-1} \leqslant ct^{-\alpha^{-1}-2+\delta_1}, \qquad r \geqslant 2,\ \delta_1 > 0.$$

Substituting this into (2.6) for $n = N$ and recalling (1.11), we get that

$$\psi_N(t) \leqslant ct^{-1-\alpha^{-1}-\delta}, \qquad t \geqslant T,\ 0 < \delta < 1.$$

It is not hard to derive from (2.5) and (2.6) that $\inf_{t\in[T,T_1]} M_N(t) > 0$ for all sufficiently large T and all $T_1 \geqslant T$ if $N > 1$. In view of this the equation

$$M_N(t) = \psi_N(t) + f'(F(t,0)) M_N * G(t)$$

satisfies all the conditions of Theorem 5. Consequently, $M_N(t) \sim t^{-1-\alpha^{-1}} L_N(t)$ as $t \to \infty$. The proof of Theorem 2 is complete.

§8. Concluding remarks

In this section it will be shown how to get relation (1.4) while imposing weaker restrictions on $G(t)$ than in Theorem B.

Theorem 7. *Suppose that* $f'(1) = 1$, $f(0) > 0$, $\sum_{k=1}^{\infty} k^{2+\delta} p_k < \infty$, $G(t) \in V$, *and*

$$\int_0^{\infty} u^3\,dG(u) < \infty. \tag{8.1}$$

Then there exists a sequence of positive constants c_n $(n = 1, 2, \ldots)$ *such that*

$$P_n(t) \sim c_n t^{-2}, \qquad t \to \infty. \tag{8.2}$$

PROOF. Let $n = 1$. Setting $b = 2$, we easily get from (7.1) that

$$\frac{\hat{M}_1'''(\lambda)}{\hat{M}_1''(\lambda)} = -\frac{3\hat{G}_1(\lambda) - 2\mu}{1 - \hat{G}(\lambda)} - \frac{\eta(\lambda)}{1 - \hat{G}(\lambda)} - \frac{\hat{\xi}_3(\lambda)}{\hat{M}_1''(\lambda)(1 - \hat{G}(\lambda))}, \tag{8.3}$$

where

$$\eta(\lambda) = \left(\hat{M}_1''(\lambda)\right)^{-1}\left[\int_0^\infty e^{-\lambda t} t^2 \sigma(t) m_1 * G(t)\, dt + 3\hat{M}_1'(\lambda)\hat{G}_2(\lambda) + \hat{M}_1(\lambda)\hat{G}_3(\lambda)\right], \tag{8.4}$$

and

$$\hat{\xi}_3(\lambda) = \int_0^\infty e^{-\lambda t}(1 - G(t)) t^2 (t + 2\mu) f'(F(t, 0))\, dt. \tag{8.5}$$

By (8.5),

$$\frac{\hat{M}_1'''(\lambda)}{\hat{M}_1''(\lambda)} \leqslant -\frac{3\hat{G}_1(\lambda) - 2\mu}{1 - \hat{G}(\lambda)} - \frac{\eta(\lambda)}{1 - \hat{G}(\lambda)}. \tag{8.6}$$

It is known (see [11], Proposition 1) that if the conditions of Theorem 7 hold, then $\sigma(t) = O(t^{-\delta})$ as $t \to \infty$, and this, together with (8.1), (8.4), (4.8), and Lemmas 4 and 5 with $b = 2$, yields the estimate

$$\eta(\lambda) \leqslant c\lambda^{\delta/2}. \tag{8.7}$$

Taking this fact into account and integrating (8.6) from λ to λ_0, we find that

$$\log\left(\hat{M}_1''(\lambda_0)/\hat{M}_1''(\lambda)\right) \leqslant c - \log(\lambda_0/\lambda).$$

Hence,

$$\hat{M}_1''(\lambda) \geqslant c_1\lambda^{-1}, \qquad 0 < \lambda < \lambda_0. \tag{8.8}$$

If we now integrate (8.3) and recall (8.1) and (8.7), we get

$$\log\frac{\hat{M}_1''(\lambda_0)}{\hat{M}_1''(\lambda)} = c(1 + o(1)) - \int_\lambda^{\lambda_0} \frac{\hat{\xi}_3(z)}{\hat{M}_1''(z)(1 - \hat{G}(z))}\, dz - \log\frac{\lambda_0}{\lambda}, \qquad \lambda \downarrow 0.$$

According to (8.7), (4.8) and (8.1),

$$\begin{aligned} 0 < \int_\lambda^{\lambda_0} \frac{\hat{\xi}_3(z)}{\hat{M}_1''(z)(1 - \hat{G}(z))}\, dz &\leqslant c\int_\lambda^{\lambda_0} \int_0^\infty e^{-tz} t^3 (1 - G(t))\, dt\, dz \\ &\leqslant c\int_0^\infty t^3(1 - G(t)) \int_\lambda^\infty e^{-tz}\, dz\, dt \leqslant c\int_0^\infty t^2(1 - G(t))\, dt < \infty. \end{aligned}$$

Consequently, $\hat{M}_1''(\lambda) \sim c_1\lambda^{-1}$ as $\lambda \downarrow 0$, and, hence,

$$\int_0^t u^2 M_1(u)\, du \sim c_1 t, \qquad t \to \infty.$$

From this, by repeating the arguments in the proof of Theorem 6, we conclude that $M_1(t) \sim c_1 t^{-2}$ as $t \to \infty$, which proves Theorem 7 for $n = 1$. The case $n > 1$ is handled similarly.

BIBLIOGRAPHY

1. V. A. Vatutin, *Limit theorems for a critical Bellman-Harris branching process with infinite variance*, Teor. Veroyatnost. i Primenen. **21** (1976), 861–863; English transl. in Theor. Probab. Appl. **21** (1976).

2. ______, *Discrete limit distributions of the number of particles in critical Bellman-Harris branching processes*, Teor. Veroyatnost. i Primenen. **22** (1977), 150–155; English transl. in Theor. Probab. Appl. **22** (1977).

3. ______, *A new limit theorem for the critical Bellman-Harris branching process*, Mat. Sb. **109 (151)** (1979), 440–452; English transl. in Math. USSR Sbornik **37** (1980).

4. I. S. Gradshteĭn and I. M. Ryzhik, *Tables of integrals, series and products*, 5th ed., "Nauka", Moscow, 1971; English transl. of 4th ed., Academic Press, 1965.

5. V. M. Zolotarev, *More exact statements of several theorems in the theory of branching processes*, Teor. Veroyatnost. i Primenen. **2** (1957), 256–266; English transl. in Theor. Probab. Appl. **2** (1957).

6. B. A. Rogozin, *Asymptotic analysis of the renewal function*, Teor. Veroyatnost. i Primenen. **21** (1976), 689–706; English transl. in Theor. Probab. Appl. **21** (1976).

7. William Feller, *An introduction to probability theory and its applications*, vol. II, Wiley, 1966.

8. Theodore E. Harris, *The theory of branching processes*, Springer-Verlag, Berlin, and Prentice-Hall, Englewood Cliffs, N.J., 1963.

9. Martin I. Goldstein, *Critical age-dependent branching processes: Single and multitype*, Z. Wahrsch. Verw. Gebiete **17** (1971), 74–88.

10. David L. Quigg, *On an integral equation arising in age-dependent branching processes*, J. Math. Anal. Appl. **65** (1978), 344–360.

11. ______, *A local limit theorem for the critical age-dependent branching process*, J. Appl. Probab. **15** (1978), 46–53.

12. Eugene Seneta, *Regularly varying functions*, Lecture Notes in Math., vol. 508, Springer-Verlag, 1976.

13. R. S. Slack, *A branching process with mean one and possibly infinite variance*, Z. Wahrsch. Verw. Gebiete **9** (1968), 139–145.

14. Howard J. Weiner, *Asymptotic probabilities in a critical age-dependent branching process*, J. Appl. Probab. **13** (1976), 476–485; Correction, ibid. **15** (1978), 661.

Translated by H. H. McFADEN

Труды Матем. Инст. Стеклов.
158(1981)

Proceedings of the Steklov
Institute of Mathematics
1983, Issue 4

ON DIRICHLET SERIES ASSOCIATED WITH THE DEFINING EQUATIONS AND CONTINUED FRACTIONS IN THE THEORY OF AUTOMORPHIC FUNCTIONS

UDC 517.862 + 517.544 + 517.522.6

A. B. VENKOV

ABSTRACT. The paper studies Dirichlet series related to automorphic functions for general Fuchsian groups of the first kind. New principles are formulated that lie at the basis of a nonspectral method for investigating such series. These principles include the ability to solve the word problem affirmatively in a given Fuchsian group and the ability to solve defining equations for matrix elements of a faithful representation of the group. On the basis of the indicated principles, and with the help of the theory of continued fractions, a new formula is derived for the main Dirichlet series, the automorphic scattering matrix for the family of Hecke groups.

Bibliography: 15 titles.

§1. Introduction

In the spectral theory of automorphic functions for Fuchsian groups of the first kind there appear Dirichlet series of a special type. One can prove many important properties of these Dirichlet series by general spectral methods. A part of the set of all such series is related to nonanalytic Eisenstein series. Dirichlet series appear in the coefficients of Fourier expansions of Eisenstein series with respect to parabolic subgroups of a given Fuchsian group Γ. These Dirichlet series we call *series of the type* $L(s; \Gamma; \chi)$, where χ is a finite-dimensional unitary representation of Γ which is involved in the definition of the general Eisenstein series. A particular feature of the series of the type $L(s; \Gamma; \chi)$ is that they are defined, in general, not on the set of positive integers but on the set of matrix elements of the transformations from Γ. The series $L(s; \Gamma; \chi)$ constitute the object of investigation in this paper.

Spectral theory yields the assertion that a meromorphic continuation of the series $L(s; \Gamma; \chi)$ to the whole complex plane $\mathbf{C}$ exists and a statement on certain functional equations for these series. These properties justify the definition of series of the type $L(s; \Gamma; \chi)$ and attract attention to their structure.

Series of the type $L(s; \Gamma; \chi)$ in the context of the spectral theory of automorphic functions were first defined by Maass (see [1]). Then, in his groundbreaking lectures [2] Selberg found all the properties of such series for a general Fuchsian group of the first kind and a one-dimensional representation χ that could be proved by spectral methods. Later the contributions [3]–[5] considered an even more general case.

1980 *Mathematics Subject Classification*. Primary 30B50, 30F35; Secondary 32N15.

0081-5438/84 $1.00 + $.25 per page

Because of the desire to obtain concrete arithmetic applications of the spectral theory of automorphic functions on the one hand, and because of the relative simplicity of the definitions of some particular arithmetic groups Γ (congruence groups) on the other, the series of the type $L(s; \Gamma; \chi)$ for such groups Γ have been thoroughly investigated, at least for one-dimensional representations χ. In these cases it has been possible to express some series of the type $L(s; \Gamma; \chi)$ in purely arithmetic terms independently of the spectral theory, and to see what the analytic continuation of these series by the spectral method yields in number theory. We will not discuss these applications here (the most important known application of spectral theory in number theory is connected with groups Γ acting in a product of Lobachevsky spaces (see [6] and [7])). We will only indicate some examples of calculation, by Hejhal, of basic series of the type $L(s; \Gamma; \chi)$, that is, scattering matrices for congruence groups in terms of classical special functions of analytic number theory (see [8] and §4 of this paper; χ is one-dimensional and trivial).

Arithmetic groups Γ allow a number-theoretical definition in terms of the matrix elements of their faithful representation; this definition distinguishes these groups from other Fuchsian groups of the first kind. Nonarithmetic groups, incomparably more numerous, are defined, according to the classical theory of Poincaré, Klein and Fricke, only as groups with a finite number of generators and relations. So far as we know, there are no examples of calculation of series of the type $L(s; \Gamma; \chi)$ for nonarithmetic groups Γ in the literature. The possibility of defining and studying such series independently of the spectral theory would be very important not only for new applications, many of which we cannot even imagine now, but also for development of the spectral and classical theories of automorphic functions.

In the present paper we formulate the principles which, in our opinion, form the basis for a direct method of investigating series of the type $L(s; \Gamma; \chi)$ in the general case. These are two such principles:

1) ability to solve the word problem affirmatively in a given Fuchsian group Γ; and

2) ability to solve constructively the defining equations for the matrix elements of the transformations from Γ.

Then we calculate upon these principles the basic series $L(s; \Gamma; \chi)$ (the scattering matrix, χ being one-dimensional and trivial) for a family of nonarithmetic groups of genus zero (Hecke groups Γ_q) in terms of Dirichlet series in which the sums are now taken over all rational integers. Note that our method applies in principle to calculation of the basic sums occurring in the definition of the classical analytic automorphic forms (see [9] and [10]) for Hecke groups. The algebraic part of the method is closely related to the technique of reduced continued fractions developed by Rosen (see [11] and [12]) for a solution of the word problem and for investigation of the set of cusp points of Hecke groups.

The author will continue his detailed study of the main formula of Theorem 8 below, as well as the general situation in subsequent papers.

For the reader's convenience we include in the present paper a summary of the main results on series of the type $L(s; \Gamma; \chi)$ defined in the general case of an arbitrary Fuchsian group Γ of the first kind with noncompact fundamental domain and of an arbitrary finite-dimensional unitary representation χ of Γ, namely results which are proved within the framework of spectral theory (§§2 and 3). We also include some arithmetic examples of the series of the type $L(s;\Gamma;\chi)$ (§4). The main results are contained in §§5 and 6.

§2. Definition of the Dirichlet series $L(s;\Gamma;\chi)$

Let $H=\{z=x+iy\in\mathbf{C}\,|\,y>0\}$ be the upper complex half-plane, one of the Poincaré models of Lobachevsky plane. H is a homogeneous symmetric space $H=G/K$, $G=PSL(2,\mathbf{R})$ and $K=SO(2)$. The group G acts on H by linear-fractional transformations $z\to gq$, $z\in H$, $g\in G$, leaving the Poincaré metric ds,

$$ds^2=(dx^2+dy^2)/y^2,$$

invariant.

Furthermore, let Γ be a Fuchsian group of the first kind with a noncompact fundamental domain F. Let us choose F in the form of a polygon in H bounded by a finite even number of geodesics. Then F has only a finite number of zero inner angles and, by the definition of Γ, this number is not zero. The vertices of these angles are fixed points of parabolic transformations from the group Γ and are called *parabolic vertices* (or *cusps*) of the domain F. Assume that F is chosen in such a way that all its cusps are pairwise nonequivalent under the group Γ. Recall that two points z, $z'\in H$ are called *equivalent under* Γ if there exists a transformation $\gamma\in\Gamma$ such that $z=\gamma z'$. A similar definition is given for the points of the absolute in H. Existence of a fundamental domain F with the above properties follows from the classical theory of Fuchsian groups (see [5], §1.2, where further references are given).

Denote by $z_1,\dots,z_n$ all the cusps of the domain F. For each z_j consider the maximal parabolic subgroup $\Gamma_j\subset\Gamma$ stabilizing z_j. The group Γ_j is generated by one parabolic element, which we denote by S_j. For each $j=1,\dots,n$ there exists a transformation $g_j\in G$ with the properties

$$g_j\infty=z_j,\qquad g_j^{-1}S_jg_jz=S_\infty z=z+1,\qquad z\in H.$$

Let V be an h-dimensional complex linear space, $V=\mathbf{C}^k$. Elements of V will be denoted by v. The inner product of elements v, $v'\in V$, antilinear with respect to the second argument, and the norm of an element v will be denoted by $\langle v,v'\rangle$ and $|v|_V$, respectively. Let χ be an arbitrary unitary (with respect to $\langle v,v'\rangle$) representation of the group Γ acting in V. For each $j=1,\dots,n$ define the subspace $V_j\subset V$ of the vectors v fixed under the operator $\chi(S_j)$; that is, $V_j=\{v\in V\,|\,\chi(S_j)v=v\}$. Denote by P_j the orthogonal projection of V onto V_j, and by $k_j=\dim V_j$ the dimension of the subspace V_j, $j=1,\dots,n$. Put $k=k(\Gamma;\chi)=\Sigma_1^n k_j$. The numbers k_j and k characterize the singularity of the representation χ with respect to Γ and strongly influence the spectral properties of the corresponding automorphic Laplacian (see [5], §2.1).

For every integer α, $1\leqslant\alpha\leqslant n$, and for every element $v\in V_\alpha\subset V$ we define Eisenstein series by the formula

$$E(z;s;\alpha;v;\Gamma;\chi)=E(z;s;\alpha;v)=\sum_{\gamma\in\Gamma_\alpha\backslash\Gamma}y^s(g_\alpha^{-1}\gamma z)\chi^*(\gamma)v,\tag{1}$$

$s\in\mathbf{C}$, $y(z)=\operatorname{Im}z$; here * means taking the adjoint operator, and γ runs over all cosets $\Gamma_\alpha\gamma$.

THEOREM 1 (see [5], Theorem 3.1.1). *The series* (1) *is absolutely convergent in the domain* $\operatorname{Re} s > 1$ *for every* $z \in H$ *and is an analytic function of* s *there. As a function of* $z \in H$ *the series is infinitely differentiable with respect to* x *and* y *and has the property that*

$$E(\gamma z; s; \alpha; v) = \chi(\gamma) E(z; s; \alpha; v)$$

for all $\gamma \in \Gamma$.

Let $e_l(\alpha)$ be an arbitrary element from a basis of the subspace $V_\alpha \subset V$.

THEOREM 2 (see [5], Theorem 3.1.2]). *The following Fourier expansion is valid for* $1 \leqslant \alpha \leqslant n$, $1 \leqslant l \leqslant k_\alpha$ *and* $1 \leqslant \beta \leqslant n$:

$$P_\beta E(g_\beta z; s; \alpha; e_l(\alpha)) = \sum_{j \in Z} a_j(y; s) \exp\left(2\pi\sqrt{-1}\, jx\right),$$

where the coefficients of the expansion are given by

$$a_j(y; s) = 2\pi^s |j|^{s-1/2} \Gamma(s)^{-1} \sqrt{y}\, K_{s-1/2}(2\pi |j| y) \eta_j(s), \qquad j \neq 0,$$

$$a_0(y; s) = \delta_{\alpha\beta} e_l(\alpha) y^2 + \sqrt{\pi}\, \Gamma(s - \tfrac{1}{2}) \Gamma(s)^{-1} y^{1-s} \eta_0(s),$$

$$\eta_j(s) = \sum_{\substack{\gamma \in \Gamma_\alpha \backslash \Gamma / \Gamma_\beta, \\ \text{except } \gamma = (\Gamma_\alpha E \Gamma_\beta)\delta_{\alpha\beta}}} \frac{P_\beta \chi^*(\gamma) e_l(\alpha)}{|c(g_\alpha^{-1} \gamma g_\beta)|^{2s}} \exp\left[2\pi\sqrt{-1}\, j \frac{d(g_\alpha^{-1} \gamma g_\beta)}{c(g_\alpha^{-1} \gamma g_\beta)}\right], \tag{2}$$

where by definition $\Gamma(s)$ *is the gamma function of Euler*, $K_s(z)$ *is the modified Bessel function and*

$$(\Gamma_\alpha E \Gamma_\beta)\delta_{\alpha\beta} = \begin{cases} \varnothing, & \alpha \neq \beta, \\ \Gamma_\alpha E \Gamma_\alpha, & \alpha = \beta, \end{cases}$$

$\varnothing$ *is an empty set*, E *is the identity of the group* Γ, $j \in \mathbf{Z}$; $c = c(g_\alpha^{-1} \gamma g_\beta)$ *and* $d = d(g_\alpha^{-1} \gamma g_\beta)$ *are matrix elements of the transformation* $g_\alpha^{-1} \gamma g_\beta \colon H \to H$, *and*

$$g_\alpha^{-1} \gamma g_\beta z = (az + b)/(cz + d).$$

The series (2) *is absolutely convergent in the norm of* V *if* $\operatorname{Re} s > 1$.

To formulate the next theorem we introduce the following notation. For every α, $1 \leqslant \alpha \leqslant n$,

$$V = V_\alpha \oplus V_\alpha^\perp, \qquad 1_V - P_\alpha = \sum_{k=k_\alpha+1}^{h} [e_k(\alpha) \oplus e_k(\alpha)], \tag{3}$$

whre 1_V is the identity operator in the space V, $h = \dim V$, and $e_k(\alpha)$ is an element of a basis in V compatible with the decomposition (3). Let the basis $\{e_l(\alpha)\}$ be such that the matrix of the operator $\chi(S_\alpha)$ in it is diagonal. Then

$$\chi(S_\alpha) e_l(\alpha) = \nu_{l\alpha} e_l(\alpha),$$

$$\nu_{l\alpha} = \exp\left(2\pi\sqrt{-1}\, \theta_{l\alpha}\right), \qquad 0 < \theta_{l\alpha} < 1,\ k_\alpha + 1 \leqslant l \leqslant h.$$

THEOREM 3 (see [5], Theorem 3.1.3). *The following Fourier expansion is valid for* $1 \leqslant \alpha \leqslant n$, $1 \leqslant l \leqslant k_\alpha$ *and* $1 \leqslant \beta \leqslant n$:

$$(1_V - P_\beta)E(g_\beta z; s; \alpha; e_l(\alpha)) = \sum_{k=k_\beta+1}^{h} [e_k(\beta) \otimes e_k(\beta)] E(g_\beta z; s; \alpha; e_l(\alpha))$$
$$= \sum_{k=k_\beta+1}^{h} \sum_{j\in\mathbf{Z}} b_j(y; s) \exp\left(2\pi\sqrt{-1}\,(j + \theta_{k\beta})x\right),$$

where the coefficients of the expansion are given by

$$b_j(y; s) = 2\pi^s |\theta_{k\beta} + j|^{s-1/2} \Gamma(s)^{-1} \sqrt{y}\, K_{s-1/2}(2\pi|\theta_{k\beta} + j|y)\xi_j(s), \qquad j \in \mathbf{Z},$$
$$\xi_j(S) = \sum_{\substack{\gamma\in\Gamma_\alpha\backslash\Gamma/\Gamma_\beta, \\ \text{except } \gamma=(\Gamma_\alpha E \Gamma_\beta)\delta_{\alpha\beta}}} [e_k(\beta) \otimes e_k(\beta)] \chi^*(\gamma) e_l(\alpha) \left(|c(g_\alpha^{-1}\gamma g_\beta)|^{-2s}\right)$$
$$\times \exp\left[2\pi\sqrt{-1}\,(\theta_{k\beta} + j) d(g_\alpha^{-1}\gamma g_\beta)\left(c(g_\alpha^{-1}\gamma g_\beta)^{-1}\right)\right]. \tag{4}$$

The series (4) *is absolutely convergent in the norm of* V *if* $\operatorname{Re} s > 1$ (*the notation of Theorems* 2 *and* 3 *is compatible*).

DEFINITION 1. Each series $\eta_j(s)$, $\xi_j(s)$, $j \in \mathbf{Z}$, appearing in Theorems 2 and 3 will be called a *Dirichlet series of the type* $L(s; \Gamma; \chi)$.

The spectral theory of automorphic functions allows one to prove the existence of a meromorphic continuation of the Dirichlet series of the type $L(s; \Gamma; \chi)$ to all of $\mathbf{C}$ and to answer some questions concerning the distribution of zeros and poles of these series. A summary of general results in this area will be given in §3; here we define the so-called *scattering matrix* $\mathfrak{S}(s)$, which determines, in essence, the analytic properties of series of the type $L(s; \Gamma; \chi)$.

For $1 \leqslant \alpha \leqslant n$, $1 \leqslant \beta \leqslant n$, $1 \leqslant l \leqslant k_\alpha$ and $1 \leqslant k \leqslant k_\beta$ we define $\mathfrak{S}_{\alpha l, \beta k}(s)$ by the formula

$$\sqrt{\pi}\,(s - \tfrac{1}{2})\Gamma(s)^{-1} \left\langle e_k(\beta), \overline{\eta_0(s; \alpha; l; \beta)} \right\rangle = \mathfrak{S}_{\alpha l, \beta k}(s), \tag{5}$$

where $\eta_0(s) = \eta_0(s; \alpha; l; \beta)$ is defined in Theorem 2 and the bar means complex conjugation. Instead of the four indices αl and βk we introduce two:

$$a = k_1 + \cdots + k_{\alpha-1} + l(\alpha), \qquad b = k_1 + \cdots + k_{\beta-1} + k(\beta).$$

In this notation $\mathfrak{S}(s) = \{\mathfrak{S}_{a,b}(s)\}$ is a matrix of order $k(\Gamma, \chi) = \sum_1^n k_\alpha$. Denote by $\mathfrak{f}(s) = \mathfrak{f}(s; \Gamma; \chi)$ the determinant of the matrix $\mathfrak{S}(s) = \mathfrak{S}(s; \Gamma; \chi)$.

§3. The series $L(s; \Gamma; \chi)$ and spectral theory

Here we formulate theorems describing the properties of the series $L(s; \Gamma; \chi)$ that can be proved by the methods of the spectral theory of automorphic functions. The proofs of these theorems are given in [5].

THEOREM 4 (see [5], Theorem 3.1.5). *Dirichlet series of the type* $L(s; \Gamma; \chi)$ *admit meromorphic continuations to the whole plane* $\mathbf{C}$.

Using Maass-Selberg relations, quantitative and qualitative description of the poles of $L(s; \Gamma; \chi)$ reduces to the description of the poles of the scattering matrix

$\mathfrak{S}(s)$. We do not restate these relations here, but we mention that they express the integral of the square of the absolute value of a regularized Eisenstein series over a fundamental domain in terms of matrix elements $\mathfrak{S}_{a,b}(s)$ of the scattering matrix (see [5], Theorem 3.2.1). Now we indicate the main properties of the scattering matrix and of its determinant.

THEOREM 5 (see [5], Theorems 2.3.5, 3.1.5, 3.5.1, 3.5.5 and 4.4.3).

1) $\mathfrak{S}(s)$ *satisfies the functional equations*

$$\mathfrak{S}(s) = \mathfrak{S}^*(s), \qquad \mathfrak{S}(s)\mathfrak{S}(1-s) = I_k, \qquad s \in \mathbf{C},$$

where I_k *is the identity matrix of order* $k(\Gamma; \chi)$ *and* * *means conjugate transposition of a matrix.*

2) *The function* $\mathfrak{f}(s)$ *is meromorphic on* $\mathbf{C}$ *of order at most two.*

3) *The function* $\mathfrak{f}(s)$ *is regular in the half-plane* $\operatorname{Re} s > \frac{1}{2}$ *except at finitely many poles of multiplicity at most* $k(\Gamma; \chi)$, *each pole located on the interval* $(\frac{1}{2}, 1]$.

4) *For every real number* $r \in \mathbf{R}$

$$\mathfrak{f}(\tfrac{1}{2} + ir) \neq 0.$$

5) $\mathfrak{f}(s)$ *satisfies the functional equations*

$$\mathfrak{f}(s) = \overline{\mathfrak{f}(s)}, \qquad \mathfrak{f}(s)\mathfrak{f}(1-s) = 1,$$

in particular $|\mathfrak{f}(\frac{1}{2} + ir)| = 1$, $r \in \mathbf{R}$.

6) *In the half-plane* $\operatorname{Re} s > 1$

$$\mathfrak{f}(s) = \left(\sqrt{\pi}\,\Gamma(s - \tfrac{1}{2})\Gamma(s)^{-1}\right)^{k(\Gamma;\chi)} l(s),$$

where $l(s)$ *is an absolutely convergent Dirichlet series* ($\operatorname{Re} s > 1$):

$$l(s) = \sum_{m=1}^{\infty} \frac{p_m}{q_m^{2s}}, \qquad p_m \in \mathbf{C}, p_1 \neq 0, 0 < q_1 < q_2 < \cdots.$$

7) $l(s)$ *may be expanded into the product*

$$\mathfrak{f}(s) = \mathfrak{f}_{\text{reg}}\left(\frac{1}{2}\right) q_1^{1-2s} \prod_{j=1}^{M} \frac{s - 1 + \sigma_j}{s - \sigma_j} \prod_{\rho} \frac{s - 1 + \tilde{\rho}}{s - \rho},$$

where ρ *runs over the set of all poles of* $\mathfrak{f}(s)$ *in the half-plane* $\operatorname{Re} s < \frac{1}{2}$ *and* σ_j, $j = 1, \ldots, M$, *runs over the set of all poles of* $\mathfrak{f}(s)$ *described in* 3).

Such are the main properties of the series $L(s; \Gamma; \chi)$ obtained by general methods of spectral theory. These methods, however, cannot give answers to a number of more delicate questions. For example, what is the precise order of meromorphicity of the function $\mathfrak{f}(s; \Gamma; \chi)$? (That is, is it equal to one or to two?) In order to try to answer this and similar questions, one must study the Dirichlet series defined by (2) and (4) using more direct methods. Such methods must be related to a precise description of the matrix elements of the transformations from the group Γ.

§4. Some arithmetic examples of the series $L(s; \Gamma; \chi)$

Generally, the only way to define an arbitrary Fuchsian group of the first kind is to indicate its generators and defining relations (this will be discussed in §5). However, examples of groups Γ defined in a different and a more constructive way have long been known; these were the first examples of Fuchsian groups. We are speaking about arithmetic groups Γ. In our situation Γ is an arithmetic group if and only if it is commensurable with the modular group $\Gamma(\mathbf{Z}) = PSL(2, \mathbf{Z})$. In particular, every subgroup of $\Gamma(\mathbf{Z})$ of finite index is arithmetic. The following are important examples of such groups:

$$\Gamma_0(m) = \left\{ \begin{Vmatrix} a & b \\ c & d \end{Vmatrix} \in \Gamma(\mathbf{Z}), c \equiv 0 \ (\mathrm{mod}\ m) \right\},$$

$$\Gamma_1(m) = \left\{ \begin{Vmatrix} a & b \\ c & d \end{Vmatrix} \in \Gamma_0(m), a \equiv 1, d \equiv 1 \ (\mathrm{mod}\ m) \right\},$$

$$\Gamma_2(m) = \left\{ \begin{Vmatrix} a & b \\ c & d \end{Vmatrix} \in \Gamma_1(m), b \equiv 0 \ (\mathrm{mod}\ m) \right\},$$

m any positive integer. It is important for us that in all these examples the matrix elements are given explicitly. This lets us calculate $\mathfrak{f}(s; \Gamma; \chi)$ for such groups Γ and at least their trivial one-dimensional representations χ, in terms of familiar functions. We will denote a trivial one-dimensional representation by $\chi = 1$. We give here the results of these calculations for $\Gamma = \Gamma(\mathbf{Z})$ and $\Gamma = \Gamma_0(m)$, where $m \geqslant 3$ is odd and square-free. The formulas are due to Selberg and Hejhal (see [8]). For the modular group such a formula, though in a different context, has long been known in the theory of the zeta function of a positive quadratic form. We have

$$\mathfrak{f}(s; \Gamma(\mathbf{Z}); 1) = \sqrt{\pi}\, \frac{\Gamma(s - \frac{1}{2})\zeta(2s - 1)}{\Gamma(S)\zeta(2s)},$$

$$\mathfrak{f}(s; \Gamma_0(m); 1) = \left[\sqrt{\pi}\, \frac{\Gamma(s - \frac{1}{2})\zeta(2s - 1)}{\Gamma(s)\zeta(2s)} \right]^{n(m)} \Pi(s; m),$$

where

$$\Pi(s; m) = \det\{a_{p_1}(s) \otimes \cdots \otimes a_{p_r}(s)\},$$

$$a_p(s) = \frac{1}{p^{2s} - 1} \begin{Vmatrix} p - 1 & p^s - p^{1-s} \\ p^s - p^{1-s} & p - 1 \end{Vmatrix},$$

$\zeta(s)$ is the Riemann zeta function, $m = p_1 \cdots p_r$ is the factorization of the number m into primes, and $n(m) = 2^r$. Similar but considerably more complex formulas are known for the groups $\Gamma_1(m)$ and $\Gamma_2(m)$ (see [8]).

Note that in all these examples $\mathfrak{f}(s; \Gamma; 1)$ is meromorphic of order one; this is very important for spectral theory.

Now we leave arithmetic groups, which are in a sense exceptional, and proceed to a consideration of the general case.

§5. The series $L(s; \Gamma; \chi)$ and defining equations

The problem considered here is how to describe the set of the matrix elements of the transformations from an arbitrary Fuchsian group Γ of the first kind.

Recall that the series $L(s; \Gamma; \chi)$ can be written in terms of these elements. Indeed, considering for simplicity only the case of the trivial one-dimensional representation χ, we note that an arbitrary series $L(s; \Gamma; 1)$ can be represented in the form

$$\sum_{\gamma} f\left(c\left(g_\alpha^{-1}\gamma g_\beta\right); d\left(g_\alpha^{-1}\gamma g_\beta\right)\right)\left(\left| c\left(g_\alpha^{-1}\gamma g_\beta\right)\right|^{-2s}\right), \tag{6}$$

where the summation is over a set of representatives of the double cosets $\Gamma_\alpha \gamma \Gamma_\beta$, f is a coefficient of the series, and

$$g_\alpha^{-1}\gamma g_\beta z = (az+b)/(cz+d), \qquad c = c\big(g_\alpha^{-1}\gamma g_\beta\big), d = d\big(g_\alpha^{-1}\gamma g_\beta\big).$$

Having determined the matrix elements of the transformations g_α, $1 \leqslant \alpha \leqslant n$, and of all transformations $\gamma \in \Gamma$, we can, in principle, calculate the series (6); that is, express it in terms of some familiar special functions.

We state now a classical theorem characterizing an arbitrary Fuchsian group Γ of the first kind.

THEOREM 6 (FRICKE). *Every Fuchsian group Γ of the first kind has a finite set of generators $A_1, B_1, \ldots, A_g, B_g, S_1, \ldots, S_r, R_1, \ldots, R_l$ with a finite system of defining relations*

$$[A_1, B_1] \cdots [A_g, B_g] S_1 \cdots S_r R_1 \cdots R_l = E, \qquad (7)$$
$$R_1^{m_1} = E, \ldots, R_l^{m_l} = E,$$

where $[\ ,\ ]$ denotes the commutator of elements and $m_j \in \mathbf{Z}$, $m_j \geqslant 1$, $j = 1, \ldots, l$. The elements $R_1, \ldots, R_l$ are elliptic; $S_1, \ldots, S_r$ parabolic; $A_1, B_1, \ldots, A_g, B_g$ hyperbolic; and E is the identity of Γ.

By definition, the group Γ from Theorem 6 has signature $\Gamma = \Gamma(g; m_1, \ldots, m_l; r)$, where g is the genus of the group Γ.

In the remaining part of the section we use terminology of combinatorial group theory (see [13]).

The set of all elements of Γ as a group given by generators and defining relations is in a natural correspondence with the set of classes of equivalent words composed from the letters that are generators of Γ and their inverses. Denote by $\{\omega\}$ the class of equivalent words with a representative ω. We index the generators of Γ in the order in which they appear in (7):

$$\gamma(1) = A_1, \gamma(2) = B_1, \ldots, \qquad \gamma(2g-1) = A_g, \gamma(2g) = B_g, \gamma(2g+1) = S_1,$$
$$\ldots, \gamma(2g+r) = S_r, \gamma(2g+r+1) = R_1, \ldots, \qquad \gamma(2g+r+l) = R_l,$$

and add to these their inverses:

$$\gamma(2g+r+l+1) = A_1^{-1}, \gamma(2g+r+l+2) = B_1^{-1}, \ldots, \gamma(4g+2r+2l) = R_l^{-1}.$$

An arbitrary element $\omega \in \Gamma$ is represented by a word

$$\omega = \gamma(j_1) \cdots \gamma(j_m), \qquad (8)$$

where $1 \leqslant j_k \leqslant 4g+2r+2l$, $1 \leqslant k \leqslant m$. Denote by

$$a(j_1, \ldots, j_m; m) = a_m, \qquad b(j_1, \ldots, j_m; m) = b_m,$$
$$c(j_1, \ldots, j_m; m) = c_m, \qquad d(j_1, \ldots, j_m; m) = d_m$$

the matrix elements of the transformation ω from (8),

$$\omega z = (a_m z + b_m)/(c_m z + d_m).$$

Introducing similar notation for the matrix elements of the transformation $\omega\gamma(j_{m+1})$ with some j_{m+1}, $1 \leqslant j_{m+1} \leqslant 4g + 2r + 2l$, we obtain the following recursion relations which we will call the defining equations for the matrix elements of the transformations from the group Γ:

$$\begin{Vmatrix} a(j_1,\ldots,j_{m+1}; m+1), & b(j_1,\ldots,j_{m+1}; m+1) \\ c(j_1,\ldots,j_{m+1}; m+1), & d(j_1,\ldots,j_{m+1}; m+1) \end{Vmatrix}$$

$$= \begin{Vmatrix} a(j_1,\ldots,j_m; m), & b(j_1,\ldots,j_m; m) \\ c(j_1,\ldots,j_m; m), & d(j_1,\ldots,j_m; m) \end{Vmatrix} \begin{Vmatrix} a(j_{m+1}; 1), & b(j_{m+1}; 1) \\ c(j_{m+1}; 1), & d(j_{m+1}; 1) \end{Vmatrix}, \quad (9)$$

where on the right side of (9) we take the usual product of matrices and all the matrices are understood as elements of $PSL(2, \mathbf{R})$. Besides,

$$\gamma(j_k)z = \frac{a(j_k; 1)z + b(j_k; 1)}{c(j_k; 1)z + d(j_k; 1)}$$

for every j_k, $1 \leqslant j_k \leqslant 4g + 2r + 2l$.

Return now to the series (6). We introduce the following notation:

$$c_{\alpha\beta}(j_1,\ldots,j_m) = c\left(g_\alpha^{-1}\omega g_\beta\right), \qquad d_{\alpha\beta}(j_1,\ldots,j_m) = d\left(g_\alpha^{-1}\omega g_\beta\right),$$

where ω is from (8). Let Ω_1 mean the set of representatives, one from each equivalence class $\{\omega\}$; let Ω_2 mean the set of representatives, one from each double coset $\Gamma_\alpha\omega\Gamma_\beta$. Introduce the characteristic functions $\Phi_k(j_1,\ldots,j_m; m)$ of the sets Ω_k, $k = 1, 2$:

$$\Phi_k(j_1,\ldots,j_m; m) = \begin{cases} 1, & \omega \in \Omega_k, \\ 0 & \text{otherwise}, \end{cases}$$

where $\omega = \gamma(j_1)\cdots\gamma(j_m)$ is an arbitrary element $\omega \in \Gamma$ written in the form (8). Thus, in this notation we have

$$\begin{aligned} L(s; \Gamma; \chi) = \sum_{m=1}^{\infty} \sum_{j_1=1}^{4g+2r+2l} \cdots \sum_{j_m=1}^{4g+2r+2l} & f(c_{\alpha\beta}(j_1,\ldots,j_m); d_{\alpha\beta}(j_1,\ldots,j_m)) \\ & \times \Phi_1(j_1,\ldots,j_m; m)\Phi_2(j_1,\ldots,j_m; m) \\ & \times \left(|c_{\alpha\beta}(j_1,\ldots,j_m)|^{-2s}\right); \qquad \chi = 1. \end{aligned} \quad (10)$$

We have proved the following theorem:

THEOREM 7. *Every Dirichlet series of the type $L(s; \Gamma; \chi)$ defined for any Fuchsian group Γ of the first kind and its trivial one-dimensional representation χ can be represented in the domain* $\operatorname{Re} s > 1$ *by the absolutely convergent Dirichlet series* (10).

Thus, Theorem 7 shows that to study analytic properties of the series of the type $L(s; \Gamma; \chi)$ by a direct method (that is, by a method not based on spectral theory) it is necessary, first, to be able to solve constructively the defining equations (9) and, second, to be able to describe constructively the characteristic functions Φ_1 and Φ_2 from (10). Note here that the description of Φ_1 is closely related to the first Dehn problem (the word problem) of combinatorial group theory; it is natural, therefore, to consider first those examples of nonarithmetic groups Γ for which it is known that the word problem has a simple positive solution.

§6. Derivation of the main formula for the scattering matrix $\mathfrak{S}(s)$ for Hecke groups

By definition, the Hecke group Γ_q is isomorphic to a free product of two cyclic groups:

$$\Gamma_q = \{T, U; T^2 = E, U^q = E\},$$

where $q \in \mathbf{Z}$, $q \geqslant 3$ (see [14]). As a discrete group of linear-fractional transformations of H, Γ_q is determined by the following matrices:

$$T = \begin{Vmatrix} 0 & 1 \\ -1 & 0 \end{Vmatrix}, \qquad S = \begin{Vmatrix} 1 & a \\ 0 & 1 \end{Vmatrix} \quad (\text{mod} \pm 1),$$

where $\alpha = 2\cos(\pi/q)$. The matrix $U = TS$ has the property $U^q = E$. In the canonical notation of the theory of Fuchsian groups, $\Gamma_q = \Gamma(0; 2; q; 1)$. It is known that Γ_q is arithmetic only in the following cases: $q = 3, 4, 6, \infty$; besides, $\Gamma_3 = \Gamma(\mathbf{Z})$. Below we assume that q is fixed and $4 \leqslant q < \infty$.

From Γ_q we proceed to a similar group, $\tilde{\Gamma}_q$:

$$\tilde{\Gamma}_q = g_\lambda \Gamma_q g_\lambda^{-1}, \qquad g_\lambda = \begin{Vmatrix} \lambda & 0 \\ 0 & \lambda^{-1} \end{Vmatrix} \quad (\text{mod} \pm 1),$$

where $\lambda = \alpha^{-1/2}$, $\alpha = 2\cos(\pi/q)$. Now we introduce the notation $\tilde{T} = g_\lambda T g_\lambda^{-1}$, $\tilde{S} = g_\lambda S g_\lambda^{-1}$, and $\tilde{U} = g_\lambda U g_\lambda^{-1}$, where

$$\tilde{T} = \begin{Vmatrix} 0 & \alpha^{-1} \\ -\alpha & 0 \end{Vmatrix}, \qquad \tilde{S} = \begin{Vmatrix} 1 & 1 \\ 0 & 1 \end{Vmatrix} \quad (\text{mod} \pm 1).$$

To simplify the notation we put $\Gamma = \tilde{\Gamma}_q$, $\gamma = \tilde{T}$ and $\mu = \tilde{S}$. In addition, we denote by $\Gamma_\infty \subset \Gamma$ the infinite cyclic subgroup generated by the element μ.

The canonical fundamental domain of the group Γ in H has only one cusp; therefore, Theorem 2 and (5) imply the following expression for the scattering matrix (or for its determinant, which is the same in this case):

$$\mathfrak{S}(s) = \sqrt{\pi}\, \frac{\Gamma(s - \frac{1}{2})}{\Gamma(s)} \sum_{\substack{g \in \Gamma_\infty \backslash \Gamma / \Gamma_\infty, \\ \text{except } g = \Gamma_\infty E \Gamma_\infty}} \frac{1}{|c(g)|^{2s}}, \tag{11}$$

where $c = c(g)$, $gz = (az + b)(cz + d)^{-1}$, the representation χ from Theorem 2 is assumed trivial and one-dimensional, and $\operatorname{Re} s > 1$. To transform (11) we parametrize the set $\Gamma_\infty \backslash \Gamma / \Gamma_\infty$ and the matrix elements $c(g)$ on the basis of the definition of Γ and according to the method of §5. It is known (see [13], §2.6 or [11], §3) that for any word in Γ there exists a reduction process that, generally speaking, shortens it and in finitely many steps reduces this word to minimal length and special (reduced) form. Assume that a word is written in the letters γ, μ and μ^{-1}. The reduced form of the word does not contain powers γ^k, $k \geqslant 2$. The exponents of the products $(\gamma\mu)$ (or $(\mu^{-1}\gamma)$) contained in the word, taking adjacent factors into account, do not exceed l, where $q = 2l + 1$ or $q = 2l$ (for more details see [11], §3). A reduced word of minimal length is called *irreducible*. Every class of equivalent words contains only one irreducible representative. Assume now that the representatives ω of the equivalence classes $\{\omega\}$ are chosen irreducible.

To transform (11), the following words in Γ will be of interest:

$$g = g_m(n_1,\dots,n_m) = \gamma\mu^{n_1}\cdots\gamma\mu^{n_m}\gamma. \tag{12}$$

The set of all such words g with $n_j \in \mathbf{Z}$, $j = 1,\dots,m$, $m \geqslant 1$ fixed, we denote by $\Pi(m)$. Let Ω be the set of all classes of equivalent words in Γ. Denote by Φ the characteristic function in $\Pi(m)$ of the intersection $\Omega_1(m) = \Pi(m) \cap \Omega$, where Ω, as above, denotes the set of irreducible representatives ω from the classes $\{\omega\}$ (one from each class). In other words,

$$\Phi(g) = \Phi(n_1,\dots,n_m;\, m) = \begin{cases} 1, & g \in \Omega_1(m), \\ 0 & \text{otherwise}. \end{cases} \tag{13}$$

Now (11) can be rewritten in the following form:

$$\mathfrak{S}(s) = \sqrt{\pi}\,\frac{\Gamma(s-\frac{1}{2})}{\Gamma(s)}\left\{\alpha^{-2s} + \sum_{m=1}^{\infty}\sum_{n_1=-\infty}^{\infty}\cdots\sum_{n_m=-\infty}^{\infty}\frac{\Phi(n_1,\dots,n_m;\, m)}{|c_m(n_1,\dots,n_m)|^{2s}}\right\}, \tag{14}$$

where $c = c(g) = c_m(n_1,\dots,n_m)$. The term α^{-2s} in the braces in (14) is due to the element $g = g_0 = \gamma$, which is not contained in the intersection $\Omega_1(m)$ for any $m \geqslant 1$. Indeed, this set contains irreducible elements g_m; therefore, $n_j \neq 0$ for every j, $1 \leqslant j \leqslant m$.

Let us find the recursion relation for the element c_m; that is, its defining equation. Let

$$g_m = \begin{Vmatrix} a_m & b_m \\ c_m & d_m \end{Vmatrix} \quad (\text{mod} \pm 1).$$

At the $(m+1)$th step we have

$$g_{m+1} = \begin{Vmatrix} a_{m+1} & b_{m+1} \\ d_{m+1} & d_{m+1} \end{Vmatrix} = \begin{Vmatrix} a_m & b_m \\ c_m & d_m \end{Vmatrix}\begin{Vmatrix} 1 & n_{m+1} \\ 0 & 1 \end{Vmatrix}\begin{Vmatrix} 0 & \alpha^{-1} \\ -\alpha & 0 \end{Vmatrix} \quad (\text{mod} \pm 1),$$

whence

$$c_{m+1} = -\alpha d_m - \alpha c_m n_{m+1}, \qquad d_{m+1} = \alpha^{-1} c_m,$$

which finally leads to the defining equation

$$c_{m+1} = -\alpha n_{m+1} c_m - c_{m-1} \tag{15}$$

with the initial conditions $c_0 = -\alpha$ and $c_1 = \alpha^2 n_1$.

We note the following. It is well known that c_m in (15) coincides in its domain of definition with the numerator of the mth convergent for a certain continued fraction, namely, for

$$P_m/Q_m = [e_0, f_1/e_1,\dots,f_m/e_m] = e_0 + \cfrac{f_1}{e_1 + \cfrac{f_2}{e_2 + \cdots + f_m/e_m}},$$

where $c_{m+1} = P_m$, $-\alpha n_{m+1} = e_m$ and $-1 = f_m$, $m = 1, 2, 3, \dots$.

The meaning of the problem demands that the solution c_m of (15) be chosen different from zero. We put

$$h_m = c_m/c_{m-1}.$$

The following relations are valid:

$$h_{m+1} = -\alpha n_{m+1} - h_m^{-1}, \qquad m = 1, 2, 3, \ldots, \tag{16}$$

with the initial condition $h_1 = -\alpha n_1$, and

$$c_m = h_m \cdots h_1 c_0. \tag{17}$$

But (16) implies that $h_m = h_m(n_1, \ldots, n_m)$ is represented by the finite continued fraction

$$h_m = [-\alpha n_m, -1/-\alpha n_{m-1}, \ldots, -1/-\alpha n_1]. \tag{18}$$

Now we rewrite (14) in terms of the h_m. From (15) and (17) we have

$$\mathfrak{S}(s) = \sqrt{\pi}\,\frac{\Gamma(s-\frac{1}{2})}{\Gamma(s)a^{2s}}\left\{1 + \sum_{m=1}^{\infty}\sum_{k=1}^{\infty}\sum_{n_k=-\infty}^{\infty} \frac{\Phi(n_1, \ldots, n_m; m)}{|h_m \cdots h_1|^{2s}}\right\}. \tag{19}$$

In subsequent transformations of (19) we assume that s is real and $s > 1$. The following are some auxiliary observations for investigation of (19). Let $M \subset \mathbf{Z}$ and $0 \notin M$. Consider the series

$$\sum_{n_m \in M} |\alpha n_m + h_{m-1}^{-1}|^{-2s}. \tag{20}$$

Consider the alternative: $|n_m| > \alpha^{-1}|h_{m-1}|^{-1}$, $|n_m| < \alpha^{-1}|h_{m-1}|^{-1}$, or $|n_m| = \alpha^{-1}|h_{m-1}|^{-1}$ (the last is impossible, since $c_m \neq 0$). Using the binomial expansion, we obtain

$$\begin{aligned}
\sum_{n_m \in M} |\alpha n_m + h_{m-1}^{-1}|^{-2s} &= \sum_{\substack{n_m \in M \\ |n_m| > \alpha^{-1}|h_{m-1}|^{-1}}} (\alpha n_m)^{-2s} \sum_{k=0}^{\infty} \binom{-2s}{k} (h_{m-1}\alpha n_m)^{-k} \\
&\quad + \sum_{\substack{n_m \in M \\ |n_m| < \alpha^{-1}|h_{m-1}|^{-1}}} h_{m-1}^{2s} \sum_{k=0}^{\infty} \binom{-2s}{k} (\alpha n_m h_{m-1})^{k} \\
&= \sum_{k=0}^{\infty} \binom{-2s}{k} \alpha^{-2s-k} h_{m-1}^{-k} \sum_{\substack{n_m \in M \\ |n_m| > \alpha^{-1}|h_{m-1}|^{-1}}} n_m^{-2s-k} \\
&\quad + \sum_{k=0}^{\infty} \binom{-2s}{k} \alpha^{k} h_{m-1}^{2s+k} \sum_{\substack{n_m \in M \\ |n_m| < \alpha^{-1}|h_{m-1}|^{-1}}} n_m^{k},
\end{aligned} \tag{21}$$

where, as usual,

$$\binom{a}{k} = \frac{a(a-1)\cdots(a-k+1)}{k!} = \frac{\Gamma(-a+k)(-1)^k}{\Gamma(-a)\Gamma(k+1)}.$$

We now study the set of solutions of the system of inequalities

$$|n_{m-k}| < \alpha^{-1}|h_{m-1-k}(n_1, \ldots, n_{m-1-k})|^{-1}, \tag{22}$$

where $0 \leqslant k \leqslant m-2$, $m \geqslant 3$. We are seeking solutions on the set of integer vectors $(n_1, \ldots, n_{m-k}, \ldots, n_m)$ such that

$$g_m = \gamma\mu^{n_1} \cdots \gamma\mu^{n_m}\gamma \in \Omega_1(m). \tag{23}$$

It suffices to study (22) for $k = 0$. For this, in turn, it suffices to answer the question of the minimum of $|h_m(n_1,\dots,n_m)|$ on the set g_m defined by (23). To the irreducible word g_m from (23) we associate the inverse word

$$g_m^{-1} = \gamma\mu^{n_m}\cdots\gamma\mu^{n_1}\gamma.$$

Obviously, the word g_m^{-1} has minimal length but is reduced with respect to the product $(\mu^{-1}\gamma)$. The minimization problem for absolute values of the fractions (18) on the words of minimal length and reduced with respect to the product $(\mu^{-1}\gamma)$ was considered in Lemma 2 of [11]. The result of that lemma is the following:

$$\min|h_m| \geqslant 2/\alpha. \tag{24}$$

This implies that none of the inequalities (22) has solutions in the domain which we are considering. This implies that for the set M, which we will presently describe, equality (21) becomes

$$\sum_{n_m}|\alpha n_m + h_{m-1}^{-1}(n_1,\dots,n_{m-1})|^{-2s}$$
$$= \sum_{k_1=0}^{\infty}\frac{\Gamma(2s+k_1)(-1)^{k_1}}{\Gamma(2s)\Gamma(k_1+1)}\alpha^{-2s-k_1}h_{m-1}^{-k}(n_1,\dots,n_{m-1})\sum_{n_m}n_m^{-2s-k_1}, \tag{25}$$

where $n_1,\dots,n_{m-1}$ are fixed and $g_m = g_m(n_1,\dots,n_m)$ is from (23).

We return to (19). To transform the Dirichlet series on the right side of this formula, we first obtain, by analogy with (25),

$$\sum_{n_{m-1}} h_{m-1}^{-2s-k_1} = \sum_{k_2=0}^{\infty}\frac{\Gamma(2s+k_1+k_2)(-1)^{k_1+k_2}h_{m-2}^{-k_2}}{\Gamma(2s+k_1)\Gamma(k_2+1)\alpha^{2s+k_1+k_2}}\sum_{n_{m-1}} n_{m-1}^{-2s-k_1-k_2}.$$

Continuing by induction, at the $(m-2)$th and $(m-1)$th steps we obtain

$$\sum_{n_2} h_2^{-2s-k_{m-2}} = \sum_{k_{m-1}=0}^{\infty}\frac{\Gamma(2s+k_{m-2}+k_{m-1})(-1)^{k_{m-1}+k_{m-2}}}{\Gamma(2s+k_{m-2})\Gamma(k_{m-1}+1)}$$
$$\times\alpha^{-2s-k_{m-1}-k_{m-1}}h_1^{-k_{m-1}}\sum_{n_2} n_2^{-2s-k_{m-1}-k_{m-1}};$$

$$\sum_{n_1} h_1^{-2s-k_{m-1}} = \sum_{n_1}(-\alpha n_1)^{-2s-k_{m-1}} = (-1)^{k_{m-1}}\alpha^{-2s-k_{m-1}}\sum_{n_1} n_1^{-2s-k_{m-1}}.$$

This yields the following desired expression for the Dirichlet series:

$$\sum_{m=1}^{\infty}\sum_{k=1}^{m}\sum_{n_k=-\infty}^{\infty}\Phi(n_1,\dots,n_m;m)\big(|h_m\cdots h_1|^{-2s}\big)$$
$$= \sum_{m=1}^{\infty}\sum_{k_1=0}^{\infty}\cdots\sum_{k_{m-1}=0}^{\infty}\alpha^{-2sm-2k_1-\cdots-2k_{m-1}}$$
$$\times\frac{\Gamma(2s+k_1)\Gamma(2s+k_1+k_2)\Gamma(2s+k_2+k_3)\cdots\Gamma(2s+k_{m-2}+k_{m-1})}{\Gamma(2s)\Gamma(k_1+1)\Gamma(2s+k_1)\Gamma(k_2+1)\Gamma(2s+k_2)\Gamma(k_3+1)\cdots\Gamma(2s+k_{m-2})\Gamma(k_{m-1}+1)}$$
$$\times\sum_{n_1=-\infty}^{\infty}\cdots\sum_{n_m=-\infty}^{\infty}\frac{\Phi(n_1,\dots,n_m;m)}{n_m^{2s+k_1}n_{m-1}^{2s+k_1+k_2}\cdots n_2^{2s+k_{m-2}+k_{m-1}}n_1^{2s+k_{m-1}}}, \tag{26}$$

where we assume $k_0 = 0$ identically.

We introduce the notation

$$\zeta(2s; k_1,\dots,k_{m-1}; m) = \sum_{n_1=-\infty}^{\infty} \cdots \sum_{n_m=-\infty}^{\infty} n_m^{-2s-k_1}$$

$$\times n_{m-1}^{-2s-k_1-k_2} \cdots n_2^{-2s-k_{m-2}-k_{m-1}} n_1^{-2s-k_{m-1}} \Phi(n_1,\dots,n_m; m). \quad (27)$$

Finally, from (26) and (27) we obtain the desired expression for the scattering matrix:

$$\mathfrak{S}(s) = \sqrt{\pi}\frac{\Gamma(s-\frac{1}{2})}{\Gamma(s)\alpha^{2s}}\left\{1 + \sum_{m=1}^{\infty}\sum_{l=1}^{m-1}\sum_{k_l=0}^{\infty}\zeta(2s; k_1,\dots,k_{m-1}; m)\right.$$

$$\left.\times\alpha^{-2ms-2k_1-\cdots-2k_{m-1}}\prod_{j=1}^{m-1}\frac{\Gamma(2s+k_{j-1}+k_j)}{\Gamma(2s+k_{j-1})\Gamma(k_j+1)}\right\} \quad (28)$$

($k_0 = 0$ identically).

We have deduced (28) under the assumption $s > 1$; it is possible to show, however, that the right side of this formula is well defined for other values of s, too. This requires a detailed study of the zeta function $\zeta(s; k_1,\dots,k_{m-1}; m)$, which will be the subject of another paper.

We formulate the result of this section in the form of a theorem.

THEOREM 8. *The scattering matrix $\mathfrak{S}(s)$ defined for a group Γ conjugate to the Hecke group Γ_q, $q \geqslant 4$, satisfies* (28).

BIBLIOGRAPHY

1. Hans Maass, *Über eine neue Art von nichtanalytischen automorphen Funktionen und die Bestimmung Dirischletscher Reihen durch Funktionalgleichungen*, Math. Ann. **121** (1949), 141–183.

2. Atle Selberg, *Harmonic analysis*. Vol. 2, Lecture notes, Göttingen, 1954.

3. Walter Roelcke, *Das Eigenwertproblem der automorphen Formen in der hyperbolischen Ebene*. I, II, Math. Ann. **167** (1966), 292–337; ibid. **168** (1967), 261–324.

4. Robert P. Langlands, *On the functional equations satisfied by Eisenstein series*, Lecture Notes in Math., vol. 544, Springer-Verlag, 1976.

5. A. B. Venkov, *Spectral theory of automorphic functions*, Trudy Mat. Inst. Steklov. **153** (1981); English transl., Proc. Steklov Inst. Math. **1982**, no. 4 (153).

6. Tomio Kubota, *Some results concerning reciprocity and functional analysis*, Proc. Internat. Congr. Math. (Nice, 1970), vol. 1, Gauthier-Villars, Paris, 1971, pp. 395–399.

7. D. R. Heath-Brown and S. J. Patterson, *The distribution of Kummer sums at prime arguments*, J. Reine Angew. Math. **310** (1979), 111–130.

8. Dennis A. Hejhal, *The Selberg trace formula and the Riemann zeta function*, Duke Math. J. **43** (1976), 441–482.

9. H. Poincaré, *Mémoire sur les fonctions fuchsiennes*, Acta Math. **1** (1882), 193–294.

10. Hans Petersson, *Konstruktion der Modulformen und der zu gewissen Grenzkreisgruppen gehörigen automorphen Formen von positiver reeller Dimension und die vollständige Bestimmung ihrer Fourierkoeffizienten*, S.-B. Heidelberger Akad. Wiss. Math.-Nat. Kl. **1950**, 417–494.

11. David Rosen, *A class of continued fractions associated with certain properly discontinuous groups*, Duke Math. J. **21** (1954), 549–563.

12. ______, *An arithmetic characterization of the parabolic points of* $G(2\cos\pi/5)$, Proc. Glasgow Math. Assoc. **6** (1963), 88–96.

13. Kurt Reidemeister, *Einführung in die kombinatorische Topologie*, Vieweg, Braunschweig, 1932; reprint, Chelsea, New York, 1950.

14. E. Hecke, *Über die Bestimmung Dirichletscher Reihen durch ihre Funktionalgleichung*, Math. Ann. **112** (1935/36), 664–699.

15. O. Perron, *Die Lehre von den Kettinbrüchen*, 2nd ed., Teubner, Leipzig, 1929; reprint, Chelsea, New York, 1950.

Translated by M. E. BENJAMIN

Труды Матем. Инст. Стеклов.
158(1981)

Proceedings of the Steklov
Institute of Mathematics
1983, Issue 4

ANALOGUES OF THE VINOGRADOV-GAUSS FORMULA IN THE CRITICAL STRIP

UDC 511.33 + 517.862

A. I. VINOGRADOV AND L. A. TAKHTADZHYAN

ABSTRACT. In this paper an analytic generalization of the classical Vinogradov-Gauss formula for the mean value of the class number of purely radical forms of negative determinant is given. A formula with two asymptotic terms for the mean value of the zeta-functions of orders of imaginary quadratic fields in the critical strip is proved.

Bibliography: 18 titles.

Introduction

In his famous *Disquisitiones Arithmeticae* [1] Gauss gave without proof an asymptotic formula for the mean value of the class number of purely radical quadratic forms of negative determinant. In a series of papers [2]–[5], the first of which appeared in 1917 and the last in 1963, I. M. Vinogradov proved, with increasing precision, a formula suggested by Gauss:

$$\sum_{D=1}^{X} H(-D) = \frac{4\pi}{21\zeta(3)} X^{3/2} - \frac{2}{\pi^2} X + O(X^{2/3}\ln^6 X), \tag{1}$$

where $H(-D)$ is the class number of purely radical quadratic forms of negative determinant $-D$ and

$$\zeta(3) = \sum_{n=1}^{\infty} \frac{1}{n^3}.$$

We call (1) the *Vinogradov-Gauss formula.*

In this paper we show that it is possible, using rather simple ideas from the spectral theory of automorphic functions, to give an analytic generalization of the Vinogradov-Gauss formula. The basis of this generalization is the classical formula of Dirichlet:

$$2\pi h(-D) = w_D \lim_{s\to 1} \left(\sqrt{D}\right)^s \zeta_D(s)(s-1), \tag{2}$$

where $h(-D)$ is the class number of primitive quadratic forms of discriminant $-D<0$, $\zeta_D(s)$ is the zeta-function of the order $\mathfrak{O}_D$ of the discriminant $-D$ in the imaginary quadratic field $\mathbf{Q}(\sqrt{-D})$, and w_D is the number of units of the order $\mathfrak{O}_D$. Formula (2) suggests that we study, instead of the sum on the left-hand side of (1), more general sums of the form

$$V_s(X) = \sum_{D=1}^{X} \left(\sqrt{D}\right)^s \zeta_D(s). \tag{3}$$

1980 *Mathematics Subject Classification.* Primary 10C07; Secondary 12A25, 12A70.

0081-5438/84 $1.00 + $.25 per page

Using a representation of $\zeta_D(s)$ as a sum of values of the Eisenstein-Maass series at special points, the imaginary quadratic irrationalities, we can reduce (3) to an overall sum involving all lattice points of some hyperbolic region. To this overall sum is directly applicable the method of Vinogradov [2]–[4], augmented by certain analytic properties of the Eisenstein-Maass series. Our main result is proved in this way.

THEOREM 1. *As $X \to \infty$, for $0 \leqslant \sigma \leqslant 1$, $\sigma = \operatorname{Re} s$, we have*

$$V_s(X) = 2^s\zeta(2s)\Phi(s)X^{1+s/2} + 2^s\zeta(2s)c(s)\Phi(1-s)X^{1+(1-s)/2} + R_s(X), \quad (4)$$

where

$$\Phi(s) = \frac{2^{-s}\zeta(s)}{(s+2)\zeta(s+2)}, \qquad c(s) = \sqrt{\pi}\,\frac{\Gamma(s-\frac{1}{2})\zeta(2s-1)}{\Gamma(S)\zeta(2s)}, \quad (5)$$

$\zeta(s)$ is Riemann's zeta-function, and $\Gamma(s)$ is Euler's gamma-function. Here $R_s(X)$ is the remainder in the asymptotic formula (4), *i.e. an analytic function of the variable s for a fixed X having a first-order pole at $s = 1$ and whose order with respect to X for s in compact subsets of the critical strip is less than the order of the two principal terms on the right-hand side of* (4). *One estimate of the remainder $R_s(X)$ has the form*

$$R_s(X) = O\left(\frac{X^{3/4}\ln^3 X}{|s-1|\,(|s-\frac{1}{2}| + \ln^{-1} X)}\right). \quad (6)$$

Estimate (6) *is uniform in s on compact subsets of the critical strip.*

Note that we do not offer new methods of estimating the remainder $R_s(X)$ here; rather we use Vinogradov's methods [2]–[5] (more precisely, in [3]). This is related to the fact that our main purpose is to obtain formula (4) with two asymptotic terms for values of s in the critical strip, in particular at $s = \frac{1}{2}$. Therefore, in order to focus the attention of the reader on the behavior of the principal terms of (4) we have deliberately shortened the calculations in estimating $R_s(X)$, using the simple method of [3] instead of [4] and [5]. Using Vinogradov's methods in [4] and [5], we could improve the remainder estimates in our theorems.

For the function $\zeta_D(s)$ we have the factorization

$$\zeta_D(s) = \zeta(s)L_D(s), \quad (7)$$

where $L_D(s)$ is the Dirichlet series corresponding to the order $\mathfrak{O}_D$. It coincides with the usual Dirichlet L-series with quadratic character of the field $\mathbf{Q}(\sqrt{-D})$ only if $\mathfrak{O}_D$ is a maximal order of $\mathbf{Q}(\sqrt{-D})$. The explicit form of the series $L_D(s)$ is given in Lemma 1.8.

Using Theorem 1 and (7), we can calculate the mean values of Dirichlet L-series with quadratic characters. We have the following assertion.

THEOREM 2. *As $X \to \infty$, for $0 \leqslant \sigma \leqslant 1$*

$$\sum_{D=1}^{X} L_D(s) = \Phi_1(s)X + \Phi_2(s)X^{1+(1-2s)/2} + O\left(\frac{X^{1/2+(1-2\sigma)/4}\ln^3 X}{|s-\frac{1}{2}| + \ln^{-1} X}\right), \quad (8)$$

where

$$\Phi_1(s) = \frac{\zeta(2s)}{2\zeta(s+2)}, \qquad \Phi_2(s) = \frac{\pi^{s-1/2}\zeta(2-2s)\Gamma((2-s)/2)}{(3-2s)\zeta(3-s)\Gamma((s+1)2)}, \quad (9)$$

and the estimate of the remainder is uniform in s on compact subsets of the critical strip.

From Theorem 2 we at once obtain

COROLLARY 1. *As* $X \to \infty$,

$$\sum_{D=1}^{X} L_D\left(\frac{1}{2}\right) = c_0 X(\ln X + c_1) + O\left(\sqrt{X}\ln^4 X\right), \tag{10}$$

where

$$c_0 = \frac{1}{4\zeta(2.5)}, \qquad c_1 = -2\frac{\zeta'}{\zeta}(2.5) - 3\ln 2 - \ln\pi + 3\gamma + \frac{\pi}{2} - 1, \tag{11}$$

and γ *is Euler's constant.*

Formulas of type (10) for primitive characters where conjectured in [6]. We will devote a separate paper to the proof of these conjectures.

We also mention the following fact.

COROLLARY 2. *For the function* $\Sigma_{1 \leq D \leq X} L_D(s)$ *there is an "analogue of the Riemann hypothesis with respect to X," i.e. for any* $T > 0$ *and for a sufficiently large X, this function has no zeros in the region* $|\operatorname{Im} s| \leq T$, $\sigma > \frac{1}{2}$.

As a consequence of (5)–(7), both sides of (4) have a simple pole at $s = 1$. Equating the results of both sides of (4) at $s = 1$, we obtain from Theorem 1 an asymptotic formula for the mean value of the number of classes of primitive quadratic forms of negative discriminant.

COROLLARY 3. *As* $X \to \infty$.

$$\sum_{D=1}^{X} H(-D) = \frac{\pi}{18\zeta(3)} X^{3/2} - \frac{3}{2\pi^2} X + O(X^{3/4}\ln^3 X). \tag{12}$$

In order to obtain from Theorem 1 the classical Vinogradov-Gauss formula, i.e. (1), we must take account of the distinction between Gauss's classical definition for purely radical forms and the modern definition of primitive forms. Taking account of this distinction, we come to consider the parametric sum

$$V_s^*(X) = \sum_{\substack{D=4 \\ D \equiv 0 \ (\mathrm{mod}\ 4)}}^{4X} \left(\sqrt{D}\right)^s \zeta_D(s), \tag{13}$$

a direct generalization of the sum on the left-hand side of (1). To reduce $V_s^*(X)$ to an overall sum we use a simple generalization of a well-known formula of Lipschitz in the theory of quadratic forms. The sum $V_s^*(X)$ reduces to an overall sum of the same type as $V_s(X)$, so that their asymptotic expansions as $X \to \infty$ can be calculated in virtually the same way. As a result, we arrive at the following asymptotic formula.

THEOREM 3. *As* $X \to \infty$, *for* $0 \leq \sigma \leq 1$

$$V_s^*(X) = 2^{s+1}\zeta(2s)\Phi^*(S)X^{1+s/2} + 2^{s+1}\zeta(2s)c(s)\Phi^*(1-s)X^{1+(1-s)/2}$$
$$+ O\left(\frac{X^{3/4}\ln^3 X}{|s-1|\left(|s-\frac{1}{2}| + \ln^{-1} X\right)}\right), \tag{14}$$

where the estimate of the remainder is uniform in s on compact subsets of the critical strip. Here

$$\Phi^*(s)=\left(1-\frac{1}{2^{s+1}}\right)\left(1-\frac{1}{2^{s+2}}\right)^{-1}\frac{\zeta(s)}{(s+2)\zeta(s+2)}. \tag{15}$$

The classical Vinogradov-Gauss formula is the equality of the residues of both sides of (14) at $s=1$.

This paper is a detailed exposition of the results announced in [7]. We will make a few remarks about its content. In the first section we discuss in considerable detail certain known facts concerning zeta-functions of orders in imaginary quadratic fields. Even though these classical results go back essentially to Dedekind, we have decided to include them here because they are not easy to locate in the modern literature. In this same section we discuss the connection between zeta-functions of orders in imaginary quadratic fields and the Eisenstein-Maass series for the full modular group, and we consider some properties of the Eisenstein-Maass series that will be needed later. For the convenience of the reader who is a specialist in analytic number theory we present in considerable detail elementary facts from the theory of automorphic functions on the Lobachevsky plane, while at the same time we freely use the standard methods of number theory. In §2 the main sum $V_s(x)$ is reduced to an overall sum $\Psi_s(X)$, which is interpreted by a method of I. M. Vinogradov, and the two asymptotic terms in the expansion of $\Psi_s(X)$ as $X\to\infty$ are obtained. The resulting asymptotic formula will be called the *generalized Vinogradov-Gauss formula*. It is from this formula that we derive the asymptotic expansions for $V_s(X)$ and $V_s^*(X)$. The third section is devoted to estimating the remainder in the generalized Vinogradov-Gauss formula by the method of [3]. Theorems 1 and 2 of the Introduction are also proved there. In §4 we prove a generalized Lipschitz formula, which is used to reduce $V_s^*(X)$ to an overall sum. Finally, in the Conclusion we briefly discuss further generalizations of our results. We use a double indexing of formulas, lemmas and theorems: the first number is that of the section, and the second is the number within the section.

§1. Zeta-functions of orders in imaginary quadratic fields

Suppose $d<0$ is the discriminant of the imaginary quadratic field $\mathbf{Q}(\sqrt{d})$, i.e. either $d\equiv 1 \pmod 4$ and is square-free or else $d=4d_1$, where $d_1\equiv 2$ or $3 \pmod 4$ and is square-free. The ring of integers $\mathfrak{O}_d$ of the field $\mathbf{Q}(\sqrt{d})$ has a basis $1,\ \omega$ over the ring of rational integers $\mathbf{Z}$, where $\omega=(1+\sqrt{d})/2$ in the first case and $\omega=\sqrt{d}/2$ in the second. The ring $\mathfrak{O}_d=\{1,\omega\}$ is a maximal order of the field $\mathbf{Q}(\sqrt{d})$. An arbitrary order $\mathfrak{O}$ of $\mathbf{Q}(\sqrt{d})$ has the form $\{1, f\omega\}$, where the natural number f is called the *conductor* of $\mathfrak{O}$. The discriminant D of the order $\mathfrak{O}$ is equal to df^2, and in connection with this we introduce the notation $\mathfrak{O}_D=\{1, f\omega\}$, where $D=df^2$. Thus, the orders in imaginary quadratic fields are in one-to-one correspondence with the negative discriminants, i.e. with all negative rational integers D such that $D\equiv 0$ or $1 \pmod 4$. These elementary facts are well known (see, for example, [8]).

Suppose $\mathfrak{a}$ is an ideal of the order $\mathfrak{O}_D$, i.e. a full module in the field $\mathbf{Q}(\sqrt{d})$ for which $\mathfrak{O}_D$ is the coefficient ring. The ideal $\mathfrak{a}$ is called *integral* if it is contained in the order $\mathfrak{O}_D$. The zeta-function $\zeta_D(s)$ of $\mathfrak{O}_D$ is defined for $\operatorname{Re} s > 1$ by the absolutely convergent series

$$\zeta_D(s) = \sum_{\mathfrak{a}} \frac{1}{N(\mathfrak{a})^s}, \tag{1.1}$$

where $N(\mathfrak{a})$ denotes the norm of $\mathfrak{a}$ and the summation in (1.1) extends over all integral ideals of $\mathfrak{O}_D$. If D is the fundamental discriminant, i.e. the discriminant of $\mathbf{Q}(\sqrt{D})$, then $\zeta_D(s)$ coincides with the Dedekind zeta-function of $\mathbf{Q}(\sqrt{D})$. Like the Dedekind zeta-function, $\zeta_D(s)$ admits an analytic continuation to the whole plane of the complex variable s, except for the point $s = 1$, where it has a simple pole.

It will be convenient in the sequel to have another representation of $\zeta_D(s)$, from which, in particular, we can obtain the analytic continuation of $\zeta_D(s)$. For this we use the well-known correspondence between ideals in orders of quadratic fields and binary quadratic forms (see, for example, [8]).

LEMMA 1.1. *Suppose $\mathfrak{a}$ is an ideal of the order $\mathfrak{O}_D$ and α, β is a basis of this ideal such that*

$$\frac{1}{i}\begin{vmatrix} \alpha & \beta \\ \bar{\alpha} & \bar{\beta} \end{vmatrix} > 0, \tag{1.2}$$

where the bar denotes conjugation in $\mathbf{Q}(\sqrt{d})$. Consider the binary quadratic form

$$f(x, y) = N(\alpha x + \beta y)/N(\mathfrak{a}) = ax^2 + bxy + cy^2, \tag{1.3}$$

where x and y are rational integers. Then a, b and c are rational integers, and

$$(a, b, c) = 1, \qquad b^2 - 4ac = D. \tag{1.4}$$

The correspondence $\mathfrak{a} \mapsto f(x, y)$ effects an isomorphism of the ideal class group of the order $\mathfrak{O}_D$ and the group of classes of properly equivalent primitive binary quadratic forms of discriminant D.

Let $h(D)$ be the ideal class number of $\mathfrak{O}_D$. We represent (1.1) in the form

$$\zeta_D(s) = \sum_{i=1}^{h(D)} \zeta_{D,i}(s), \tag{1.5}$$

where

$$\zeta_{D,i}(s) = \sum_{\mathfrak{a} \in C_i} \frac{1}{N(\mathfrak{a})^s} \tag{1.6}$$

and $\mathfrak{a}$ ranges over all integral ideals of a given ideal class C_i of the order $\mathfrak{O}_D$. Suppose $\mathfrak{a}'$ is an integral ideal of the inverse class C_i^{-1}. Then for a fixed $\mathfrak{a}'$ the mapping $\mathfrak{a} \mapsto \mathfrak{a}\mathfrak{a}' = (\xi)$ establishes a one-to-one correspondence between all integral ideals $\mathfrak{a}$ in C_i and the principal ideals (ξ) divisible by $\mathfrak{a}'$. Since $N(\mathfrak{a})N(\mathfrak{a}') = N(\xi)$, we can rewrite (1.6) in the form

$$\zeta_{D,i}(s) = N(\mathfrak{a}')^s \sum_{(\xi)} \frac{1}{N(\xi)^s}, \qquad \xi \equiv 0 \pmod{\mathfrak{a}'} \tag{1.7}$$

where the summation extends over all principal integral ideals (ξ) of $\mathfrak{O}_D$ that are divisible by $\mathfrak{a}'$. In view of the correspondence (1.3) and formulas (1.5) and (1.7), we see that $\zeta_D(s)$ can be represented in the form

$$\zeta_D(s) = \frac{1}{w_D} \sum_{i=1}^{h(D)} \sum_{x,y=-\infty}^{\infty}{}' \frac{1}{f_i(x,y)^s}, \tag{1.8}$$

where w_D is the number of units in $\mathfrak{O}_D$, $w_D = 2$ when $D < -4$, and the $f_i(x, y)$ range over a complete system of representatives of the classes of properly equivalent primitive quadratic forms of discriminant D; the prime on the inner summation symbol in (1.8) signifies that the summand with $x = y = 0$ is omitted.

To convert (1.8) into a more convenient form we introduce a number of definitions. Let $\mathcal{H} = \{z = x + iy \in \mathbf{C}^1,\ y > 0\}$ be the upper half-plane of the variable z equipped with the Poincaré metric $|dz|^2 = (dx^2 + dy^2)/y^2$, i.e. the Poincaré model of the Lobachevsky plane. The group of proper motions of the Lobachevsky plane is the group $G = \mathrm{PSL}(2, \mathbf{R})$ consisting of the real 2×2 matrices with determinant 1, where matrices differing only in sign are considered equivalent. The group G acts on $\mathcal{H}$ by linear-fractional transformations, i.e. for $g \in G$ and $z \in \mathcal{H}$ we have

$$g \cdot z = \frac{az + b}{cz + d}, \qquad g = \begin{Vmatrix} a & b \\ c & d \end{Vmatrix}. \tag{1.9}$$

Consider the discrete subgroup $\Gamma = \mathrm{PSL}(2, \mathbf{Z})$ of G, which is called the full modular group. Let $\Gamma_\infty \subset \Gamma$ be the stabilizer of the apex $i\infty$, i.e. the cyclic subgroup generated by the translation $z \mapsto z + 1$. For $\operatorname{Re} s > 1$ we define the Eisenstein-Maass series $E(z, s)$ for the group Γ by the following absolutely convergent series:

$$E(z, s) = \sum_{\gamma \in \Gamma_\infty \backslash \Gamma} y^s(\gamma \cdot z), \qquad z \in \mathcal{H}. \tag{1.10}$$

This can be rewritten as

$$E(z, s) = \frac{1}{2} \sum_{\substack{m,n=-\infty \\ (m,n)=1}}^{\infty} \frac{y^s}{|-nz + m|^{2s}}. \tag{1.11}$$

Now suppose $f(x, y) = ax^2 + bxy + cy^2$ is a primitive form of discriminant $D < 0$. Let ω denote its root belonging to $\mathcal{H}$, i.e.

$$\omega = \left(-b + i\sqrt{|D|}\right)/2a. \tag{1.12}$$

We now return to (1.8) and denote by ω_i the roots of the forms f_i, $i = 1, \ldots, h(D)$. From (1.11) and (1.12) we at once obtain

$$\sum_{x,y=-\infty}^{\infty}{}' \frac{1}{f_i(x,y)^s} = 2\zeta(2s)\left(\frac{2}{\sqrt{|D|}}\right)^s E(\omega_i, s). \tag{1.13}$$

Thus, we arrive at the following assertion.

LEMMA 1.2. *For* $\operatorname{Re} s > 1$,

$$\zeta_D(s) = \frac{2}{w_D} \sum_{i=1}^{h(D)} \left(\frac{2}{\sqrt{|D|}}\right)^s E^*(\omega_i, s), \tag{1.14}$$

where $E^*(z, s) = \zeta(2s)E(z, s)$.

In the general case of an arbitrary Fuchsian group of the first kind, Eisenstein series possess many remarkable properties. For the discrete group Γ we mention only the simplest of these that will be needed later.

LEMMA 1.3. *The Eisenstein-Maass series $E(z,s)$ admits an analytic continuation to the whole plane of the complex variable s, except for the point $s=1$, where it has a simple pole with residue $3/\pi$, and poles at the points $s=\frac{1}{2}\rho$, where ρ ranges over the nontrivial zeros of Riemann's zeta-function. For all values of s we have a Fourier expansion*

$$E(z,s)=y^s+c(s)y^{1-s}+\frac{2\pi^s\sqrt{y}}{\Gamma(s)\zeta(2s)}\sum_{n=-\infty}^{\infty}{}'\sigma_s(n)$$
$$\times K_{s-1/2}(2\pi|n|y)e^{2\pi inx}, \tag{1.15}$$

where

$$c(s)=\sqrt{\pi}\,\frac{\Gamma(s-1/2)\zeta(2s-1)}{\Gamma(s)\zeta(2s)},\qquad \sigma_s(n)=|n|^{s-1/2}\sum_{\substack{d|n\\ d>0}}d^{1-2s}, \tag{1.16}$$

and $K_\nu(y)$ is the Macdonald function having the integral representation

$$K_\nu(y)=\frac{1}{2}\int_0^\infty e^{-y(t+1/t)/2}t^\nu\frac{dt}{t},\qquad y>0. \tag{1.17}$$

This lemma is well known in the elementary theory of Eisenstein series (see, for example, [9]).

Lemmas 1.2 and 1.3 give use, in particular, an analytic continuation of the zeta-function $\zeta_D(s)$ to the whole plane of the variable s, except for the point $s=1$, where it has a simple pole with residue $2\pi h(D)/w_D\sqrt{|D|}$, which is equivalent to (2).

This method of proving Dirichlet's classical formula goes back to Kronecker.

Let $\mathfrak{L}=-y^2(\partial^2/\partial x^2+\partial^2/\partial y^2)$ be the Laplace operator of the Poincaré metric. We have the following assertion.

LEMMA 1.4. *For all values of s, the Eisenstein-Maass series is an automorphic function relative to the discrete group Γ, i.e., for any $\gamma\in\Gamma$,*

$$E(\gamma\cdot z,s)=E(z,s). \tag{1.18}$$

It is also an even function of x, i.e.,

$$E(-\bar{z},s)=E(z,s). \tag{1.19}$$

Moreover, $E(z,s)$ is an eigenfunction of the invariant Laplace operator:

$$\mathfrak{L}E(z,s)=s(1-s)E(z,s). \tag{1.20}$$

For $\operatorname{Re}s>1$, Lemma 1.4 follows from the commutativity of the operator $\mathfrak{L}$ with the action of G, the definition of the series $E(z,s)$ (see (1.10)), and (1.15). For the other values of s, Lemma 1.4 follows from the principle of analytic continuation.

Let $\mathfrak{F}=\Gamma\backslash\mathfrak{H}$ be the canonical fundamental domain of the discrete group Γ. It has the form $\mathfrak{F}=\{z=x+iy\colon y>0;\ |z|>1,\ -\frac{1}{2}\leqslant x<\frac{1}{2}\text{ or }|z|=1,\ -\frac{1}{2}\leqslant x\leqslant 0\}$.

The domain $\mathfrak{F}$ is characterized by the property that it contains the roots of the reduced (in the sense of Lagrange) binary quadratic forms of negative discriminant. Let $d\mu(z)$ denote the invariant measure on the Lobachevsky plane

$$d\mu(z) = \frac{1}{y^2} dx\, dy. \tag{1.21}$$

We have the following assertion.

LEMMA 1.5. *Let $\mathfrak{F}_D$ be the part of the fundamental domain $\mathfrak{F}$ with $y \leqslant D$. Then*

$$\iint_{\mathfrak{F}_D} E(z, s) d\mu(z) = \frac{D^{s-1}}{s-1} - \frac{c(s)}{s} D^{-s}. \tag{1.22}$$

PROOF. It follows from (1.20) and Green's integral formula that

$$\begin{aligned} \iint_{\mathfrak{F}_D} E(z, s) d\mu(z) &= \frac{1}{s(s-1)} \iint_{\mathfrak{F}_D} \Delta E(z, s)\, dx\, dy \\ &= \frac{1}{s(s-1)} \int_{\partial \mathfrak{F}_D} \frac{\partial}{\partial n} E(z, s)\, dl, \end{aligned} \tag{1.23}$$

where $\Delta = \partial^2/\partial x^2 + \partial^2/\partial y^2$ is the usual Laplace operator, $\partial\mathfrak{F}_D$ is the boundary of $\mathfrak{F}_D$, $\partial/\partial n$ denotes the outer normal derivative, and dl is the measure on $\partial\mathfrak{F}_D$. It follows from the invariance of the Eisenstein-Maass series under the transformations $z \mapsto z+1$ and $z \mapsto -1/z$ that the integral over $\partial\mathfrak{F}_D$ in (1.23) is equal to zero, except for the integral over the segment $y = D$, $-\frac{1}{2} \leqslant x \leqslant \frac{1}{2}$. Thus,

$$\iint_{\mathfrak{F}_D} E(z, s)\, d\mu(z) = \frac{1}{s(s-1)} \int_{-1/2}^{1/2} \frac{\partial}{\partial y} E(z, s)\Big|_{y=D} dx. \tag{1.24}$$

Substituting the Fourier expansion (1.15) into (1.24), we obtain (1.22). Note that Lemma 1.5 is a special case of the well-known Maass-Selberg relation in the theory of Eisenstein series.

LEMMA 1.6. *As $y \to \infty$ the asymptotic formula*

$$E(z, s) = y^s + c(s) y^{1-s} + O\left(\frac{e^{-\pi y}}{|\Gamma(s)\zeta(2s)|} \right), \tag{1.25}$$

holds uniformly in s on compact subsets of the complex plane. Formula (1.25) *admits term-by-term differentiation.*

To prove (1.25) it suffices to use the Fourier expansion (1.15) of the series $E(z, s)$ and the asymptotic behavior of the Macdonald function as $y \to \infty$:

$$K_\nu(y) = \sqrt{\frac{\pi}{2y}}\, e^{-y} \left(1 + O\left(\frac{1}{y} \right) \right), \tag{1.26}$$

which is given, for example, in [10].

We now return to the function $\zeta_D(s)$ and show that it has the factorization (7). The following assertion is true.

LEMMA 1.7 (main lemma of the theory of binary quadratic forms). *Let $r(a, D)$ denote the number of primitive representations of the natural number a by a complete* *system of representatives of the discriminant D. Then $r(a, D)$ is equal to the number of solutions* mod $2a$ *of the congruence*

$$b^2 \equiv D \pmod{4a}. \tag{1.27}$$

The proof can be found in [8] or [11].

It follows from (1.8) that for $\operatorname{Re} s > 1$ we have

$$\zeta_D(s) = \zeta(2s) \sum_{n=1}^{\infty} \frac{r^*(n, D)}{n^s}, \tag{1.28}$$

where $r^*(n, D)$ is the number of primitive representations of n by a complete system of representatives of the classes of primitive forms of discriminant D. From the expansion

$$r(n, D) = \sum_{\delta \mid (n, f)} r^*\left(\frac{n}{\delta}, \frac{D}{\delta^2}\right), \tag{1.29}$$

where f is the conductor of the order $\mathfrak{O}_D$, we obtain

$$r^*(n, D) = \sum_{\delta \mid (n, f)} \mu(\delta) r\left(\frac{n}{\delta}, \frac{D}{\delta^2}\right), \tag{1.30}$$

where $\mu(n)$ is the Möbius function. It follows from (1.30) that

$$\zeta_D(s) = \zeta(2s) \sum_{\delta \mid f} \sum_{n=1}^{\infty} \frac{\mu(\delta)}{\delta^s} \frac{r(n, D/\delta^2)}{n^s}. \tag{1.31}$$

Next, it is easy to deduce from Lemma 1.7 (see, for example, [12]) that

$$\zeta(2s) \sum_{n=1}^{\infty} \frac{r(n, D)}{n^s} = \zeta(s) L(s, \chi_d) \Phi(s, D), \tag{1.32}$$

where $L(s, \chi_d)$ is the Dirichlet L-series with character χ_d, $\chi_d(n) = (d/n)$ is the Kronecker symbol, and

$$\Phi(s, D) = \sum_{n \mid f} \frac{\mu(n)'(d/n)}{n^s} \sum_{\delta \mid (f/n)} \frac{1}{\delta^{2s-1}}. \tag{1.33}$$

It now follows from (1.31)–(1.33) that

$$\zeta_D(s) = \zeta(s) L_D(s), \tag{1.34}$$

where

$$L_D(s) = L(s, \chi_d) \sum_{\delta \mid f} \frac{\mu(\delta)}{\delta^s} \Phi\left(s, \frac{D}{\delta^2}\right). \tag{1.35}$$

Thus, we arrive at the following assertion.

LEMMA 1.8. *The zeta-function $\zeta_D(s)$ of the order $\mathfrak{O}_D$ in an imaginary quadratic field* $\mathbf{Q}(\sqrt{D})$ *has the factorization* (1.34), *where $L_D(s)$ is the Dirichlet series of $\mathfrak{O}_D$. It differs from the L-series $L(s, \chi_d)$ of the field* $\mathbf{Q}(\sqrt{D})$ *only by a finite Dirichlet series over the divisors of the conductor f of $\mathfrak{O}_D$.*

This concludes our discussion of the properties of zeta-functions of orders in imaginary quadratic fields and the Eisenstein-Maass series that will be needed in the sequel.

§2. The generalized Vinogradov-Gauss formula

Consider the sum

$$V_s(X) = \sum_{D=1}^{X} \frac{w_{-D}}{2} \left(\sqrt{D}\right)^s \zeta_{-D}(s). \tag{2.1}$$

Using (1.14), we can write it in the form

$$V_s(X) = 2^s \zeta(2s) \sum_{D=1}^{X} \sum_{i=1}^{h(-D)} E(\omega_i, s), \tag{2.2}$$

where the inner sum in (2.2) extends over the roots in $\mathcal{H}$ of all pairwise nonequivalent primitive forms of discriminant $-D$. Eliminating the primitivity condition, we obtain

$$V_s(X) = 2^s \sum_{n=1}^{\sqrt{X}} \mu(n) \Psi_s\left(\frac{X}{n^2}\right), \tag{2.3}$$

where

$$\Psi_s(X) = \sum_{D=1}^{X} \sum_{i=1}^{h^*(-D)} E^*(\tilde{\omega}_i, s), \tag{2.4}$$

and $\tilde{\omega}_i$ ranges over the roots in $\mathcal{H}$ of all $h^*(-D)$ pairwise nonequivalent quadratic forms of discriminant $-D$.

We will now apply Vinogradov's method [2], [3] to the overall sum (2.4). For this purpose we choose as representatives of the classes of forms those that are reduced in the sense of Lagrange, i.e. forms $f(x, y) = ax^2 + bxy + cy^2$ whose coefficients a, b, c lie in the classical reduction domain $|b| < a < c$ or $0 \leqslant b < a = c$ or $0 < b = a \leqslant c$. As noted in §1, the roots of reduced (in the sense of Lagrange) quadratic forms lie in the fundamental domain $\mathcal{F}$.

To calculate the sum (2.4) we need the following well-known assertion.

LEMMA 2.1 (N. Ya. Sonin's summation formula). *Suppose the complex-valued function $f(x)$ is twice continuously differentiable in the interval $P \leqslant x \leqslant Q$. Then*

$$\sum_{P<x\leqslant Q} f(x) = \int_P^Q f(x)\,dx + \rho(Q)f(Q) - \rho(P)f(P) + \rho_1(P)f'(P) - \rho_1(Q)f'(Q) + \int_P^Q \rho_1(x) f''(x)\,dx, \tag{2.5}$$

where

$$\rho(x) = \frac{1}{2} - \{x\}, \qquad \rho_1(x) = \int_0^x \rho(x')\,dx', \tag{2.6}$$

and $\{x\} = x - [x]$ is the fractional part of x.

A proof can be found in [13].

We now rewrite (2.4) in the form

$$\Psi_s(X) = \Psi_s^{(1)}(X) + \Psi_s^{(2)}(X), \tag{2.7}$$

where

$$\Psi_s^{(1)}(X) = \sum_{\substack{0<4ac-b^2=D\leqslant X \\ 0<a<c,\, -a<b\leqslant a}} E^*\left(\frac{-b+i\sqrt{D}}{2a}, s\right), \tag{2.8}$$

$$\Psi_s^{(2)}(X) = \sum_{\substack{0<4ac-b^2=D\leqslant X \\ a=c,\, 0\leqslant b\leqslant a}} E^*\left(\frac{-b+i\sqrt{D}}{2a}, s\right). \tag{2.9}$$

Let us first consider $\Psi_s^{(1)}(X)$. We rewrite its definition (2.8) as a triple sum

$$\Psi_s^{(1)}(X) \subset \sum_{a=1}^{\sqrt{X/3}} \sum_{-a<b\leqslant a} \sum_{a<c\leqslant(x+b^2)/4a} E^*\left(\frac{-b+i\sqrt{4ac-b^2}}{2a}, s\right). \tag{2.10}$$

The inner sum over c in (2.10) can be replaced, according to Sonin's formula, by the integral

$$\begin{aligned}\sum_{a<c\leqslant(X+b^2)/4a} E^*\left(\frac{-b+i\sqrt{4ac-b^2}}{2a}, s\right) &= \int_a^{(X+b^2)/4a} E^*\left(\frac{-b+i\sqrt{4ac-b^2}}{2a}, s\right) dc \\ &+ \rho\left(\frac{X+b^2}{4a}\right) E^*\left(\frac{-b+i\sqrt{X}}{2a}, s\right) \\ &- \frac{1}{2} E^*\left(\frac{-b+i\sqrt{4a^2-b^2}}{2a}, s\right) \\ &- \rho_1\left(\frac{X+b^2}{4a}\right) \frac{\partial}{\partial c} E^*\left(\frac{-b+i\sqrt{4ac-b^2}}{2a}, s\right)\Bigg|_{c=(X+b^2)/4a} \\ &+ \rho_1(a) \frac{\partial}{\partial c} E^*\left(\frac{-b+i\sqrt{4ac-b^2}}{2a}, s\right)\Bigg|_{c=a} \\ &+ \int_a^{(X+b^2)/4a} \rho_1(c) \frac{\partial^2}{\partial c^2} E^*\left(\frac{-b+i\sqrt{4ac-b^2}}{2a}, s\right) dc.\end{aligned} \tag{2.11}$$

The first integral in the right-hand side of (2.11) can obviously be brought to the form

$$2a \int_{\sqrt{1-(b/2a)^2}}^{\sqrt{X}/2a} E^*\left(\frac{-b}{2a} + iy, s\right) y\, dy. \tag{2.12}$$

Let us now consider the last integral in (2.11). We rewrite it as

$$\frac{1}{2a} \int_{\sqrt{1-(b/2a)^2}}^{\sqrt{X}/2a} \rho_1\left(ay^2 + \frac{b^2}{4a}\right) \frac{\partial}{\partial y} \left(\frac{1}{y} \frac{\partial}{\partial y} E^*\left(\frac{-b}{2a} + iy, s\right)\right) dy. \tag{2.13}$$

Using Lemma 1.6, we can represent the integral (2.13) in the form

$$\frac{\zeta(2s)}{2a}\int_{\sqrt{1-(b/2a)^2}}^{\sqrt{X}/2a}\rho_1\left(ay^2+\frac{b^2}{4a}\right)\left(s(s-2)y^{s-3}\right.$$
$$\left.+(s+1)(s-1)c(s)y^{-s-2}\right)dy+O\left(\frac{1}{a}\right), \quad (2.14)$$

from which it follows that the order of magnitude of the integral (2.13) for $0 \leqslant \sigma \leqslant 1$, where $\sigma = \operatorname{Re} s$, does not exceed

$$\frac{1}{a}\left(\left(\frac{\sqrt{X}}{2a}\right)^{\sigma-2}+\left(\frac{\sqrt{X}}{2a}\right)^{-\sigma-1}\right)\frac{1}{\left(|s-\frac{1}{2}|+\ln^{-1}X\right)} \ll \frac{1}{a\left(|s-\frac{1}{2}|+\ln^{-1}X\right)}. \quad (2.15)$$

Thus, the contribution of the integral (2.13) to the sum (2.10) can be estimated by

$$O\left(\frac{\sqrt{X}}{|s-\frac{1}{2}|+\ln^{-1}X}\right).$$

Analogously, the order of magnitude of the contribution of the fourth and fifth terms of (2.11) to the sum (2.10) does not exceed

$$\frac{\sqrt{X}}{|s-\frac{1}{2}|+\ln^{-1}X}.$$

Therefore,

$$\begin{aligned}\Psi_s^{(1)}(X) = &\sum_{a=1}^{\sqrt{X/3}}\sum_{\substack{-a<b\leqslant a\\4a^2-b^2\leqslant X}} a\int_{\sqrt{1-(b/2a)^2}}^{\sqrt{X}/2a}E^*\left(\frac{-b}{2a}+iy, s\right)y\,dy\\ &-\frac{1}{2}\sum_{a=1}^{\sqrt{X/3}}\sum_{\substack{-a<b\leqslant a\\4a^2-b^2\leqslant X}} E^*\left(\frac{-b+i\sqrt{4a^2-b^2}}{2a}, s\right)\\ &+\sum_{a=1}^{\sqrt{X/3}}\sum_{\substack{-a<b\leqslant a\\4a^2-b^2\leqslant X}} \rho\left(\frac{X+b^2}{4a}\right)E^*\left(\frac{-b+i\sqrt{X}}{2a}, s\right)\\ &+O\left(\frac{\sqrt{X}}{|s-\frac{1}{2}|+\ln^{-1}X}\right), \end{aligned} \quad (2.16)$$

where $0 \leqslant \sigma \leqslant 1$.

Let us now consider the second summand in (2.16). In the sum $\Psi_s(X)$ there is a counterterm for it with the opposite sign, the function $\Psi_s^{(2)}(X)$. Thus, we arrive at the expression

$$\sum_{a=1}^{\sqrt{X/3}}\left(\sum_{\substack{0\leqslant b\leqslant a\\4a^2-b^2\leqslant X}} E^*\left(\frac{-b+i\sqrt{4a^2-b^2}}{2a}, s\right)\right.$$
$$\left.-\frac{1}{2}\sum_{\substack{-a<b\leqslant a\\4a^2-b^2\leqslant X}} E^*\left(\frac{-b+i\sqrt{4a^2-b^2}}{2a}, s\right)\right). \quad (2.17)$$

Suppose first that $a \leqslant \sqrt{X}/2$ in (2.17). Then, using (1.19), we obtain

$$\sum_{-a<b\leqslant a} E^*\left(\frac{-b+i\sqrt{4a^2-b^2}}{2a}, s\right) = -E^*(i,s) - E^*(\omega, s) + 2\sum_{0\leqslant b\leqslant a} E^*\left(\frac{-b+i\sqrt{4a^2-b^2}}{2a}, s\right), \quad (2.18)$$

where $\omega = e^{2\pi i/3}$. Thereby the part of (2.17) that extends over $a \leqslant \sqrt{X}/2$ is equal to

$$\tfrac{1}{2}(E^*(i,s) + E^*(\omega, s))\left[\sqrt{X}/2\right]. \quad (2.19)$$

Analogously, the part of (2.17) with $\sqrt{X}/2 < a \leqslant \sqrt{X/3}$ is equal to

$$\tfrac{1}{2}E^*(\omega, s)\left(\left[\sqrt{X/3}\right] - \left[\sqrt{X}/2\right]\right). \quad (2.20)$$

It follows from (2.19) and (2.20) that (2.17) can be estimated by $\sqrt{X}/|s-1|$. Thus, for $0 \leqslant \sigma \leqslant 1$ we have

$$\begin{aligned}\Psi_s(X) = 2\sum_{a=1}^{\sqrt{X/3}} \sum_{\substack{-a<b\leqslant a\\ 4a^2-b^2\leqslant X}} a\int_{\sqrt{1-(b/2a)^2}}^{\sqrt{X}/2a} E^*\left(\frac{-b}{2a} + iy, s\right) y\,dy \\ + \sum_{a=1}^{\sqrt{X/3}} \sum_{\substack{-a<b\leqslant a\\ 4a^2-b^2\leqslant X}} \rho\left(\frac{X+b^2}{4a}\right) E^*\left(\frac{-b+i\sqrt{X}}{2a}, s\right) \\ + O\left(\frac{\sqrt{X}}{|s-1|\left(|s-\frac{1}{2}| + \ln^{-1} X\right)}\right). \end{aligned} \quad (2.21)$$

Now let us consider the first summand in (2.21). When $a \leqslant \sqrt{X}/2$ we can, according to Sonin's formula, replace the inner sum over b by the integral

$$\begin{aligned}\sum_{-a<b\leqslant a} \int_{\sqrt{1-(b/2a)^2}}^{\sqrt{X}/2a} E^*\left(\frac{-b}{2a} + iy, s\right) y\,dy \\ = 2a\int_{-1/2}^{1/2}\int_{\sqrt{1-x^2}}^{\sqrt{X}/2a} E^*(z,s)\,y\,dx\,dy \\ - \int_{-1/2}^{1/2}\int_{\sqrt{1-x^2}}^{\sqrt{X}/2a} \frac{\partial}{\partial x} E^*(z,s)\rho(2ax)\,y\,dx\,dy \\ - \int_{-1/2}^{1/2} E^*\left(x + i\sqrt{1-x^2}, s\right)\frac{x}{\sqrt{1-x^2}}\rho(2ax)\,dx, \end{aligned} \quad (2.22)$$

where $z = x + iy$. Using Lemma 1.3, we see that the contribution of the last two terms of (2.22) to the sum (2.21) can be estimated by $\sqrt{X}\,|s-1|$.

Analogously, for $\sqrt{X}/2 < a \leqslant \sqrt{X/3}$ we have

$$\sum_{\substack{-a<b\leqslant a\\ 4a^2-b^2\leqslant X}} \int_{\sqrt{1-(b/2a)^2}}^{\sqrt{X}/2a} E^*\left(\frac{-b}{2a}+iy, s\right) y\,dy$$

$$= 4a\int_{\sqrt{1-X/4a^2}}^{\frac{1}{2}} \int_{\sqrt{1-x^2}}^{\sqrt{X}/2a} E^*(z,s)\,y\,dx\,dy + O\left(\frac{1}{a|s-1|}\right). \quad (2.23)$$

Using (2.22), (2.23) and the notation of Lemma 1.5, we obtain

$$\begin{aligned}\Psi_s(X) &= 4\sum_{a=1}^{\sqrt{X/3}} \int\int_{\mathfrak{F}\sqrt{X}/2a} E^*(z,s)\,y\,dx\,dy \\ &\quad + \sum_{a=1}^{\sqrt{X/3}} \sum_{\substack{-a<b\leqslant a\\ 4a^2-b^2\leqslant X}} \rho\left(\frac{X+b^2}{4a}\right) E^*\left(\frac{-b+i\sqrt{X}}{2a}, s\right) \\ &\quad + O\left(\frac{\sqrt{X}}{|s-1|(|s-\frac{1}{2}|+\ln^{-1} X)}\right) \\ &= 4\int\int_{\mathfrak{F}\sqrt{X}/2} E^*(z,s) \sum_{a=1}^{\sqrt{X}/2y} a^2\,y\,dx\,dy \\ &\quad + \tilde{R}_s(X) + O\left(\frac{\sqrt{X}}{|s-1|(|s-\frac{1}{2}|+\ln^{-1} X)}\right),\end{aligned} \quad (2.24)$$

where in the second equality we have interchanged the order of summation and integration.

Now consider the double integral in (2.24). Note first that by contour integration it is easy to obtain

$$\sum_{1\leqslant a\leqslant \sqrt{X}/2} a^2 = \frac{1}{2\pi i}\int_{\alpha-iT}^{\alpha+iT} \left(\frac{\sqrt{X}}{2y}\right)^w \frac{\zeta(w-2)}{w}\,dw + O\left(\left(\frac{\sqrt{X}}{2y}\right)^\alpha \frac{\ln X}{T}\right), \quad (2.25)$$

where $\alpha = 3 + \ln^{-1} X$, $T = X^3$, and the prime on the summation sign signifies that for an integral $\sqrt{X}/2y$ the summand $X/4y^2$ is taken with factor $\frac{1}{2}$. We now move the contour of integration in (2.25) to the straight line $\operatorname{Re} w = 1$. We obtain

$$\begin{aligned}\sum_{1\leqslant a\leqslant \sqrt{X}/2}{}' a^2 &= \frac{1}{3}\left(\frac{\sqrt{X}}{2y}\right)^3 + \frac{1}{2\pi i}\int_{1-iT}^{1+iT} \left(\frac{\sqrt{X}}{2y}\right)^\omega \frac{\zeta(w-2)}{w}\,dw \\ &\quad + \frac{1}{2\pi i}\int_{1+iT}^{3+iT} \left(\frac{\sqrt{X}}{2y}\right)^w \frac{\zeta(w-2)}{w}\,dw \\ &\quad - \frac{1}{2\pi i}\int_{1-iT}^{3-iT} \left(\frac{\sqrt{X}}{2y}\right)^w \frac{\zeta(w-2)}{w}\,dw + O\left(\left(\frac{\sqrt{X}}{2y}\right)^\alpha \frac{\ln X}{T}\right).\end{aligned} \quad (2.26)$$

Next, substitute (2.26) into the double integral in (2.24). For $0 \leqslant \sigma \leqslant 1$ we obtain the following expression:

$$\frac{4}{3}\left(\frac{\sqrt{X}}{2}\right)^3 \iint_{\mathfrak{F}_{\sqrt{X}/2}} E^*(z,s)\,d\mu(z)$$

$$+\frac{2}{\pi i}\int_{1-iT}^{1+iT}\left(\frac{\sqrt{X}}{2}\right)^w \frac{\zeta(w-2)}{w}\iint_{\mathfrak{F}_{\sqrt{X}/2}} E^*(z,s)\,y^{1-w}\,dx\,dy\,dw$$

$$+\frac{2}{\pi i}\int_{1+iT}^{3+iT}\left(\frac{\sqrt{X}}{2}\right)^w \frac{\zeta(w-2)}{w}\iint_{\mathfrak{F}_{\sqrt{X}/2}} E^*(z,s)\,y^{1-w}\,dx\,dy\,dw$$

$$-\frac{2}{\pi i}\int_{1-iT}^{3-iT}\left(\frac{\sqrt{X}}{2}\right)^w \frac{\zeta(w-2)}{w}\iint_{\mathfrak{F}_{\sqrt{X}/2}} E^*(z,s)\,y^{1-w}\,dx\,dy\,dw$$

$$+O\left(\frac{\ln^2 X}{|s-1|}\right). \tag{2.27}$$

Let us consider the first summand in (2.27). By Lemma 1.5, it has the form

$$\frac{4}{3}\zeta(2s)\left(\frac{1}{s-1}\left(\frac{\sqrt{X}}{2}\right)^{s+2} - \frac{c(s)}{s}\left(\frac{\sqrt{X}}{2}\right)^{3-s}\right). \tag{2.28}$$

Now consider the second summand in (2.27). Let $E_0^*(z,s)\zeta(2s)y^s + \zeta(2s)c(s)y^{1-s}$ denote the constant term of the Fourier series for $E^*(z,s)$, and put $E^*(z,s) = E_0^*(z,s) + E_1^*(z,s)$. Then

$$\iint_{\mathfrak{F}_{\sqrt{X}/2}} E_0^*(z,s)y^{1-w}\,dx\,dy = \frac{\zeta(2s)}{s-w+2}\left(\frac{\sqrt{X}}{2}\right)^{s-w+2} + \frac{\zeta(2s)c(s)}{3-s-w}\left(\frac{\sqrt{X}}{2}\right)^{3-s-w}$$
$$-\frac{\zeta(2s)}{s-w+2}\int_{-1/2}^{1/2}\left(\sqrt{1-x^2}\right)^{s-w+2}dx$$
$$-\frac{\zeta(2s)c(s)}{3-s-w}\int_{-1/2}^{1/2}\left(\sqrt{1-x^2}\right)^{3-s-w}dx. \tag{2.29}$$

Thus, the contribution of $E_0^*(z,s)$ to the second summand of (2.27) can be represented as a sum of four integrals:

$$\left(\frac{\sqrt{X}}{2}\right)^{s+2}\frac{2\zeta(2s)}{\pi i}\int_{1-iT}^{1+iT}\frac{\zeta(w-2)}{w(s-w+2)}\,dw$$

$$+\left(\frac{\sqrt{X}}{2}\right)^{3-s}\frac{2\zeta(2s)c(s)}{\pi i}\int_{1-iT}^{1+iT}\frac{\zeta(w-2)}{2(3-s-w)}\,dw$$

$$-\frac{2\zeta(2s)}{\pi i}\int_{1-iT}^{1+iT}\int_{-1/2}^{1/2}\frac{\zeta(w-2)\left(\sqrt{1-x^2}\right)^{s-w+2}}{w(s-w+2)}\left(\frac{\sqrt{X}}{2}\right)^w dx\,dw$$

$$-\frac{2\zeta(2s)c(s)}{\pi i}\int_{1-iT}^{1+iT}\int_{-1/2}^{1/2}\frac{\zeta(w-2)\left(\sqrt{1-x^2}\right)^{3-s-w}}{w(3-s-w)}\left(\frac{\sqrt{X}}{2}\right)^w dx\,dw. \tag{2.30}$$

We shift the contour of integration in the first two integrals in (2.30) to the right to infinity. Since $\zeta(-1+it) = O(t^{3/2})$, these integrals are equal to

$$-\frac{4}{3}\frac{\zeta(2s)}{s-1}\left(\frac{\sqrt{X}}{2}\right)^{s+2} + \frac{4}{3}\frac{\zeta(2s)c(s)}{s}\left(\frac{\sqrt{X}}{2}\right)^{3-s}$$

$$+ 4\zeta(2s)\left(\frac{\zeta(s)}{s+2}\left(\frac{\sqrt{X}}{2}\right)^{s+2} + \frac{\zeta(1-s)c(s)}{3-s}\left(\frac{\sqrt{X}}{2}\right)^{3-s}\right) + O\left(\frac{\ln X}{|s-1|}\right), \quad (2.31)$$

where $0 \leqslant \sigma \leqslant 1$. To estimate the last two integrals in (2.30), note that as $\operatorname{Im} w \to \infty$ and for s in compact subsets of the critical strip we have the estimates

$$\int_{-1/2}^{1/2}\left(\sqrt{1-x^2}\right)^{s-w+2} dx = O\left(\frac{1}{\sqrt{|\operatorname{Im} w|}}\right),$$
$$\int_{-1/2}^{1/2}\left(\sqrt{1-x^2}\right)^{3-s-w} dx = O\left(\frac{1}{\sqrt{|\operatorname{Im} w|}}\right). \quad (2.32)$$

It follows from (2.32) and the estimate for $\zeta(-1+it)$ that the order of magnitude of the last two integrals in (2.30) does not exceed $(\sqrt{X}\ln X)/|s-1|$. We now consider the contribution of $E_1^*(z, s)$ to the second integral in (2.27). Note that

$$\iint_{\mathfrak{F}_{\sqrt{X}/2}} E_1^*(z, s) y^{1-w}\, dx\, dy = \int_{-1/2}^{1/2}\int_{\sqrt{1-x^2}}^{\sqrt{X}/2} E_1^*(z, s) y^{1-w}\, dx\, dy$$
$$= \frac{1}{w-2}\int_{-1/2}^{1/2}\int_{\sqrt{1-x^2}}^{\sqrt{X}/2} \frac{\partial}{\partial y} E_1^*(z, s) y^{2-w}\, dx\, dy$$
$$+ \frac{1}{w-2}\int_{-1/2}^{1/2} E_1^*\left(x + i\sqrt{1-x^2}, s\right)\left(\sqrt{1-x^2}\right)^{2-w} dx, \quad (2.33)$$

where we have used the fact that

$$\int_{-1/2}^{1/2} E_1^*(z, s)\, dx = 0. \quad (2.34)$$

Integrating by parts repeatedly in the first integral in (2.33), we see that for s in compact subsets of the critical strip the order of magnitude of this integral does not exceed $1/|w|^2$. At the same time, the second integral in (2.33) can be estimated by $1/|w|^{3/2}$. Thus, the order of magnitude of the contribution of $E_1^*(z, s)$ to the second integral in (2.27) does not exceed $\sqrt{X}\ln X$. Combining these results, we see that the first two integrals in (2.27) are equal to

$$4\zeta(2s)\left(\frac{\zeta(s)}{s+2}\left(\frac{\sqrt{X}}{2}\right)^{s+2} + \frac{\zeta(1-s)c(s)}{3-s}\left(\frac{\sqrt{X}}{2}\right)^{3-s}\right) + O\left(\frac{\sqrt{X}\ln X}{|s-1|}\right). \quad (2.35)$$

To estimate the last two integrals in (2.27) it suffices to use the following estimate, which is a consequence of (2.33):

$$\iint_{\mathfrak{F}_{\sqrt{X}/2}} E_1^*(z, s) y^{1-w}\, dx\, dy = O\left(\frac{1}{T}\right), \quad (2.36)$$

where $|\operatorname{Im} w| = T$. As a result we obtain that the contribution of these integrals to (2.27) is $O(\ln X/|s-1|)$.

Thus, we have proved

LEMMA 2.2 (the generalized Vinogradov-Gauss formula). *For $0 \leqslant \sigma \leqslant 1$ and uniformly in s on compact subsets of the critical strip the asymptotic formula*

$$\Psi_s(X) = 4\zeta(2s)\left(\frac{\zeta(s)}{s+2}\left(\frac{\sqrt{X}}{2}\right)^{s+2} + \frac{\zeta(1-s)c(s)}{3-s}\left(\frac{\sqrt{X}}{2}\right)^{3-s}\right)$$
$$+\tilde{R}_s(X) + O\left(\frac{\sqrt{X}\ln X}{|s-1|}\right), \tag{2.37}$$

holds, where

$$\tilde{R}_s(X) = \sum_{a=1}^{\sqrt{X/3}} \sum_{\substack{-a<b\leqslant a\\ 4a^2-b^2\leqslant X}} \rho\left(\frac{X+b^2}{4a}\right)E^*\left(\frac{-b+i\sqrt{X}}{2a}, s\right). \tag{2.38}$$

Now observe that if we substitute the principal term of (2.37) into (2.3), we obtain two terms of the asymptotic relation in Theorem 1. Thus, to complete the proof of Theorem 1 it remains to estimate the remainder $\tilde{R}_s(X)$. It is to this problem that we devote the next section.

§3. Estimate of the remainder $\tilde{R}_s(X)$ by a method of I. M. Vinogradov

We will follow the idea of Vinogradov [3], i.e. the reduction of the estimate of the remainder $\tilde{R}_s(X)$ to estimates of Gaussian sums. We first mention some classical results.

LEMMA 3.1 (I. M. Vinogradov's "little cups" lemma). *Suppose r is an integer, $r > 0$, α and β are real numbers, $0 < \Delta < 0.25$ and $\Delta \leqslant \beta - \alpha \leqslant 1 - \Delta$. Then there exists a periodic function $\psi(x)$ with period 1 such that:*

1. $\psi(x) = 1$ *in the interval* $\alpha + 0.5\Delta \leqslant x \leqslant \beta - 0.5\Delta$;
2. $0 \leqslant \psi(x) \leqslant 1$ *in the intervals* $\alpha - 0.5\Delta \leqslant x \leqslant \alpha + 0.5\Delta$ *and* $\beta - 0.5\Delta \leqslant x \leqslant \beta + 0.5\Delta$;
3. $\psi(x) = 0$ *in the interval* $\beta + 0.5\Delta \leqslant x \leqslant 1 + \alpha - 0.5\Delta$; *and*
4. $\psi(x)$ *can be expanded into a Fourier series of the form*

$$\psi(x) = \beta - \alpha + \sum_{m=-\infty}^{\infty}{}' a_m e^{2\pi i m x}, \tag{3.1}$$

where

$$|a_m| \leqslant \frac{2}{|\pi m|}, \quad |a_m| \leqslant 2(\beta-\alpha), \quad |a_m| \leqslant \frac{2}{|\pi m|}\left(\frac{r}{|\pi m|\Delta}\right)^r \tag{3.2}$$

and the prime on the summation symbol in (3.1) *signifies that the term with $m = 0$ is omitted.*

A proof can be found in [14].

LEMMA 3.2. *Suppose $(a, n) = \delta$ and*

$$G_\alpha(n, m) = \sum_{b=1}^{a} e^{2\pi i(nb^2+mb)/a}. \tag{3.3}$$

Then (*see* [13])

$$|G_a(n, m)| \leqslant \sqrt{2a\delta}. \tag{3.4}$$

We now estimate the remainder $\tilde{R}_s(X)$. Let $\chi(z)$ denote the characteristic function of the fundamental domain $\mathfrak{F}$. Then the sum (2.38) can be written without the constraint $4a^2 - b^2 \leqslant X$ in the form

$$\sum_{a=1}^{\sqrt{X/3}} \sum_{-a<b\leqslant a} \chi(z)\rho\left(\frac{X+b^2}{4a}\right)E^*(z, s), \tag{3.5}$$

where $z = (-b + i\sqrt{X})/2a$. The inner sum over b in (3.5) can be brought by partial summation to the form

$$E^*\left(\frac{1}{2} + i\frac{\sqrt{X}}{2}, s\right)S_a\left(\frac{1}{2}, \chi\right) - \int_{-1/2}^{1/2} \frac{\partial}{\partial x} E^*(z, s) S_a(x, \chi)\, dx, \tag{3.6}$$

where now $z = x + i\sqrt{X}/2a$ and

$$S_a(x, \chi) = \sum_{-a<b\leqslant 2ax} \chi\left(\frac{-b+i\sqrt{X}}{2a}\right)\rho\left(\frac{X+b^2}{4a}\right). \tag{3.7}$$

In order to reduce the summation over b in (3.7) to the complete interval $-a < b \leqslant a$ we introduce the characteristic function χ_1 of the interval $(-\frac{1}{2}, x]$. Then the sum (3.7) assumes the form

$$S_a(x, \chi) = \sum_{-a<b\leqslant a} \chi\left(\frac{-b+i\sqrt{X}}{2a}\right)\chi_1\left(\frac{b}{2a}\right)\rho\left(\frac{X+b^2}{4a}\right). \tag{3.8}$$

Now observe that $\chi((-b + i\sqrt{X})/2a) = 1$ whenever $a \leqslant \sqrt{X}/2$. Thus, for $a \leqslant \sqrt{X}/2$ we have

$$S_a(x, \chi) = \sum_{-a<b\leqslant a} \chi_1\left(\frac{b}{2a}\right)\rho\left(\frac{X+b^2}{4a}\right), \tag{3.9}$$

and for $\sqrt{X}/2 < a \leqslant \sqrt{X/3}$ we have

$$S_a(x, \chi) = \sum_{-a<b\leqslant a} \chi_1\left(\frac{b}{2a}\right)\chi_2\left(\frac{b}{2a}\right)\rho\left(\frac{X+b^2}{4a}\right), \tag{3.10}$$

where χ_2 is the characteristic function of the union of the intervals

$$\left(-\tfrac{1}{2}, -\sqrt{1 - X/4a^2}\right) \quad \text{and} \quad \left(\sqrt{1 - X/4a^2}, \tfrac{1}{2}\right].$$

We first consider the sum (3.9). In order to eliminate subtle questions connected with the conditional convergence of the Fourier series of χ_1 we approximate it by a "little cup" of Vinogradov. Let $\tilde{\chi}_1$ be the function of Lemma 3.1, where $\alpha = -\frac{1}{2}$, $\beta = x$, $r = 1$ and Δ is to be chosen later. We replace (3.9) by the sum

$$\tilde{S}_a(x, \chi) = \sum_{-a<b\leqslant a} \tilde{\chi}_1\left(\frac{b}{2a}\right)\rho\left(\frac{X+b^2}{4a}\right). \tag{3.11}$$

Using (3.1), we write $\tilde{S}_a(x, \chi)$ as follows:

$$\tilde{S}_a(x, \chi) = \left(x + \frac{1}{2}\right)S(a, 0) + \sum_{m=-\infty}^{\infty}{}' a_m S(a, m), \tag{3.12}$$

where

$$S(a, m) = \sum_{-a<b\leqslant a} \rho\left(\frac{X + b^2}{4a}\right) e^{\pi imb/a}. \tag{3.13}$$

We now estimate $S(a, m)$. To do this it suffices to expand $\rho(x)$ into a Fourier series and reduce the problem to that of estimating Gaussian sums. The only subtlety is that the Fourier series of $\rho(x)$ converges conditionally. To get around this difficulty we can replace the whole Fourier series by a sufficiently long (finite) partial sum, as was done in [3], or observe that X can be assumed to be an integer and therefore the rational numbers $\rho((X + b^2)/4a)$ are also values of a continuous periodic function $\tilde{\rho}(x)$ with period 1. The function $\tilde{\rho}(X)$ is defined as follows:

$$\tilde{\rho}(x) = \begin{cases} \frac{1}{2} - x & \text{if } 0 \leqslant x \leqslant 1 - 1/4a, \\ \frac{1}{2} + (4a - 1)(x - 1) & \text{if } 1 - 1/4a \leqslant x \leqslant 1. \end{cases} \tag{3.14}$$

Therefore, $\tilde{\rho}(x)$ can be expanded into an absolutely convergent Fourier series, and it follows that

$$\frac{1}{2} - \left\{\frac{X + b^2}{4a}\right\} = \frac{1}{8a} + \sum_{n=-\infty}^{\infty}{}' b_n e^{\pi in(X+b^1)/2a}, \tag{3.15}$$

where

$$b_n = \frac{a}{\pi^2 n^2}(1 - e^{\pi in/2a}), \qquad n \neq 0. \tag{3.16}$$

Thus, for $m \neq 0$ we have

$$S(a, m) = \frac{1}{2} \sum_{n=-\infty}^{\infty}{}' b_n G_{4a}(n, 2m) e^{\pi inX/2a}, \tag{3.17}$$

where $G_{4a}(n, 2m)$ is the Gaussian sum (3.3). It follows from (3.4) and (3.16) that

$$\begin{aligned} S(a, m) &\ll \sqrt{a} \sum_{1\leqslant n\leqslant a} \frac{\sqrt{(n, a)}}{n} + a\sqrt{a} \sum_{a+1\leqslant n<\infty} \frac{\sqrt{(n, a)}}{n^2} \\ &\ll \sqrt{a} \sum_{\delta|a} \frac{1}{\sqrt{\delta}}\left(\ln\frac{a}{\delta} + 2\right). \end{aligned} \tag{3.18}$$

The same estimate holds for $m = 0$. Substituting (3.18) into (3.12) and using the estimates $a_m = O(1/|m|)$ for $|m| \leqslant \Delta^{-1}$ and $a_m = O(1/m^2\Delta)$ for $|m| \geqslant \Delta^{-1}$, which are consequences of (3.2), we obtain

$$\tilde{S}_a(x, \chi) \ll \sqrt{a} \ln\frac{1}{\Delta} \sum_{\delta|a} \frac{1}{\sqrt{\delta}}\left(\ln\frac{a}{\delta} + 2\right). \tag{3.19}$$

Next, it follows from the definition of $\tilde{\chi}_1$ that

$$S_a(x, \chi) - \tilde{S}_a(x, \chi) \ll a\Delta; \tag{3.20}$$

hence for $a \leqslant \sqrt{X}/2$ we have

$$S_a(x, \chi) \ll \sqrt{a} \ln\frac{1}{\Delta} \sum_{\delta|a} \frac{1}{\sqrt{\delta}}\left(\ln\frac{a}{\delta} + 2\right) + a\Delta. \tag{3.21}$$

We now substitute (3.21) into (3.6) and sum over a from 1 to $\sqrt{X}/2$. Using the estimates

$$\frac{\partial}{\partial x}E^*(z, s) = O(e^{-\pi y}), \tag{3.22}$$

$$E^*(z, s) \ll \left(y^\sigma + \frac{1}{|s-1|}y^{1-\sigma}\right)\frac{1}{\left(|s-\frac{1}{2}|+\ln^{-1}y\right)}, \tag{3.23}$$

where $0 \leqslant \sigma \leqslant 1$, which follow from Lemma 1.6, we obtain

$$\sum_{1 \leqslant a \leqslant \sqrt{X}/2} \sum_{-a<b\leqslant a} \rho\left(\frac{X+b^2}{4a}\right)E^*\left(\frac{-b+i\sqrt{X}}{2a}, s\right)$$
$$\ll X^{3/4}\ln X \ln\frac{1}{\Delta}\frac{1}{|s-1|\left(|s-\frac{1}{2}|+\ln^{-1}X\right)} + \Delta X. \tag{3.24}$$

If we now choose $\Delta = X^{-1/4}$, for the sum in (3.24) we obtain

$$\frac{X^{3/4}\ln^2 X}{|s-1|\left(|s-\frac{1}{2}|+\ln^{-1}X\right)}.$$

Analogously, it can be shown that for $\sqrt{X}/2 < a \leqslant \sqrt{X/3}$

$$S_a(x, \chi) \ll \sqrt{a}\ln^2 X \sum_{\delta|a}\frac{1}{\sqrt{\delta}}\left(\ln\frac{a}{\delta}+2\right) + \frac{a}{\sqrt[4]{X}}, \tag{3.25}$$

where the superfluous factor $\ln X$ is connected with the double use of Vinogradov's "little cups." Thus, we have proved the following assertion.

LEMMA 3.3. *For $0 \leqslant \sigma \leqslant 1$, and uniformly in s on compact subsets of the critical strip,*

$$\tilde{R}_s(X) = O\left(\frac{X^{3/4}\ln^3 X}{|s-1|\left(|s-\frac{1}{2}|+\ln^{-1}X\right)}\right). \tag{3.26}$$

Theorem 1 follows immediately from Lemma 2.2 and 3.3 and formula (2.3).

We now turn to the proof of Theorem 2. It follows from Lemma 1.8 that the left-hand side of (4) is divisible by $\zeta(s)$; also, we see from the explicit form of the two principal terms of (4) and the functional equation for Riemann's zeta-function that these terms are also divisible by $\zeta(s)$. Thus, the remainder $R_s(X)$ in (4) has the form

$$R_s(X) = \zeta(s)M(s, X), \tag{3.27}$$

where $M(s, X)$ is an analytic function of s in the critical strip, $0 \leqslant \sigma \leqslant 1$. For any compact subdomain Ω of the critical strip we have

$$|(s-1)R_s(X)| \leqslant c(\Omega)X^{3/4}\ln^3 X, \tag{3.28}$$

where we have assume that Ω does not contain the point $s = \frac{1}{2}$.

It follows from (4) that for a fixed X the function $(s-1)R_s(X)$ is analytic in the critical strip. Thus, the maximum of the left-hand side of (3.28) is attained on the boundary $\partial\Omega$ of Ω. We may assume without loss of generality that $\zeta(s)$ has no zeros on $\partial\Omega$. Then it follows from (3.27) that on $\partial\Omega$ we have

$$|M(s, X)| \leqslant mc(\Omega)X^{3/4}\ln^3 X, \tag{3.29}$$

where m is the maximum of the function $1/|(s-1)\zeta(s)|$ on $\partial\Omega$. By the maximum modulus principle, (3.29) holds for all $s \in \Omega$. In the case $s = \frac{1}{2}$, the corresponding estimate contains a superfluous factor $\ln X$ in comparison with (3.29). Thus, we obtain from (4) an asymptotic formula for the sum $\Sigma_{1 \leqslant D \leqslant X}(\sqrt{D})^s L_D(s)$, where the remainder has the same order with respect to X as has the remainder in (4); now (8) can be obtained by partial summation. Thereby we have also proved Theorem 2. By elementary use of the theory of functions we could get rid of the superfluous factor $1/\zeta(s)$ in the estimate

$$O\left(\frac{X^{3/4}\ln^3 X}{|(s-1)\zeta(s)|}\right),$$

which follows immediately from (4).

Thus, we have completely proved Theorems 1 and 2 stated in the Introduction. The next section of this paper deals with the connection between the generalized and classical Vinogradov-Gauss formulas, and with the proof of Theorem 3.

§4. The generalized Lipschitz formula

Recall that in accordance with a tradition stemming from Gauss and Dirichlet, Russian classics in number theory have considered binary quadratic forms of the type (1.3) with even middle coefficient, i.e. forms of the type

$$f(x, y) = ax^2 + 2bxy + cy^2 \tag{4.1}$$

with determinant

$$|f| = b^2 - ac. \tag{4.2}$$

Following Gauss, a quadratic form (4.1) with rational integral coefficients is called *purely radical* if $(a, 2b, c) = 1$, and *not purely radical* if $|f| \equiv 1 \pmod 4$ and $(a, 2b, c) = 2$.

Now suppose $g(x, y)$ is a primitive quadratic form of discriminant D, i.e.

$$g(x, y) = ax^2 + bxy + cy^2 \tag{4.3}$$

and

$$D = b^2 - 4ac. \tag{4.4}$$

Then in the case $D \equiv 0 \pmod 4$ we must have $b \equiv 0 \pmod 2$, so that $g(x, y)$ is also a purely radical form of determinant $D/4$. In the case $D \equiv 1 \pmod 4$, there corresponds to $g(x, y)$ a not purely radical form $f(x, y) = 2g(x, y)$ of determinant D. Thus, the sum in (1) is the residue at $s = 1$ of the parametric sum

$$V_s^*(X) = \sum_{\substack{D=4 \\ D \equiv 0 \pmod 4}}^{4X} \frac{w_{-D}}{2}\left(\sqrt{D}\right)^s \zeta_{-D}(s). \tag{4.5}$$

Reduction of $V_s^*(X)$ to an overall sum can be effected by a simple generalization of the well-known Lipschitz formula, to whose derivation we now turn. The

Lipschitz formula itself can be found, for example, in [15]. Suppose $\varphi(z)$ is an arbitrary continuous function on $\mathcal{H}$. Consider the sums

$$F(n) = \sum_{\omega \in \mathcal{F}} \varphi(\omega), \tag{4.6}$$

where ω ranges over the roots of all forms (4.3) of discriminant $-n$, and the sums

$$F^*(n) = \sum_{\omega^* \in \mathcal{F}} \varphi(\omega^*), \tag{4.7}$$

where ω^* ranges over the roots of all forms (4.1) of determinant $-n$, $n > 0$. We have the following decompositions:

$$F(n) = \sum_{n=4km^2} \tilde{F}^*(k) + \sum_{\substack{n=km^2 \\ -k\equiv 1 \pmod 4}} \tilde{F}(k) \tag{4.8}$$

and

$$F^*(n) = \sum_{n=km^2} \tilde{F}^*(k) + \sum_{\substack{n=km^2 \\ -k\equiv 1 \pmod 4}} \tilde{F}(k), \tag{4.9}$$

where $\tilde{F}(k)$ denotes the part of the sum $F(k)$ that extends over the roots of primitive forms, and $\tilde{F}^*(k)$ the part of $F^*(k)$ that extends over the roots of purely radical forms. The second summands in (4.8) and (4.9) corresponds to not purely radical forms. It follows from (4.8) and (4.9) that

$$F^*(n) - F(n) = \sum_{n=km^2} \tilde{F}^*(k) - \sum_{n=4km^2} \tilde{F}^*(k), \tag{4.10}$$

hence, by the Möbius inversion formula,

$$\tilde{F}^*(n) = \sum_{\substack{\delta^2 | n \\ \delta\equiv 1 \pmod 2}} \mu(\delta)\left(F^*\left(\frac{n}{\delta^2}\right) - F\left(\frac{n}{\delta^2}\right)\right). \tag{4.11}$$

Thus, we arrive at the following assertion.

LEMMA 4.1 (the generalized Lipschitz formula).

$$\sum_{n=1}^{X} \tilde{F}^*(n) = \sum_{\substack{n=1 \\ n\equiv 1 \pmod 2}}^{\sqrt{X}} \mu(n) \sum_{m=1}^{X/n^2} (F^*(m) - F(m)). \tag{4.12}$$

In the case where $\varphi(z)$ is identically equal to 1, (4.12) becomes the usual Lipschitz formula. It was this formula that was used by Vinogradov [2], [3] in proving (1).

If we choose as $\varphi(z)$ the Eisenstein-Maass series $E(z, s)$, it follows from Lemma 4.1 that

$$V_s^*(X) = 2^s \sum_{\substack{n=1 \\ n\equiv 1 \pmod 2}}^{\sqrt{X}} \mu(n)\left(\Psi_s^*\left(\frac{4X}{n^2}\right) - \Psi_s\left(\frac{X}{n^2}\right)\right), \tag{4.13}$$

where $\Psi_s^*(4X)$ is the part of the sum $\Psi_s(4X)$ that extends over those $D \equiv 0 \pmod 4$. For $\Psi_s^*(4X)$ we obviously have the following analogue of Lemma 2.2.

LEMMA 4.2. *As* $X \to \infty$, *for* $0 \leqslant \sigma \leqslant 1$,

$$\Psi_s^*(4X) = 2\zeta(2s)\left(\frac{\zeta(s)}{s+2}\left(\sqrt{X}\right)^{s+2} + \frac{\zeta(1-s)c(s)}{3-s}\left(\sqrt{X}\right)^{3-s}\right) + O\left(\frac{X^{3/4}\ln^3 X}{|s-1|\left(|s-\frac{1}{2}|+\ln^{-1} X\right)}\right), \tag{4.14}$$

where the estimate of the remainder is uniform in s on compact subsets of the critical strip.

Lemma 4.2 and (4.13) prove Theorem 3 stated in the Introduction. This theorem is a direct generalization of the classical Vinogradov-Gauss formula.

Conclusion

The scheme discussed in this paper admits a number of generalizations.

First, it is possible to prove an analogue of Theorem 1 for zeta-functions of orders in real quadratic fields, but this requires a much more complicated technical exposition. The results thus obtained generalize the Vinogradov-Gauss formula for quadratic forms of positive determinant, which was proved by I. M. Vinogradov [16]. To do this, instead of (1.14), for the zeta-function $\zeta_D(s)$ of the order $\mathfrak{O}_D$ of the discriminant D in the real quadratic field $\mathbf{Q}(\sqrt{D})$ one should use the following representation, which can be found in another form in Hecke [17]:

$$\zeta_D(s) = 2\zeta(2s)\frac{\Gamma(s)}{\Gamma^2(s/2)}\left(\sqrt{D}\right)^{-s}\sum_{j=1}^{h(D)}\int_{l_j} E(z,s)\frac{dl}{y},$$

where $h(D)$ is the class number of primitive quadratic forms of discriminant D and $l_j \subset \mathcal{H}$ is an arc of the circle $|z - b_j/2a_j|^2 = D/4a_j$, $z = x+iy$, with initial point at $y = \sqrt{D}/(2a_j\cosh 2R_D)$ and terminal point at $y = \sqrt{D}/2a_j$. Here a_j, b_j and c_j are the coefficients of forms $f_j(x,y) = a_jx^2 + b_jxy + c_jy^2$ constituting a complete system of representatives of the classes of primitive quadratic forms of discriminant D, $R_D = \ln \varepsilon_D$, where ε_D is the fundamental unit of the order $\mathfrak{O}_D$, and dl is the measure on the arc l_j. It is convenient to take the $f_j(x,y)$ to be reduced forms. By partitioning the l_j into several arcs consisting of the roots of the forms of given period, and using the fact that the series $E(z,s)$ is automorphic relative to the group Γ, we can, with good accuracy, replace the Hecke representation by a sum over the roots of all primitive forms in the classical reduction domain. After this, the problem does not differ essentially from the one considered in the main text. Because the formulas are somewhat cumbersome and the space here is insufficient, we will not give the detailed calculations.

Second, instead of the zeta-functions $\zeta_D(s)$ we can consider series $L_D(s,\lambda^k)$, Hecke L-series with Grössencharakters, for imaginary quadratic fields. In Hecke's classical definition,

$$L_D(s,\lambda^k) = \sum_{\alpha}\frac{\lambda^k(\alpha)}{N(\alpha)^s}, \qquad \lambda(\alpha) = e^{iw_D\arg\alpha}, \quad \operatorname{Re} s > 1,$$

where $k \in \mathbf{Z}$, α ranges over the ring of ideal numbers of the order $\mathfrak{O}_D$, and $N(\alpha)$ is the norm of α. Instead of the Eisenstein-Maass series $E(z,s)$, $z \in \mathcal{H}$, there appears a family of Eisenstein series $E_k(g, s)$, where $g \in \mathrm{PSL}(2, \mathbf{R})$ and $k \in \mathbf{Z}$. The result is a two-parameter generalization of the Vinogradov-Gauss formula.

Third, instead of a quadratic extension of the field of rational numbers $\mathbf{Q}$ we can pose the analogous problem for zeta-functions of orders in a quadratic extension of an arbitrary totally real field of algebraic numbers $\mathcal{K}$. In the case of an imaginary extension, the zeta-functions of the orders will have a representation analogous to (1.14) in which $\zeta(s)$ is replaced by $\zeta_{\mathcal{K}}(s)$, the Dedekind zeta-function of $\mathcal{K}$, and $E(z, s)$ by the Eisenstein series corresponding to the full Hilbert modular group of $\mathcal{K}$. The essential difference between this problem and the one considered in the main text is the complicated structure of the fundamental domain of the Hilbert modular group, which significantly impedes estimation of the remainder in this generalization of the Vinogradov-Gauss formula. Thus, our method utilizing the explicit form of the fundamental domain $\mathcal{F}$ and Sonin's summation formula needs to be modified. However, it is suitable for distinguishing the corresponding principal terms. Indeed, it follows from the arguments in §2 that the principal terms of the asymptotics are formed from the "top" of the domain $\mathcal{F}_{\sqrt{X}}$, i.e. from the points $z = x + iy$ with y "close" to $\sqrt{X}$, and the "top" of the fundamental domain of the Hilbert modular group has a very simple structure.

We also mention here the connection between Theorem 1 and Linnik's asymptotics, the uniform distribution of the roots of primitive quadratic forms of discriminant $D < 0$ in the fundamental domain $\mathcal{F}$. It was shown in [18] that for "almost all" D we have

$$\frac{N(\Omega)}{h(D)} = \frac{3}{\pi}\mu(\Omega) + O\left(|D|^{-1/4+\varepsilon}\right),$$

where $\Omega \subset \mathcal{F}_{\sqrt{D}/2}$ and $N(\Omega)$ is the number of corresponding roots in Ω. It is clear that for domains Ω lying at the very "top" of $\mathcal{F}_{\sqrt{D}/2}$ the indicated asymptotic behavior loses its meaning. Roughly speaking, Theorem 1 shows that this must be so. That is to say, if we assume that Linnik's asymptotics hold for all D up to the very "top" of the domain $\mathcal{F}_{\sqrt{D}/2}$, then we can easily calculate the sum $V_s(X)$. It turns out that the orders with respect to X of the two principal terms will be correct, but not the coefficients; they will have a pole at $s = 1$, but will not contain the factors $\zeta(s)$ and $\zeta(1-s)$. Thus, it follows from the Vinogradov-Gauss formula in the critical strip that if Linnik's asyptotics can hold at all up to the very "top", they do only for exceptional values of D. The importance of the extension of the range of Linnik's asymptotics, even if the exceptional values of D, was indicated in [18].

Finally, we mention that for other sequences of values of D (fundamental discriminants, primes, primes in progressions) the method given here is inapplicable. This is connected with the fact that it is precisely for the sequence of all discriminants $D \leqslant X$ that we can reduce $V_s(X)$ to an overall sum $\Psi_s(X)$ that can be calculated with great accuracy. In considering other sequences of values of D we can, by using various discontinuous operators, again reduce the problem to an overall sum, but the situation is more complicated and we have not been able to distinguish a second term in the asymptotics. However, for the indicated sequences of values of D there is another approach, based on A. F. Lavrik's truncated equations and mean density theorems for L-series with quadratic characters. In this way corresponding asymptotic formulas may be obtained which, in particular, prove two conjectures of [6].

BIBLIOGRAPHY

1. Carl Friedrich Gauss, *Werke*. Band II, Königl. Ges. Wiss., Göttingen, 1863.

2. I. M. Vinogradov, *A new method obtaining asymptotic expressions of arithmetic functions*, Izv. Rossiĭsk. Akad. Nauk (6) **1917**, 1347–1378; reprinted in his *Selected works*, Izdat. Akad. Nauk SSSR, Moscow, 1952, pp. 3–28. (Russian)

3. ______, *On the mean value of the number of classes of binary quadratic forms with negative determinant*, Comm. Soc. Math. Kharkow = Soobshch. Khar'kov. Mat. Obshch. (2) **16** (1918), 10–38; reprinted in his *Selected works*, Izdat. Akad. Nauk SSSR, Moscow, 1952. (Russian)

4. ______, *The number of integer points in a ball*, Trudy Mat. Inst. Steklov. **9** (1935), 17–38. (Russian)

5. ______, *On the number of integer points in a ball*, Izv. Akad. Nauk SSSR Ser. Mat. **27** (1963), 957–968. (Russian)

6. D. Goldfeld and C. Viola, *Mean values of L-functions associated to elliptic, Fermat and other curves at the centre of the critical strip*, J. Number Theory **11** (1979), 305–320.

7. A. I. Vinogradov and L. A. Takhtadzhyan, *On analogues of the Vinogradov-Gauss formula*, Dokl. Akad. Nauk SSSR **254** (1980), 1298–1300; English transl. in Soviet Math. Dokl. **22** (1980).

8. Z. I. Borevich and I. R. Shafarevich, *Number theory*, 2nd rev. ed., "Nauka", Moscow, 1972; English transl. of 1st ed., Academic Press, 1966.

9. Atle Selberg and S. Chowla, *On Epstein's zeta-function*, J. Reine Angew. Math. **227** (1967) 86–110.

10. A. Erdélyi, et al., *Higher transcendental functions*, Vol. 2, McGraw-Hill, 1953.

11. Edmund Landau, *Vorlesungen über Zahlentheorie*. Band I: *Aus der elementaren Zahlentheorie*, Hirzel, Leipzig, 1927; English transl., Elementary number theory, Chelsea, New York, 1958.

12. F. Hirzebruch and D. Zagier, *Intersection numbers of curves on Hilbert modular surfaces and modular forms of Nebentypus*, Invent. Math. **36** (1976), 57–113.

13. I. M. Vinogradov, *Foundations of the theory of numbers*, 8th ed., "Nauka", Moscow, 1972; English transl. of sixth ed., Pergamon Press, 1955.

14. ______, *Trigonometrical sums in number theory*, rev. ed., "Nauka", Moscow, 1971; English transl., Statist. Publ. Soc., Calcutta, 1975.

15. Paul Bachmann, *Zahlentheorie*. 2 Teil: *Die analytische Zahlentheorie*, Teubner, Leipzig, 1894.

16. I. M. Vinogradov, *Sur une égalité asymptotique de la théorie des formes quadratiques*, Zh. Fiz.-Mat. Obshch. Perm. Gos. Univ. Vyp. **1** (1918) (1919), 18–28. (Russian; French summary); reprinted in his *Selected works*, Izdat. Akad. Nauk SSSR, Moscow, 1952 (Russian only).

17. Erich Hecke, *Über die Kroneckersche Grenzformel für reelle quadratische Körper und die Klassenzahl relativ-abelscher Körper*, Verh. Naturforsch. Ges. Basel **28** (1917), 363–372; reprinted in his *Mathematische Werke*, Vandenhoeck & Ruprecht, Göttingen, 1959.

18. A. I. Vinogradov and L. A. Takhtadzhyan, *On Linnik-Skubenko asymptotics*, Dokl. Akad. Nauk SSSR **253** (1980), 777–780; English transl. in Soviet Math. Dokl. **22** (1980).

Translated by G. A. KANDALL

Труды Матем. Инст. Стеклов.
158(1981)

Proceedings of the Steklov
Institute of Mathematics
1983, Issue 4

THE DISTRIBUTION OF POINTS ON TWO-SHEETED HYPERBOLOIDS

UDC 511.465 + 517.863

E. P. GOLUBEVA AND O. M. FOMENKO

ABSTRACT. Assuming a conjectural bound for the zeros of Dirichlet L-series, the author determines the asymptotic behavior of the number of lattice points in arbitrary regions on two-sheeted hyerboloids for a large class of diagonal ternary forms.

Bibliography: 22 titles.

1. We consider a ternary quadratic form with integer coefficients and the associated surfaces ($n \in \mathbf{Z}$, $n > 0$)

$$f(x_1, x_2, x_3) = 1, \tag{1}$$

$$f(x_1, x_2, x_3) = n, \tag{2}$$

where in Theorem 1 below f is a diagonal form with at least one positive coefficient, and in Theorems 2–4 the surfaces (1) and (2) are two-sheeted diagonal hyperboloids.

Here and below $\Omega, \Omega_0, \Omega_1, \Omega_2, \ldots$ are bounded rectifiable regions on (1). By the measure $\omega(\Omega)$ of Ω we mean the volume of the cone with vertex at the origin and base Ω. Let $Q(n, \Omega) = Q_f(n, \Omega)$ be the number of lattice points (i.e., points with integer coordinates x_1, x_2, x_3) in the projection from the origin of the region Ω onto the surface (2). We fix Ω_2, and take Ω_1 to be an arbitrary region on the surface (1). If as $n \to \infty$

$$Q_f(n, \Omega_1) = (\omega(\Omega_1)/\omega(\Omega_2))Q_f(n, \Omega_2)(1 + o(1))$$

we say that the lattice points are uniformly distributed on (2). If we then know the asymptotic behavior as ($n \to \infty$) of $Q_f(n, \Omega_2)$, we thereby know the asymptotic behavior of $Q_f(n, \Omega_1)$ for an arbitrary region Ω_1.

In this paper we study the asymptotic distribution of lattice points on the two-sheeted hyperboloids

$$a_1x_1^2 - a_2x_2^2 - a_3x_3^2 = n > 0 \qquad (n \to \infty) \tag{3}$$

where $a_i > 0$, $a_i \in \mathbf{Z}$ ($i = 1, 2, 3$), the determinant $d = a_1a_2a_3$ is squarefree, and $(n, 2d) = 1$. Under a certain assumption concerning the zeros of Dirichlet L-series, we find (Theorem 4) the asymptotic behavior of the number of lattice points in arbitrary regions on (3), and we give estimates for the remainders.

1980 *Mathematics Subject Classification.* Primary 10J25, 10E10, 10E25; Secondary 10B35, 10C05.

The study of representability of numbers by ternary forms started with Gauss's famous formula for the number $Q(n)$ of lattice points on the sphere $x_1^2 + x_2^2 + x_3^2 = n > 0$, which, along with Siegel's theorem, shows that, if $n \neq 4^a\,(8k+7)$, then $Q(n) \gg n^{1/2-\varepsilon}$ ($\varepsilon > 0$ arbitrarily small).

In 1939–1940, Yu. V. Linnik started studying the representation of numbers by certain positive ternary forms (or, equivalently, classes) of a genus (the question of representability by the genus itself had already been studied by Smith and Minkowski) (see [1] or the survey [2]). In 1957 ([3]; reprinted in [1], p. 209–228; reproduced in the book [4]), Linnik proved that the lattice points on a sphere are asymptotically uniformly distributed.

For positive ternary forms the most general results are due to Malyshev [5]. For a large class of forms he obtained two-sided order-sharp estimates for the number of lattice points in any sufficiently good region on the corresponding ellipsoid. A major achievement of Linnik and Malyshev was the possibility of studying positive ternary forms belonging to multi-class genera. However, the problem of the uniform distribution of lattice points on ellipsoids of general type is not completely solved (even conditionally); for diagonal ellipsoids the uniform distribution follows from a combination of the results of Golubeva [6] and the results of Malyshev [5] just mentioned. There is also the more difficult unsolved problem (not even solved conditionally) of finding the asymptotic behavior of the number of lattice points in arbitrary regions on ellipsoids, in particular, on the entire ellipsoid. To be sure, Malyshev [5] and Golubeva [7], [8] have found such asymptotic formulas for certain classes of ellipsoids (for which the total number of lattice points can be found by reducing to the case of a sphere).

The study of indefinite ternary forms began with Linnik's paper [9] (reprinted in [1], p. 141–200; reproduced in [4]), in which he proved an asymptotic formula for the number of lattice points in an arbitrary region on the two-sheeted hyperboloid

$$x_1x_2 - x_3^2 = n > 0 \tag{4}$$

(U. M. Pachev [10] generalized this result somewhat.)

In the more difficult case of the one-sheeted hyperboloid

$$x_1x_2 - x_3^2 = -n < 0 \tag{5}$$

B. F. Skubenko ([11]; reproduced in [4]) obtained uniform distribution of the lattice points.

In [6], Golubeva proved uniform distribution on one- and two-sheeted hyperboloids for a large class of diagonal forms (not including the form $x_1x_2 - x_3^2$, however). The indefinite ternary forms considered by Linnik, Skubenko and Golubeva are zero-forms, i.e., they vanish at some lattice point other than $(0, 0, 0)$.

In the results mentioned above, one needs either the existence of a prime number satisfying certain conditions or else certain unproved though rather weak conjectures on Dirichlet L-series of the H conjecture type (see below). Until now there have been no results on the distribution of lattice points on hyperboloids of general type; the only exception is the above-mentioned note by Golubeva [6].

Here we shall prove uniform distribution of the lattice points on the two-sheeted hyperboloids (3) in the case when the ternary form under consideration is not a zero-form. The ternary forms we study belong to one-class genera; however, unlike the case of positive ternary forms, this condition is not very restrictive, since a classical theorem of Maier ensures us that the genus of indefinite ternary forms has only one class under very general conditions. Here we shall assume that the following conjecture holds:

CONJECTURE H. *There exists $\varepsilon_0 > 0$ such that for any sufficiently large $|m|$ the Dirichlet L-function*

$$L(s, \chi) = \sum_{k=1}^{\infty} \left(\frac{m}{k}\right) \frac{1}{k^s} \tag{6}$$

has no zeros in the rectangle

$$1 - (\log|m|)^{\varepsilon_0 - 1} < \operatorname{Re} s < 1, \qquad |\operatorname{Im} s| < 1. \tag{7}$$

We note that our results can be generalized by letting the variables x_i take values in an arithmetic progression.

2. This section contains the formulation and proof (except for Theorem 2) of our results, along with some comments on the methods used. §3 is devoted to the proof of Theorem 2. We first introduce two definitions in the case of diagonal ternary forms.

DEFINITION 1. Let $f(x_1, x_2, x_3) = a_1x_1^2 + a_2x_2^2 + a_3x_3^2$. A region Ω on the surface (1) given by $f(x_1, x_2, x_3) = 1$ is called an *x_i-domain* (or a *standard domain*) if its points satisfy the inequalities

$$c_1 \leqslant x_i \leqslant c_2, \qquad t_1 \leqslant x_j/x_k \leqslant t_2,$$

where (i, j, k) is any permutation of $(1,2,3)$, and $c_1 = c_1(\Omega)$ and $c_2 = c_2(\Omega)$ are such that $1 - a_ix_i^2 \neq 0$ for $c_1 \leqslant x_i \leqslant c_2$. The curves $x_i = c$ on (1) will be called *standard lines*.

DEFINITION 2. Two x_i-domains Ω_1 and Ω_2 on the surface (1), where $f(x_1, x_2, x_3) = a_1x_1^2 + a_2x_2^2 + a_3x_3^2$, are said to be *equivalent* if

$$c_1(\Omega_1) = c_1(\Omega_2), \qquad c_2(\Omega_1) = c_2(\Omega_2), \qquad \omega(\Omega_1) = \omega(\Omega_2).$$

In other words, two x_i-domains on the surface (1) are equivalent if one is obtained from the other by "translating" along the standard lines $x_i = c$.

Our results on uniform distribution of integral points on two-sheeted hyperboloids are based upon the following rather general assertion.

THEOREM 1. *Suppose that $f(x_1, x_2, x_3)$ is a diagonal quadratic form*

$$f(x_1, x_2, x_3) = a_1x_1^2 + a_2x_2^2 + a_3x_3^2,$$

where $a_i \in \mathbf{Z}$, $a_k \neq 0$ $(i = 1,2,3)$, the determinant $d = a_1a_2a_3$ is squarefree, $a_2a_3 \neq -1$, and at least one of the a_i is positive. Suppose that $n \in \mathbf{Z}$, $n > 0$, $(n, 2d) = 1$, and na_1 and $-nd$ are not perfect squares. Let Ω_1 and Ω_2 be two equivalent x_1-domains on the surface (1)

$$f(x_1, x_2, x_3) = 1.$$

Then

$$Q_f(n, \Omega_1) = Q_f(n, \Omega_2) + O\left(\frac{n^{1/2}L(1,\chi)}{\log^{\eta} n}\right),$$

where

$$\eta > 0, \quad \chi(k) = \left(\frac{-nd}{k}\right), \qquad L(s, \chi) = \sum_{k=1}^{\infty} \frac{\chi(k)}{k^s} \quad (\operatorname{Re} s > 0).$$

Later Theorem 1 will be applied to the case of a two-sheeted hyperboloid. The theorem can also be used to study the distribution of lattice points on ellipsoids and one-sheeted hyperboloids (see [6]).

The proof of Theorem 1, which will not be given here, repeats the proof of Theorem 2 in [7], except that, in addition to the group characters, the argument also involves Grössencharakters of the ideal numbers of the quadratic field. We note that the proof of Theorem 1 uses Conjecture H and a result of Barban and Vekhov [12]; the latter result can be replaced by the use of the dispersion (or variance) method (see [13]; reprinted in [14], pp. 299–316) or the technique in A. I. Vinogradov's paper [15].

Before stating Theorem 2 we note that, when we study a two-sheeted hyperboloid, we shall always consider the upper sheet.

THEOREM 2. *Let*

$$f(x_1, x_2, x_3) = a_1x_1^2 - a_2x_2^2 - a_3x_3^2,$$

where $a_i \in \mathbf{Z}$, $a_i > 0$ $(i = 1, 2, 3)$, *and the determinant* $d = a_1a_2a_3$ *is squarefree. Suppose that* $n \in \mathbf{Z}$, $n > 0$ *and* $(n, 2d) = 1$; Ω_1 *and* Ω_2 *are bounded rectifiable regions on* (1). *Then*

$$Q_f(n, \Omega_1) = \frac{\omega(\Omega_1)}{\omega(\Omega_2)} Q_f(n, \Omega_2) + O\left(\frac{n^{1/2}L(1,\chi)}{\log^\theta n}\right),$$

where $\theta > 0$ *and*

$$L(s, \chi) = \sum_{k=1}^{\infty} \left(\frac{-nd}{k}\right) \frac{1}{k^s} \qquad (\operatorname{Re} s > 0).$$

The proof of this theorem, along with Lemmas 1 and 2 which it requires, are contained in the next section. We now mention only the main features of the proof.

In Linnik's discrete ergodic method one takes a lattice point L on the surface (2). Under the action of the element P_L of a certain discrete group, this point goes to the point $L' = P_L \cdot L$ on (2); we thus have a flow defined on the set of lattice points L on (2), which describes wanderings along a "trajectory of the flow", and in the overwhelming majority of cases the "sojourn time" of a trajectory in a region Λ on (2) is asymptotically proportional to the volume of the cone whose vertex is at the origin and whose base is the projection of Λ onto (1) from the origin.

In the method of the present paper, which stems from [6] and [7], the motion of a point along a surface is replaced by the "motion of an infinitely small region" along the standard lines. Theorem 1 ensures uniform distribution in the narrow strips formed by the standard lines. If we move along two "mutually perpendicular" directions, we can fill the entire region under consideration uniformly with lattice points. The only technical difficulty is the transition from one standard direction to the other, since different standard directions have different standard domains. In Lemma 1 we show that it is possible to make this transition is such a way that the measures of the corresponding standard domains are very close to one another.

In Lemma 2 we show that the number $Q(n,\Omega)$, where Ω is a sufficiently small standard subdomain in Ω_0, is proportional to the measure of Ω.

Theorem 2 can be proved immediately once we have Lemmas 1 and 2.

The plan of proof of Theorem 2 is suitable for a large class of second order surfaces (both definite and indefinite) (see [6]). But the forms for which the absolute value of the determinant d is a perfect square constitute an exception, for example, the form $x_1x_2 - x_3^2$, which was studied by Linnik and Skubenko. In that case Theorem 1 only ensures uniform distribution along one standard direction, since the quadratic form is decomposable along the second direction. When $|d| = d_1^2$, our consideration must be modified. In order that Theorem 2 not become trivial, we must impose a genus condition for f on the number n and find a region Ω_2 for which

$$Q(n,\Omega_2) \gg \frac{n^{1/2}L(1,\chi)}{\psi(n)}, \tag{8}$$

where $\psi(n) = o(\log^\theta n)$, $\theta > 0$.

Using (8), one can easily understand the meaning of Theorem 2: if there are sufficiently many lattice points in some region on the surface of our hyperboloid, then there are also sufficiently many lattice points in any other region of the hyperboloid, and the lattice points are uniformly distributed on the surface of the hyperboloid. We now proceed to Theorem 3, which gives a region Ω_2 on the two-sheeted hyperboloid (1) under consideration for which $Q(n,\Omega_2)$ has an exact expression satisfying (8).

Suppose that the form

$$f(x_1,x_2,x_3) = a_1x_1^2 - a_2x_2^2 - a_3x_3^2 = X^TAX,$$

where $A = [a_1, -a_2, -a_3]$ is a diagonal 3×3-matrix, $X^T = (x_1,x_2,x_3)$, $a_i \in \mathbf{Z}$, $a_i > 0$ $(i = 1,2,3)$, and the determinant $d = |A| = a_1a_2a_3$ is squarefree, is not a zero-form. Let $n > 0$ be an integer, $(n, 2d) = 1$, which satisfies the genera conditions for f. Let $\Gamma(A)$ be the group of units for A (with determinant $+1$). The group $\Gamma(A)$ acts discretely on the surfaces (1) and (2) according to the rule $X \to UX$, where $U \in \Gamma(A)$; let $F(A,n)$ be a $\Gamma(A)$-fundamental domain on (2), and let $F(A) = F(A,1)$ be the corresponding $\Gamma(A)$-fundamental domain on (1) ($F(A,n)$ is obtained by projecting $F(A)$ onto (2) from the origin). $F(A)$ is a compact polygon on the surface of the hyperboloid (1). Below $h(-D)$ denotes the number of proper primitive classes of positive definite binary quadratic forms $ax^2 + 2bxy + cy^2$ having determinant $D = ac - b^2 > 0$. Next, we call a point (x_1,x_2,x_3) *primitive* if g.c.d.$(x_1,x_2,x_3) = 1$.

THEOREM 3. *The number $Q'_f(n, F(A))$ of primitive lattice points in the polygon $F(A,n)$ satisfies the formula*

$$Q'_f(n, F(A)) = 2^{-\nu(d)}h(-D)\rho_D, \tag{9}$$

where $D = nd > 0$, $\nu(d)$ is the number of odd prime divisors of d; $\rho_D = \rho$ depends on D (mod 8) and is defined as follows: $\rho = 1/2$ if $D \equiv 1,2,4,5,6 \pmod 8$; $\rho = 2$, if $D \equiv 7 \pmod 8$; if $D \equiv 3 \pmod 8$, then $\rho = 1/3$, except for two cases when $\rho = 1$: i) the Hasse invariant $c_2(f) = 1$; ii) $D = 3$; $\rho = 1/4$, if $D \equiv 0 \pmod 8$. The total number of lattice points in $F(A,n)$ is equal to

$$Q_f(n, F(A)) = \sum_{t^2 | n} Q'_f(n/t^2, f(A)). \tag{10}$$

Here a lattice point $\mathfrak{g}$ *which is a vertex of* $F(A, n)$ *is taken with weight* $(1/E)(A, \mathfrak{g})$, *where* $E(A, \mathfrak{g})$ *is the order of the (finite) group* $\Gamma(A, \mathfrak{g}) = \{U \in \Gamma(A) \mid U_{\mathfrak{g}} = \mathfrak{g}\}$.

To prove Theorem 3, we recall that the form of a $\Gamma(A)$-fundamental domain for an arbitrary two-sheeted hyperboloid was studied long ago by Fricke and Klein [16], and in our day by Venkov [17]. The fact that f is not a zero-form guarantees compactness of the fundamental domain. The formula (9) is proved by directly applying the results of Jones ([18]; reproduced in [19]) or else using results of Maass [20] on computing the Fourier coefficients of theta-series associated to indefinite ternary forms, and then using Siegel's theorem [21]. One need only take into account that the genus of f has only one class, since d is squarefree; this follows from a theorem of Maier (see [19]).

The formulas (9) and (10) are still valid in the case of a two-sheeted hyperboloid associated to a ternary zero-form in a genus having one class; here the fundamental domain $F(A)$ is noncompact, or, in picturesque language, it has infinite "horns" (in this connection, see [17]). Linnik encountered an analogous situation: he considered a triangle on the surface (4) (the reduction domain for positive definite binary quadratic forms) with the angles $\pi/3$, $\pi/3$, 0. To apply our Theorem 2, which requires boundedness of the regions under consideration, one must "cut off the horns" of the fundamental domain $F(A)$ in a way similar to what Linnik did in his proof. However, this requires some additional considerations, and therefore the case of zero-forms will not be included in this paper. It should be noted that for an entire class of ternary zero-forms Golubeva [6] found finite regions Ω on one-sheeted and two-sheeted hyperboloids (1) for which one can compute the asymptotic behavior of $Q(n, \Omega)$. Here both the case of one class and the case of many classes were considered. We note that the group $\Gamma(A)$ does not act discretely on the one-sheeted hyperboloid $f(x_1, x_2, x_3) = -n < 0$, and the fundamental domain does not exist; hence, formulas of the type (9) are not obtained by our method. Skubenko examined the case of the simplest one-sheeted hyperboloid (5). On the surface (5) we take the classical reduction domain Δ for indefinite binary quadratic forms; let $T(\Delta)$ be the number of integral points in Δ. After proving an interesting "cycle theorem", Skubenko derived the asymptotic formula

$$\log T(\Delta) \sim \tfrac{1}{2} \log n \qquad (n \to \infty).$$

However the asymptotic behavior of $T(\Delta)$ itself is as yet unknown, although Takhtadzhyan [22] proved such an asymptotic formula using a certain difficult assumption connected with Linnik's conjecture on a sum of Kloosterman sums.

Theorems 2 and 3 immediately imply the main Theorem 4. We need only note that we have considered above all of the lattice points on the surface. It is not difficult to carry over our results to the primitive points.

THEOREM 4. *Suppose the following form is not a zero-form*:

$$f(x_1, x_2, x_s) = a_1 x_1^2 - a_2 x_2^2 - a_3 x_3^2,$$

where $a_i \in \mathbf{Z}$, $a_i > 0$ $(i = 1, 2, 3)$ *and* $d = a_1 a_2 a_3$ *is squarefree. Suppose that* $n > 0$ *is an integer*, $(n, 2d) = 1$, *which satisfies the genera conditions for* f, *and* Ω *is an arbitrary bounded rectifiable region on* (1). *Then, in the notation of Theorem* 3,

$$Q_f'(n,\Omega) = \frac{\omega(\Omega)}{\omega(F(A))} 2^{-\nu(d)} h(-D)\rho_D + O\left(\frac{n^{1/2}L(1,\chi)}{\log^\theta n}\right), \tag{11}$$

$$Q_f(n,\Omega) = \frac{\omega(\Omega)}{\omega(F(A))} Q_f(n, F(A)) = O\left(\frac{n^{1/2}L(1,\chi)}{\log^\theta n}\right); \tag{12}$$

here $\theta > 0$, $D = nd$,

$$L(s,\chi) = \sum_{k=1}^{\infty} \left(\frac{-D}{k}\right)\frac{1}{k^s} \qquad (\operatorname{Re} s > 0),$$

and the expression for $Q_f(n, F(A))$ *is given in* (9) *and* (10).

3. In this section we prove Theorem 2. We first prove two lemmas.

LEMMA 1. *Let* Ω *be an* x_3*-domain on the surface* (1):

$$f(x_1, x_2, x_3) = 1$$

(*where* $f(x_1, x_2, x_3)$ *satisfies the conditions of Theorem* 2) *such that*

$$0 \leqslant c_1 \leqslant x_3 \leqslant c_2, \qquad 0 \leqslant x_2/x_1 \leqslant t_1,$$

where for some constants A and B

$$c_1 - c_2 = t_2 \leqslant At, \qquad t_1 \leqslant Bt.$$

Let $\tilde{\Omega}$ *be the smallest* x_2*-domain containing* Ω, *let* $\tilde{\Omega}'$ *be an arbitrary region equivalent to* $\tilde{\Omega}$ *and such that its points have* $x_1 \leqslant C$, *where C is some constant, and let* $\tilde{\tilde{\Omega}}'$ *be the smallest* x_3*-domain containing* $\tilde{\Omega}'$. *Then as* $t \to 0$

$$\omega(\Omega) = t_1 t_2/3a_1 + O(t^3), \tag{13}$$

$$\omega(\tilde{\Omega}) = \omega(\tilde{\Omega}') = \omega(\Omega) + O(t^3), \tag{14}$$

$$\omega(\tilde{\tilde{\Omega}}') = \omega(\tilde{\Omega}') + O(t^3), \tag{15}$$

where the constants in O depend on the form f and on A, B and C.

PROOF. We make the change of variables

$$x_1 = \left(r/\sqrt{a_1}\right)\cosh\varphi\cosh\theta, \quad x_2 = \left(r/\sqrt{a_2}\right)\cosh\varphi\sinh\theta, \quad x_3 = \left(r/\sqrt{a_3}\right)\sinh\varphi.$$

Then

$$\omega(\Omega) = \frac{1}{\sqrt{d}}\int_0^1 r^2\,dr \int_{\operatorname{arcsinh}\sqrt{a_3}\,c_1}^{\operatorname{arcsinh}\sqrt{a_3}\,c_2} \cosh\varphi\,d\varphi \int_0^{\operatorname{arctanh}\sqrt{a_2/a_1}\,t_1} d\theta$$

$$= \frac{1}{\sqrt{d}}\frac{1}{3}\sqrt{a_3}\,t_2\sqrt{\frac{a_2}{a_1}}\,t_1 + O(t^3) = \frac{t_1 t_2}{3a_1} + O(t^3),$$

and (13) is proved.

The region $\tilde{\Omega}$ is defined by the inequalities

$$0 \leqslant x_2 \leqslant t_1 \sqrt{\frac{1 + a_3 c_3^2}{a_1 - a_2 t_1^2}} = l_1,$$

$$m_1 = \sqrt{\frac{(a_1 - a_2 t_1^2) c_1^2}{1 + a_3 c_1^2}} \leqslant \frac{x_3}{x_1} \leqslant \frac{\sqrt{a_1}\, c_2}{\sqrt{1 + a_3 c_2^2}} = m_2.$$

We make a change of variables

$$x_1 = \left(r/\sqrt{a_1}\right)\cosh\varphi\cosh\theta, \quad x_2 = \left(r/\sqrt{a_2}\right)\sinh\varphi, \quad x_3 = \left(r/\sqrt{a_3}\right)\cosh\varphi\sinh\theta,$$

then

$$\begin{aligned}
\omega(\Omega) &= \frac{1}{\sqrt{d}} \int_0^1 r^2\, dr \int_0^{\operatorname{arcsinh}\sqrt{a_2}} \cosh\varphi\, d\varphi \int_{\operatorname{arctanh} m_2\sqrt{a_3/a_1}}^{\operatorname{arctanh} m_1\sqrt{a_3/a_1}} d\theta \\
&= \frac{1}{3}\sqrt{\frac{a_2}{d}}\, l \left(\operatorname{arctanh} m_1 \sqrt{\frac{a_3}{a_1}} - \operatorname{arctanh} m_2 \sqrt{\frac{a_2}{a_1}} \right) + O(t^3) \\
&= \frac{1}{3} \frac{1}{\sqrt{a_1 a_3}} t_1 \frac{\sqrt{1 + a_3 c_2^2}}{\sqrt{a_1}} \sqrt{\frac{a_3}{a_1}} (m_2 - m_1) \frac{1}{1 - a_1 c_2^2 a_3/(1 + a_3 c_2^2) a_1} + O(t^3) \\
&= \frac{1}{3} \frac{t_1\left(\sqrt{1 + a_3 c_2^2}\right)^3 t_2 \sqrt{a_1}\left(\sqrt{1 + a_3 c_2^2} - c_2^2 a_3/\sqrt{1 + a_3 c_2^2}\right)}{a_1 \sqrt{a_1}\left(1 + a_3 c_2^2\right)} + O(t^3) \\
&= \frac{1}{3} \frac{t_1 t_2}{a_1} + O(t^3).
\end{aligned}$$

Thus, (14) is proved. In exactly the same way we obtain (15) by a direct computation.

LEMMA 2. *Let Ω_0 be the region on the surface* (1) (*where $f(x_1, x_2, x_3)$ satisfies the conditions of Theorem* 2) *given by the condition $0 < x_1 \leqslant C$, and let Ω be an arbitrary x_3-domain for which*

$$c_2 - c_1 = \log^{\xi} n, \qquad t_1 \leqslant x_2/x_1 \leqslant t_1 + \log^{\xi} n, \tag{16}$$

where $\xi = \frac{2}{5}\eta$, and η is the constant in Theorem 1. *Then*

$$Q(n, \Omega) = \frac{\omega(\Omega)}{\omega(\Omega_0)} Q(n, \Omega_0) + O\left(\frac{n^{1/2} L(1, \chi)}{\log^{\eta} n} \right). \tag{17}$$

PROOF. We denote $n^{1/2} L(1, \chi)/\log^{\eta} n = N$. By Theorem 1, there exists a constant A such that for any equivalent region $\Omega^{(1)}$ and $\Omega^{(2)}$

$$-AN < Q(n, \Omega^{(1)}) - Q(n, \Omega^{(2)}) < AN. \tag{18}$$

From Theorem 3 or from the results in [12] it follows that

$$Q(n, \Omega_0) = O\left(n^{1/2} L(1, gc)(\log\log \neq)^{\gamma}\right), \tag{19}$$

where $\gamma > 0$ is a constant, and the constant in O depends on f, γ and C. Thus, for any constant A_1 and for n sufficiently large

$$A_1 \frac{Q(n, \Omega_0)}{\log^{3\xi} n} \leqslant AN. \tag{20}$$

We claim that

$$-6AN < Q(N, \Omega) - \frac{\omega(\Omega)}{\omega(\Omega_0)} Q(n, \Omega_0) < 6AN. \tag{21}$$

If we suppose the contrary, then there exist n arbitrarily large for which

$$Q(n, \Omega) - \frac{\omega(\Omega)}{\omega(\Omega_0)} Q(n, \Omega_0) > 6AN, \tag{22}$$

or

$$Q(n, \Omega) - \frac{\omega(\Omega)}{\omega(\Omega_0)} Q(n, \Omega_0) < -6AN. \tag{23}$$

Suppose that (22) holds. We consider Ω', a region equivalent to Ω, whose points satisfy $0 \leqslant x_2/x_1 \leqslant t$, where t is some constant. Then, by (18),

$$Q(n, \Omega') - \frac{\omega(\Omega')}{\omega(\Omega_0)} Q(n, \Omega_0) > 5AN.$$

Let $\tilde{\Omega}'$ be the smallest x_2-domain containing Ω'; then, obviously,

$$Q(n, \tilde{\Omega}') \geqslant Q(n, \Omega').$$

By Lemma 1,

$$\omega(\tilde{\Omega}') = \omega(\Omega') + O\left(\frac{1}{\log^{3\xi} n}\right).$$

Hence, for n sufficiently large, from (20) we obtain

$$Q(n, \tilde{\Omega}') - \frac{\omega(\tilde{\Omega}')}{\omega(\Omega_0)} Q(n, \Omega_0) > 5AN - A_1 \frac{Q(n,\Omega_0)}{\log^{3\xi} n} > 4AN.$$

Similarly, for an arbitrary region $\tilde{\Omega}''$ equivalent to $\tilde{\Omega}'$,

$$Q(\tilde{\Omega}'') - \frac{\omega(\tilde{\Omega}'')}{\omega(\Omega_0)} Q(n, \Omega_0) > 3AN,$$

and for $\tilde{\tilde{\Omega}}''$ the smallest x_3-domain containing $\tilde{\Omega}''$,

$$Q(n, \tilde{\tilde{\Omega}}'') - \frac{\omega(\tilde{\tilde{\Omega}}'')}{\omega(\Omega_0)} Q(n, \Omega_0) > 2AN.$$

Finally, for an arbitrary region $\tilde{\tilde{\Omega}}'''$ equivalent to $\tilde{\tilde{\Omega}}''$,

$$Q(n, \tilde{\tilde{\Omega}}''') - \frac{\omega(\tilde{\tilde{\Omega}}''')}{\omega(\Omega_0)} Q(n, \Omega_0) > AN. \tag{24}$$

We partition Ω_0, except for a strip along the boundary of width $O(\log^{\xi} n)$, into regions $\Omega^{(k)}$ of the type $\tilde{\tilde{\Omega}}'''$ which intersect only along their boundaries; the number of regions is $K = D_1 \log^{2\xi} n$, where D_1 is some constant.

Summing (24) over the $\Omega^{(k)}$, we then have

$$\sum_{k=1}^{K}\left(Q(n,\Omega^{(k)})-\frac{\omega(\Omega^{(k)})}{\omega(\Omega_0)}Q(n,\Omega_0)\right)>D_1\log^{2\xi}n\cdot AN.$$

Since $\Omega^{(k)}\subset\Omega_0$, we have

$$Q(n,\Omega_0)\geqslant\sum_{k=1}^{K}Q(n,\Omega^{(k)})+O(\log^{2\xi+1}n).$$

In addition, obviously

$$\sum_{k=1}^{K}\omega(\Omega^{(k)})=\omega(\Omega_0)+O\left(\frac{\log^{\xi}n}{\log^{2\xi}n}\right).$$

Thus,

$$Q(n,\Omega_0)-Q(n,\Omega_0)>D_1\log^{2\xi}n\cdot AN-E\frac{n^{1/2}L(1,\chi)(\log\log n)\gamma}{\log^{\xi}n},$$

where E is some constant. Since $\xi=2\eta/5$, the last inequality cannot hold for arbitrarily large n, and the lemma is proved in this case.

The case of inequality (23) is analyzed in exactly the same way.

PROOF OF THEOREM 2. We partition Ω_0 into regions $\Omega^{(k)}$ of the type (16). By Lemma 2,

$$Q(n,\Omega^{(k)})=\frac{\omega(\Omega^{(k)})}{\omega(\Omega_0)}Q(n,\Omega_0)+O\left(\frac{n^{1/2}L(1,\chi)}{\log^{\eta}n}\right). \tag{25}$$

Summing (24) over the regions $\Omega^{(k)}$ which intersect with Ω_1, we then have

$$Q(n,\Omega_1)=\frac{\omega(\Omega_1)}{\omega(\Omega_0)}Q(n,\Omega_0)+O\left(\frac{n^{1/2}L(1,\chi)}{\log^{\eta}n}\cdot\log^{2\xi}n\right)$$
$$+O\left(\sum_{\Omega^{(k)}}{}'\frac{1}{\log^{2\xi}n}Q(n,\Omega_0)\right),$$

where the summation in Σ' is taken over the regions $\Omega^{(k)}$ along the boundary of Ω_1. Consequently, Σ' contains $O(\log^{\xi}n)$ terms, and, by Lemma 2,

$$\begin{aligned}Q(n,\Omega_1)&=\frac{\omega(\Omega_1)}{\omega(\Omega_0)}Q(n,\Omega_0)+O\left(\frac{n^{1/2}L(1,\chi)}{\log^{\eta/5}n}\right)+O\left(\frac{Q(n,\Omega_0)}{\log^{2\eta/5}n}\right)\\&=\frac{\omega(\Omega_1)}{\omega(\Omega_0)}Q(n,\Omega_0)+O\left(\frac{n^{1/2}L(1,\chi)}{\log^{\eta/5}n}\right).\end{aligned} \tag{26}$$

In exactly the same way,

$$Q(n,\Omega_2)=\frac{\omega(\Omega_2)}{\omega(\Omega_0)}Q(n,\Omega_0)+O\left(\frac{n^{1/2}L(1,\chi)}{s\log^{\eta/5}n}\right). \tag{27}$$

Comparing (26) and (27), we obtain the theorem.

BIBLIOGRAPHY

1. Yu. V. Linnik, *Selected Works*. Vol. 1: *Number theory. The ergodic method and L-functions*, "Nauka", Leningrad, 1979. (Russian)

2. Yu. V. Linnik and A. V. Malyshev, *Applications of the arithmetic of quaternions to the theory of ternary quadratic forms and to the decomposition of numbers into cubes*, Uspekhi Mat. Nauk **8** (1953), no. 5 (57), 3–71; English transl. in Amer. Math Soc. Transl. (2) **3** (1956).

3. Yu. V. Linnik, *Asymptotic-geometric and ergodic properties of sets of lattice points on a sphere*, Mat. Sb. **43 (85)** (1957), 257–276; English transl. in Amer. Math. Soc. Transl. (2) **13** (1960).

4. ______, *Ergodic propeties of algebraic fields*, Izdat. Leningrad. Univ., Leningrad, 1967; English transl., Springer-Verlag, 1968.

5. A. V. Malyshev, *On the representation of integers by positive quadratic forms*, Trudy Mat. Inst. Steklov **65** (1962). (Russian)

6. E. P. Golubeva, *On the representation of large numbers by ternary quadratic forms*, Dokl. Akad. Nauk SSSR **191** (1970), 519–521; English transl. in Soviet Math. Dokl. **11** (1970).

7. ______, *Asymptotics of the number of lattice points on certain ellipsoids*, Mat. Zametki **11** (1972), 625–634; English transl. in Math. Notes **11** (1972).

8. ______, *The asymptotic behavior of the number of representations of large numbers by certain ternary positive quadratic forms*, Zap. Nauchn. Sem. Leningrad. Otdel. Mat. Inst. Steklov. (LOMI) **76** (1978), 53–59; English transl. in J. Soviet Math. **18** (1982), no. 3.

9. Yu. V. Linnik, *The asymptotic distribution of reduced binary quadratic forms in relation to the geometry of Lobachevsky*. I, II, III, Vestnik Leningrad. Univ. **10** (1955), no. 2, 3–23; no. 5, 3–32; no. 8, 15–27. (Russian)

10. U. M. Pachev, *On the distribution of integer points on some two-sheeted hyperboloids*, Zap. Nauchn. Sem. Leningrad. Otdel. Mat. Inst. Steklov. (LOMI) **93** (1980), 87–141; English transl., to appear in J. Soviet Math.

11. B. F. Skubenko, *The asymptotic distribution of lattice points on a hyperboloid of one sheet, and ergodic theorems*, Izv. Akad. Nauk SSSR Ser. Mat. **26** (1962), 721–752. (Russian)

12. M. B. Barban and P. P. Vekhov, *Summation of multiplicative functions of polynomials*, Mat. Zametki **5** (1969), 669–680; English transl. in Math Notes **5** (1969).

13. B. M. Bredikhin and Yu. V. Linnik, *Asymptotic behavior and ergodic properties of solutions of the generalized Hardy-Littlewood equation*, Mat. Sb. **71 (113)** (1966), 145–161; English transl. in Amer. Math. Soc. Transl. (2) **82** (1969).

14. Yu. V. Linnik, *Selected Works*. Vol. 2: *Number theory. L-functions and the dispersion method*, "Nauka", Leningrad, 1980. (Russian)

15. A. I. Vinogradov, *The general Hardy-Littlewood equation*, Mat. Zametki **1** (1967), 189–197; English transl. in Math. Notes **1** (1967).

16. Robert Fricke and Felix Klein, *Vorlesungen über die Theorie der automorphen Funktionen*, Vol. I, Teubner, Leipzig, 1892; reprint, Johnson Reprint, New York, and Teubner, Stuttgart, 1965.

17. B. A. Venkov, *On the fundamental domain of an indefinite ternary quadratic form*, Uchen. Zap. Leningrad. Gos. Ped. Inst. **14** (Fiz.-Mat. Fak. **1**) (1955), 16–45. (Russian)

18. Burton W. Jones, *Representations by quadratic forms*, Ann. of Math. (2) **50** (1949), 884–899.

19. ______, *The arithmetic theory of quadratic forms*, Math. Assoc. Amer. and Wiley, New York, 1950.

20. Hans Maass, *Über die räumliche Verteilung der Punkte in gittern mit indefiniter Metrik*, Math. Ann. **138** (1959), 287–315.

21. K. G. Ramanathan, *On the analytic theory of quadratic forms*, Acta Arith. **21** (1972), 423–436.

22. L. A. Takhtadzhyan, *Asymptotic formula for the sum of the lengths of the periods of the quadratic irrationalities of discriminant D*, Zap. Nauchn. Sem. Leningrad. Otdel. Mat. Inst. Steklov. (LOMI) **91** (1979), 134–144; English transl., J. Soviet Math **17** (1981), no. 5.

Translated by N. KOBLITZ

Труды Матем. Инст. Стеклов.
158(1981)

Proceedings of the Steklov
Institute of Mathematics
1983, Issue 4

ALGEBRAIC PROPERTIES OF REGULARLY CLOSED FIELDS

UDC 516.3

YU. L. ERSHOV

ABSTRACT. This paper gives an account of the basic algebraic properties of the class of regularly closed fields. This class has arisen in the study of the elementary theories of fields. For such fields an analogue of the Chebotarev density theorem is obtained.
Bibliography: 12 titles.

The purpose of this paper is to give an account of the basic algebraic properties of regularly closed fields—fields whose importance first became apparent in the study of the elementary theories of finite fields [2], [5]. The subsequent studies [8], [10] and [12] revealed other interesting classes of regularly closed fields. In [1] the author has stated a number of general theorems concerning the elementary theories of regularly closed fields. It should be noted that different papers call this notion by different names (Σ-field, Ax field), and foreign writers have recently begun in the main to call such fields pseudo-algebraically closed [7], [9]; but in [12] this term has been used for a stronger property, and we shall therefore stick to the term proposed in [1], which more closely reflects the essence of the matter.

DEFINITION. A field F is called *regularly closed* if for any finitely generated regular extension $F(t_1,\ldots,t_n)$ there exists an F-homomorphism of the ring $F[t_1,\ldots,t_n]$ onto F.

REMARK. For the definition in [1] we used the property that every absolutely irreducible variety defined over F has an F-rational point. In the present paper we shall avoid algebraic-geometrical terminology, using only the language of field theory.

PROPOSITION 1. *For a field F the following properties are equivalent*:

1. *F is regularly closed.*

2. *For any finitely generated regular extension $F(t_1,\ldots,t_n)$ there exists an F-place p: $F(t_1,\ldots,t_n) \mapsto F$ which is finite on all elements $t_1,\ldots,t_n$.*

3. *For any absolutely irreducible polynomial $f \in F[x_1,\ldots,x_k, y]$ which, as a polynomial in y, is unitary, separable and of degree greater than* 1, *and for $0 \neq g \in F[x_1,\ldots,x_k]$, there exist elements $\alpha_1,\ldots,\alpha_k$ and β in F such that $f(\alpha_1,\ldots,\alpha_k,\beta) = 0$ and $g(\alpha_1,\ldots,\alpha_k) \neq 0$.*

$1 \Rightarrow 2$. We prove a slightly more general fact, which will also be needed later.

1980 *Mathematics Subject Classification.* Primary 12F20; Secondary 03C10, 12A75, 12F10, 12L05.

LEMMA 1. *Let F be a regularly closed field and $F \leqslant F_0 \leqslant F_1$ a tower of extensions such that F_1 is a finitely generated regular extension of F and a finite separable extension of F_0. Then for any finite subset $S \subseteq F_1$ there exist an F-place p: $F_0 \mapsto F$ and an extension π of p to F_1 such that π is finite over S, F-valued (i.e., $\pi(F_1) = F$) and unramified over p.*

In F_0 pick a separating transcendence basis $t = t_1,\ldots,t_n$; then $F \leqslant F(\bar{t}) \leqslant F_0 \leqslant F_1$, and F_1 is a finite separable extension of $F(\bar{t})$. From the statement of the lemma it is evident that we need prove it only for the case $F_0 = F(\bar{t})$. Let $\alpha \in F_1$ be an element such that α is integral over $F[\bar{t}]$ and $F_1 = F_0(\alpha)$ $(= F(\bar{t}, \alpha))$; denote by $h(y)$ the minimal polynomial for α over F. Then $h(y) \in F[\bar{t}, y]$ $(= F[\bar{t}][y])$. Let $d(h)$ be the discriminant of h; then $h(h) \in F[\bar{t}]$ and $d(h) \neq 0$, since α is separable over F_0. Any element $\beta \in S$ can be represented in the form $\beta = r_0\alpha^k + r_1\alpha^{k-1} + \cdots + r_k$, where $r_i \in F(\bar{t})$, $i = 0,\ldots,k$, $k \rightleftharpoons [F_1 ; F_0] - 1$. Let g_β be any nonzero element of $F[\bar{t}]$ which is divisible by all the denominators of the rational functions r_i, $i = 0,\ldots,k$; then $\beta g_\beta \in F[\bar{t}, \alpha]$. Put $g \rightleftharpoons d(h)\Pi_{\beta\in S} g_\beta$; then $g \neq 0$, $g \in F[\bar{t}]$, and $\beta_g \in F[\bar{t}, \alpha]$, $\beta \in F[\bar{t}, g^{-1}, \alpha]$ for $\beta \in S$.

Since F is regularly closed and $F_1 = F(\bar{t}, g^{-1}, \alpha)$ is a regular extension of F, there exists an F-homomorphism φ of the ring $F[\bar{t}, g^{-1}, \alpha]$ into F. Observe that $\varphi(g) \neq 0$ and $\varphi(d(h)) \neq 0$; consequently the discriminant of the polynomial $(\varphi h)(y)$, which is $\varphi(d(h))$, is different from zero. Let p: $F(t) \mapsto F$ be an F-place which is finite on $t_1,\ldots,t_n$ and such that $p(t_i) = \varphi(t_i)$, $i = 1,\ldots,n$. (For $n = 1$ the existence of such a place is obvious; the general case is established by a simple induction.) The place p (is finite and) coincides with φ on the ring $F[\bar{t}]$; consequently $p(g) = \varphi(g) \neq 0$, p is defined on g^{-1}, and $p(g^{-1}) = \varphi(g^{-1})$. Since the discriminant of the polynomial $[p(h)](y) = (\varphi h)(y)$ is nonzero, it follows that any extension of p to F_1 must be unramified, and the relative degree of the extension is determined by the degree of the image of α over F under the extension. Let π be an extension of p to F_1 such that $\pi(\alpha) = \varphi(\alpha)$; the existence of such an extension follows from the fact that $\varphi(\alpha)$ is a root of the polynomial $p(h) = \varphi(h)$, since φ is a homomorphism from $F[\bar{t}, g^{-1}, \alpha]$ to F. Since $\varphi(\alpha) \in F$, it follows that π is an F-valued place which is finite on $\{t_1,\ldots,t_n, g^{-1}, \alpha\}$. But then π is also finite on $F[t_1,\ldots,t_n, g^{-1}, \alpha]$ and on $S \subseteq F[t_1,\ldots,t_n, g^{-1}, \alpha]$.

$2 \Rightarrow 1$. Obvious.

$1 \Rightarrow 3$. Let F be regularly closed, and f and g as in the statement of property 3. Let $F_0 \rightleftharpoons F(t_1,\ldots,t_k)$ be a pure transcendental extension of F, and γ a root of the polynomial $f(\bar{t}, y) \in F_0[y]$; then $F_1 \rightleftharpoons F_0(\gamma)$ is a regular extension of F. Consequently, there exists an F-homomorphism φ: $F[t_1,\ldots,t_k, g(\bar{t})^{-1}, \gamma] \to F$; then for $\alpha_i \rightleftharpoons \varphi(t_i)$, $i = 1,\ldots,k$, and $\beta \rightleftharpoons \varphi(\gamma)$ we have $f(\bar{\alpha}, \beta) = 0$ and $g(\bar{\alpha}) \neq 0$.

$3 \Rightarrow 1$. Suppose F satisfies condition 3. Let F_1 be a finitely generated regular extension of F, and S a finite subset of F_1. As in the proof of Lemma 1, we can find a separating transcendence basis $t_1,\ldots,t_n$ and polynomials $f \in F[x_1,\ldots,x_n, y]$, $g \in F[x_1,\ldots,x_n]$ such that $h(y) \rightleftharpoons f(\bar{t}, y)$ is the minimal polynomial for an element $\gamma \in F_1$ such that $F_1 = F(\bar{t}, \gamma)$ and γ is integral and separable over $F(\bar{t})$, while the element $g_* \rightleftharpoons g(\bar{t})$ is such that $S \subseteq F[\bar{t}, g_*^{-1}, \gamma]$. Observe that the polynomial f is absolutely irreducible over F, since $F_1 = F(t_1,\ldots,t_n, \gamma)$ is a regular extension of F. By property 3, F contains elements $\alpha_1,\ldots,\alpha_n$ and β such that $f(\bar{\alpha}, \beta) = 0$ and

$g(\bar{\alpha}) \neq 0$. The correspondence $t_i \mapsto \alpha_i$, $\gamma \mapsto \beta$ extends to an F-homomorphism φ_* of the ring $F[\bar{t}, \gamma]$ into F. If $\mathfrak{J}$ is the kernel of φ_*, then φ_* extends uniquely to an F-homomorphism ψ from $F[\bar{t}, \gamma]\mathfrak{J}$—the ring of fractions with respect to the prime ideal $\mathfrak{J}$—to F. Since $\varphi_*(g_*) = g(\bar{\alpha}) \neq 0$, we have $g_*^{-1} \in F[\bar{t}, \gamma]_{\mathfrak{J}}$; consequently, $F[S] \leqslant F[\bar{t}, g_*^{-1}, \gamma] \leqslant F[\bar{t}, \gamma]_{\mathfrak{J}}$, and $\varphi \rightleftharpoons \psi \upharpoonright F[S]$ gives an F-homomorphism of $F[S]$ into F.

COROLLARY. *If F_i, $i \in I$, is a directed (by inclusion) family of regularly closed fields, then the field $\bigcup_{i \in I} F_i$ is likewise regularly closed.*

This is easily verified by using property 3.

We show now that the property of being regularly closed is preserved under separable algebraic extension.

REMARK. This has already been noted previously [5], [9] with reference to Weil's method of descent. We give here an independent proof, avoiding algebraic-geometrical techniques.

PROPOSITION 2. *Let F be a regularly closed field, and F_0 a separable algebraic extension of F. Then F_0 is likewise regularly closed.*

LEMMA 2. *Let F_0 be a finitely generated separable extension of a field F, and L_0 the algebraic closure of F in F_0. Then there exist extensions $F_0 \leqslant L$ and $F \leqslant L_1$ such that L_1 is a finitely generated regular extension of F and $L = L_0 L_1$.*

We shall regard F_0 as imbedded in some universal domain F_1 over F; i.e., F_1 is an algebraically closed field containing F and of infinite degree of transcendence over F. Let $n \rightleftharpoons [L_0 : F]$, and let $\varphi_1 \rightleftharpoons \mathrm{id}_{L_0}$, be all the different F-imbeddings of L_0 into F_1 (L_0 is separable over F). Extend the imbeddings $\varphi_1, \ldots, \varphi_n$ to isomorphic imbeddings $\psi_1 \rightleftharpoons \mathrm{id}_{F_0}, \ldots, \psi_n$ of F_0 into F_1 such that, for $1 \leqslant i < j \leqslant n$, the images $\psi_i(F_0)$ and $\psi_i(F_0)$ are algebraically disjoint over F. Let L_* be the composite of the fields $\varphi_1(L_0), \ldots, \varphi_n(L_0)$; then L_* is the smallest Galois extension of F that contains L_0. Let G be the Galois group $G(L_*/F)$. Let $\bar{L}$ be the composite of the fields $\psi_1(F_0), \ldots, \psi_n(F_0)$; L is a separable finitely generated extension of F. If by K_i we denote the composite $\psi_i(F_0)L_*$, $i = 1, \ldots, n$, then L_* is the algebraic closure of F in K_i and K_j is a regular extension of L_*, for each $i = 1, \ldots, n$. From the construction it follows that K_i and K_j are algebraically disjoint over L_* $(i \neq j)$; consequently they are linearly disjoint over L_*, and $L = K_1 \cdot \cdots \cdot K_n = K_1 \oplus_{L_*} K_2 \oplus_{L_*} \cdots \oplus_{L_*} K_n$. We define now an action of the group G on $\bar{L}$ which is compatible with the action of G on L_*. If $\sigma \in G$, we define the action of σ on the elements of $\psi_i(F_0)$ as follows: consider the isomorphic imbedding $\sigma\varphi_i$ of L_0 into F_1; there exists a j such that $\sigma\varphi_i = \varphi_j$; then for $\alpha \in \psi_i(F_0)$ we put $\sigma\alpha \rightleftharpoons \psi_j\psi_i^{-1}(\alpha)$. A routine verification shows that the action of σ so defined on the components $\psi_i(F_0)$ extends uniquely to an action of σ on $\bar{L}$, and on L^* this action agrees with the original. Put $L_1 \rightleftharpoons \bar{L}^G$; then $\bar{L}$ is a Galois extension of L_1 with Galois group G, F is algebraically closed in L_1, and consequently L_1 is a regular extension of F. Put $L \rightleftharpoons L_0 L_1$ $(= L_0 \otimes_F L_1)$; we show that $F_0 \leqslant L_0 L_1$. Let H be the subgroup of G such that $L_0 = L_*^H$ (i.e., $H \rightleftharpoons \{\sigma \mid \sigma\alpha = \alpha$ for every $\alpha \in L_0\}$); we claim that $L = \bar{L}^H$. Indeed, the inclusion $L \leqslant \bar{L}^H$ is obvious; and since $\bar{L} = L_* L_1 = L_* \otimes_F L_1$, we have $[\bar{L} : L] = [L_* : L_0] = |H| = [\bar{L} : \bar{L}^H]$, and therefore $L = \bar{L}^H$. It remains to show that $F_0 \leqslant \bar{L}^H$. Suppose $\sigma \in H$ and $\alpha \in F_0 = \psi_1(F_0)$; then since $\varphi_1 = \mathrm{id}_{L_0}$, we have $\sigma\varphi_1 = \varphi_1$, and therefore $\sigma\alpha = \psi_1\psi_1^{-1}(\alpha) = \alpha$. Thus, $F_0 \leqslant \bar{L}^H$.

We proceed now to the proof of the proposition. If we use the corollary of the preceding proposition, it suffices to consider only the case where F_0 is a finite extension of F.

Let $F_0(t_1,\dots,t_n)$ be a finitely generated regular extension of F_0. By Lemma 2, there exists a finitely generated regular extension F_1 of F such that $F_0(t_1,\dots,t_n) \leqslant F_0F_1 = F_0 \otimes_F F_1$. Let $\alpha_1 \rightleftharpoons 1, \alpha_2,\dots,\alpha_k$ be a basis for F_0 over F. Then any element γ of F_0F_1 is uniquely representable in the form $\gamma = \sum_{i=1}^k \alpha_i\beta_i$, $\beta_i \in F_1$. Let $\beta_{ij} \in F_1$, $i = 1,\dots,n$, $j = 1,\dots,k$, be such that $t_j = \sum_{i=1}^k \alpha_i\beta_{ij}$, $j = 1,\dots,n$. Since $F(\beta_{ij} \mid i = 1,\dots,n; j = 1,\dots,k) \leqslant F_1$ is a regular extension of F and F is regularly closed, there exists an F-homomorphism ψ: $F[\beta_{ij}] \to F$, which extends uniquely to an F_0-homomorphism φ_*: $F_0[\beta_{ij}] \to F_0$ because $F_0[\beta_{ij}] = F[\beta_{ij}] \otimes_F F_0$; and since $t_1,\dots,t_n \in F_0[\beta_{ij}]$, the mapping $\varphi \rightleftharpoons \varphi_* \upharpoonright F_0[t_1,\dots,t_n]$ is the desired f_0-homomorphism from $F_0[\bar{t}]$ to F_0.

PROPOSITION 3. *Any field F has a regular extension F_0 which is regularly closed.*

The construction of such an extension goes through in a fashion quite standard. We first indicate how to construct a regular extension F' of F such that for any pair $\langle f, g\rangle$, $f \in F[x_1,\dots,x_k, y]$, $g \in F[x_1,\dots,x_k]$, with f absolutely irreducible and, as a polynomial in y, unitary, separable and of degree greater than 1, and with $g \neq 0$, there exist in F' elements $\alpha_1,\dots,\alpha_k$ and β such that $f(\bar{\alpha}, \beta) = 0$ and $g(\bar{\alpha}) \neq 0$. Let X be the set of all such pairs of polynomials, and $\leqslant$ some well-ordering of X. For every $\xi \in X$ we define a regular extension F_ξ of F by transfinite induction, as follows:

1. If $\xi_0 = \langle f_0, g_0\rangle$ is the smallest element of X, then $F_{\xi_0} \rightleftharpoons F(t_1,\dots,t_{k_0}, \tau)$ is the field of fractions of the integral ring $F[x_1,\dots,x_{k_0}, y]/(f_0)$. Observe that F_{ξ_0} is a regular extension of F, since f_0 is absolutely irreducible.

2. Suppose that for all $\eta < \xi$ the fields F_η have been constructed so that $F_{\eta_0} \leqslant F_{\eta_1}$ whenever $\eta_0 \leqslant \eta_1 < \xi$. Put $\tilde{F}_\xi \rightleftharpoons \bigcup_{\eta<\xi} F_\eta$; $\tilde{F}_\xi$ is a regular extension of F. Let $\xi = \langle f, g\rangle$; then since F is absolutely irreducible, the field $F_\xi \rightleftharpoons F(t_1,\dots,t_k, \tau)$ of fractions of the ring $\tilde{F}_\xi[x_1,\dots,x_k, y]/(f)$ is a regular extension of $\tilde{F}_\xi$, and therefore a regular extension of F.

Putting $F' \rightleftharpoons \bigcup_{\xi\in X} F_\xi$, we obtain a regular extension of F that satisfies the above conditions.

If we now construct the tower of extensions

$$F \leqslant F' \leqslant \cdots \leqslant F^{(n)} \leqslant F^{(n+1)}(\rightleftharpoons F^{(n)'}) \leqslant \cdots,$$

then $F_0 \rightleftharpoons \bigcup_{n\in\omega} F^{(n)}$ is a regular extension of F which, as is evident from the construction, satisfies the condition of property 3 of Proposition 1, i.e., is regularly closed.

Using Propositions 2 and 3, we can characterize the Galois groups of regularly closed fields, where by the Galois group $G(F)$ of a field F we mean the Galois group $G(F_s/F)$ of the separable closure F_s of F over F.

PROPOSITION 4. *A profinite group G is isomorphic to the Galois group of a suitable regularly closed field if and only if G is isomorphic to a closed subgroup of a suitable free profinite group.*

In [3] it was shown that the property of a group G described in the proposition is equivalent to the requirement that G be projective, or, what is the same, that G have cohomological dimension $\leqslant 1$.

That the Galois group of a regularly closed field has cohomological dimension $\leqslant 1$ has already been noted in [**5**, p. 269]. This implies necessity.

Let us prove sufficiency. Let G be a closed subgroup of a free profinite group $F(X)$ over a set X of free generators. Let L be an arbitrary algebraically closed field of characteristic 0, and $L(X)$ a pure transcendental extension of L with transcendence basis X. It is known [11] that then the Galois group G_0 of $L(X)$ is isomorphic to $F(X)$. Let L_1 be a regular extension of $L(X)$ which is regularly closed. Let $G_1 \rightleftharpoons G(L_1)$. The restriction mapping gives an epimorphism $\varphi\colon G_1 \to G_0$. Since G_0 is projective, there exists a homomorphism $\psi\colon G_0 \to G_1$ such that $\varphi\psi = \mathrm{id}_{G_0}$; and then $\psi(G_0)$ is a closed subgroup of G_1 isomorphic to $F(X)$. Let $\overline{G}$ be the image of the subgroup $G \leqslant F(X)$ in $\psi(G_0)$. Then $\overline{G} \simeq G$. Consider the field $L_0 \rightleftharpoons (L_1^*)^{\overline{G}}$, where L_1^* is the algebraic closure of L_1. Then L_0 is an algebraic extension (separable, since the characteristic of L is zero) of the regularly closed field L_1; therefore L_0 is itself regularly closed, and $G(L_0) = \overline{G} \simeq G$. This proves sufficiency.

We come now to a theorem that describes the possible decomposition groups for coverings of rational points of varieties over regularly closed fields. Theorems of this type are called [6], [9] analogues of the Chebotarev density theorem, and they play an important role in theorems on elimination of quantifiers for "Galois assertions." The theorem obtained below is very general and can be used in eliminating quantifiers for many classes of regularly closed fields (e.g., fields having a universal Galois group; see [7]).

Let F be a regularly closed field; F_0 a finitely generated regular extension of F; F_1 a finite Galois extension of F_0; L_0 the algebraic closure of F in F_1; $H \rightleftharpoons G(F_1/F_0)$, $H_0 \rightleftharpoons G(L_0/F)$, and $\psi\colon H \to H_0$ the restriction epimorphism. Let $\varphi\colon G(F) \to H$ be a homomorphism such that the diagram

$$\begin{array}{ccc} & G(F) & \\ \swarrow & & \searrow{\scriptstyle\varphi} \\ H_0\,(= G(L_0/F)) & \xleftarrow{\ \psi\ } & H \end{array}$$

is commutative. Put $H_1 \rightleftharpoons \varphi(G(F))$ and $L_1 \rightleftharpoons F_8^{\ker\ \varphi}$; note that $G(L_1/F) = G(F)/\ker\varphi \simeq H_1$.

THEOREM (analogue of the Chebotarev density theorem). *For any finite subset $S \subseteq F_0$, there exist an F-place $p\colon F_0 \mapsto F$, finite on S, and an extension $\pi\colon F_1 \mapsto\!\!\!\!\!\rightarrow L_1$ of p to F_1, unramified over p, whose decomposition group over p is H, and such that the isomorphism $H_1 \xrightarrow{\sim} G(L_1/F) = G(F)/\ker\varphi$ induced by π is the inverse of the isomorphism $G(F)/\ker\varphi \xrightarrow{\sim} H_1$ induced by φ.*

Observe that the special case of this theorem when $L_0 = F$, i.e., F_1 is a regular extension of F, and φ is the trivial mapping ($H_1 = \{e\}$) follows at once from the already proved Lemma 1.

Consider the field $K \rightleftharpoons L_1F_1$. This is a Galois extension of the field F_0, and its Galois group (over F_0) is described in the following lemma.

LEMMA 3. *Let M_0 and M_1 be Galois extensions of a field M; let $G_0 \rightleftharpoons G(M_0/M)$, $G_1 \rightleftharpoons G(M_1/M)$ and $H \rightleftharpoons G(M_0 \cap M_1/M)$. Then the composite M_0M_1 is a Galois extension of M, and its Galois group $H_0 \rightleftharpoons G(M_0M_1/M)$ is isomorphic to the fiber product $G_0 \times_H G_1$ of the groups G_0 and G_1 over H, i.e., to the subgroup of $G_0 \times G_1$ consisting of all pairs $\langle g_0, g_1\rangle$ such that $g_0 \upharpoonright (M_0 \cap M_1) = g_1 \upharpoonright (M_0 \cap M_1)$ (or, what is the same, such that $\pi_0 g_0 = \pi_1 g_1$, where π_i: $G_i \to H$ is the restriction mapping).*

This lemma sharpens Theorem 5 of [4], Chapter VIII, §1. According to the latter, H_0 imbeds isomorphically into $G_0 \times G_1$; from the definition of the imbedding it is evident that H_0 in fact goes into $G_0 \times_H G_1$. It suffices to prove the lemma for finite extensions M_0 and M_1; then what remains is to demonstrate equality of orders: $|H_0| = |G_0 \times_H G_1|$. Consider the extension diagram

$$\begin{array}{ccc} \ulcorner\!\!\longrightarrow & M_0 \quad M_1 & \longleftarrow\!\!\urcorner \\ M_0 & & M_1, \\ \llcorner\!\!\uparrow & M_0 \cap M_1 & \uparrow\!\!\lrcorner \\ & \uparrow & \\ & M & \end{array}$$

and let

$$l \rightleftharpoons [M_0 \cap M_1 \colon M],\ m \rightleftharpoons [M_0 \colon M_0 \cap M_1],\ n \rightleftharpoons [M_1 \colon M_0 \cap M_1].$$

Then since $M_0 \|_{M_0 \cap M_1} M_1$, we have $[M_0M_1 \colon M_0] = n$, $[M_0M_1 \colon M_1] = m$ and $[M_0M_1 \colon M] = mnl$. Therefore

$$|H_0| = [M_0M_1 \colon M] = mnl = \frac{ml \cdot nl}{l} = \frac{|G_0| \cdot |G_1|}{|H|}.$$

If now φ: $G_0 \times_H G_1 \mapsto\!\!\!\!\twoheadrightarrow G_0$ is the natural epimorphism (projection onto the first coordinate), then

$$|G_0 \times_H G_1| = |G_0| \cdot |\operatorname{Ker} \varphi|.$$

Now, $\operatorname{Ker} \varphi \rightleftharpoons \{\langle e, g_1\rangle | \pi_0 e = e = \pi_1 g_1\}$, i.e., $\operatorname{Ker} \varphi \simeq \operatorname{Ker} \pi_1$; but $|\operatorname{Ker} \pi_1| \cdot |H| = |G_1|$, and therefore

$$|\operatorname{Ker} \varphi| = |\operatorname{Ker} \pi_1| = |G_1|/|H|$$

and

$$\left|G_0 \underset{H}{\times} G_1\right| = |G_0| \cdot |\operatorname{Ker} \varphi| = \frac{|G_0| \cdot |G_1|}{|H|} = |H_0|.$$

Since $K = (L_1F_0)F_1$, and L_1F_0 is a Galois extension of F_0 such that $G(L_1F_0/F_0) \simeq G(L_1/F)$ and $L_1F_0 \cap F_1 = L_0F_0$, it follows by Lemma 3 that the Galois group G of K over F_0 is isomorphic to $H_1 \times_{H_0} H$. In G, consider the subgroup H_* corresponding to the subgroup $\{\langle h, h\rangle | h \in H_1\}$ of $H_1 \times_{H_0} H$; then $H_* \simeq H_1$. Let $K_0 \rightleftharpoons K^{H^*}$. Then, since L_1 is the algebraic closure of F in K, and H_* acts on L_1 in the same way as (the isomorphic group) $G(L_1/F)$ $(= G(F)/\operatorname{Ker} \varphi)$, the field F is algebraically closed in K_0; and since K, and therefore K_0, is separable over F, it follows that K_0 is a regular extension of F and $K = K_0 \otimes_F L_1$. It is easily verified that $F_0 \leqslant K_0$ and K_0 is a finite separable extension of F_0. Applying Lemma 1 to $F \leqslant F_0 \leqslant K_0$, we obtain an F-place p: $F_0 \mapsto\!\!\!\!\twoheadrightarrow F$ which is finite on S, and an extension

π' of p of K_0 which is unramified over p and F-valued (i.e., π': $K_0 \mapsto\!\!\!\twoheadrightarrow F$). Now extend π' to an L_1-place π_0: $K = K_0 \otimes_F L_1 \mapsto\!\!\!\twoheadrightarrow L_1$ (such an extension obviously exists and is unique); then π_0 is unramified over π' and over p. Denoting now by π the restriction of π_0 to F_1, we claim that π is the desired extension of p. Indeed, since π_0 is unramified over p, so is its restriction π; and it remains only to observe that $\pi(F_1) = L_1$, which follows from the fact that $F_1 \cap K_0 = F_1^{H_*} = F_1^{H_1}$ and $[K: K_0] = |H_1| = [F_1: F_1 \cap K_0] = [L_1: F]$, and H_1 is the decomposition group of π over p.

BIBLIOGRAPHY

1. Yu. L. Ershov, *Regularly closed fields*, Dokl. Akad. Nauk SSSR **251** (1980), 783–785; English transl. in Soviet Math. Dokl. **21** (1980).

2. ______, *Fields with a decidable theory*, Dokl. Akad. Nauk SSSR **174** (1967), 19–20; English transl. in Soviet Math. Dokl. **8** (1967).

3. ______, *On profinite groups*. Algebra i Logika **19** (1980), 552–565; English transl. in Algebra and Logic **19** (1980).

4. Serge Lang, *Algebra*, Addison-Wesley, 1965.

5. James Ax, *The elementary theory of finite fields*, Ann. of Math. (2) **88** (1968), 239–271.

6. Michael Fried, *On Hilbert's irreducibility theorem*, J. Number Theory **6** (1974), 211–231.

7. M. Fried and G. Sacerdote, *Solving Diophantine problems over all residue class fields of a number field and all finite fields*, Ann. of Math. (2) **104** (1976), 203–233.

8. Moshe Jardin, *The elementary theory of ω-free Ax fields*, Invent. Math. **38** (1976/77), 187–206.

9. ______, *An analogue of Čebotarev density theorem for fields of finite corank*, J. Math. Kyoto Univ. **20** (1980), 141–147.

10. Moshe Jardin and Ursel Kiehne, *The elementary theory of algebraic fields of finite corank*, Invent. Math. **30** (1975), 275–294.

11. Luis Ribes, *Introduction to profinite groups and Galois cohomology*, Queen's Papers in Pure and Appl. Math., No. 24, Queen's Univ., Kingston, Ont., 1970.

12. William H. Wheeler, *Model complete theories of pseudo-algebrically closed fields*, Ann. Math. Logic **17** (1979), 205–226.

Translated by J. A. ZILBER

Труды Матем. Инст. Стеклов.
158(1981)

Proceedings of the Steklov
Institute of Mathematics
1983, Issue 4

OPTIMALITY OF PURSUIT TIME IN A DIFFERENTIAL GAME OF SEVERAL OBJECTS WITH SIMPLE MOTION

UDC 518.9

R. P. IVANOV AND YU. S. LEDYAEV

ABSTRACT. A differential game is considered in which a vector $x_0(t)$ is pursued by vectors $x_1(t),\ldots,x_k(t)$. The motion of $x_i(t)$ $(i=0,\ldots,k)$ is described by the equation $dx_i(t)/dt = u_i(t)$, where the control $u_i(t)$ satisfies a constraint of the form $\|u_i\| \leqslant \sigma_i$. The pursuit terminates at the time t if there is an i, $1 \leqslant i \leqslant k$, such that $\|x_0(t) - x_i(t)\| \leqslant l_i$. In this article pursuer strategies $\{u_i(x_0, x_i, u_0)\}_{i=1}^k$ are given which ensure termination of the pursuit at a finite time, and conditions are obtained for this time to be optimal.

Bibliography: 12 titles.

In this article we obtain conditions for an optimal pursuit time in a differential game involving one evader and several pursuers with simple motions.

The motion of a vector $x_i(t)$ in n-dimensional Euclidean space R^n is said to be *simple* if it is described by an equation

$$dx_i(t)/dt = u_i(t), \tag{1}$$

where the admissible controls $u_i(t)$ are measurable vector-valued functions satisfying the constraint

$$u_i(t) \in S_{\sigma_i}. \tag{2}$$

($S_\sigma = \{x \mid |x| \leqslant \sigma\}$ is the ball of radius σ in R^n, $|x| = (x, x)^{1/2}$ is the norm of the vector x, and (x, y) is the inner product of the vectors x and y).

Differential games of pursuit and evasion involving several objects with simple motion have attracted the attention of many authors (see [5]–[8]).

A differential game is considered here in the formalization of Pontryagin ([1]–[3]); use of the procedure of control with a guide [4] enables us to pass to the Krasovskiĭ-Subbotin formalization of a positional differential game.

1. Suppose that the motion of the vectors $x_0(t)$ and $x_1(t),\ldots,x_k(t)$ is described by equation (1). The motion $x_i(t)$ corresponding to an admissible control $u_i(t)$ and an initial position (t_0, x_i^0) is defined to be the absolutely continuous function $x_i(t)$ given by

$$x_i(t) = x_i^0 + \int_{t_0}^t u_i(\tau)\,d\tau.$$

1980 *Mathematics Subject Classification*. Primary 49A45, 90D26.

The control $u_0(t)$ is the *control of the evader*, and the controls $u_i(t)$, $i = 1,\ldots,k$, are the *controls of the pursuers*.

For each $i \in J$, $J = \{1,\ldots,k\}$, fix a quantity l_i, $l_i \geqslant 0$. We say that an *l-approach* takes place at time t for the motions $x_0(t)$ and $x_1(t),\ldots,x_k(t)$ if there exists an $i \in J$ such that $|x_i(t) - x_0(t)| \leqslant l_i$.

Further, it is assumed that the following conditions hold:

$$\sigma_i \geqslant \sigma_0, \quad \sigma_i + l_i > \sigma_0 \quad \text{for each } i \in J \tag{3}$$

(without loss of generality it is assumed that $\sigma_0 = 1$).

The pursuit problem. For a given initial state $(x_0, x_1^0,\ldots,x_k^0)$ the pursuit problem is solvable if there exist a time ϑ and functions $u_i(x_0, x_i, u_0)$, $u_i\colon R^n \times R^n \times S_1 \to S_{\sigma_i}$, $i \in J$, called *pursuer strategies*, such that for any admissible control $u_0(t)$ of the evader and any solution $(x_0(t),\ldots,x_k(t))$ of the system of equations

$$dx_0(t)/dt = u_0(t), \qquad dx_i(t)/dt = u_i(x_0(t), x_i(t), u_0(t)), \quad i \in J,$$
$$x_i(0) = x_i^0, \qquad i = 0, 1,\ldots,k, \tag{4}$$

in the sense of Carathéodory [12] an l-approach takes place at some time $t \in (0, \vartheta]$:

$$|x_i(t) - x_0(t)| \leqslant l_i \quad \text{for some } i \in J.$$

The time ϑ is called a *guaranteed pursuit time*.

If there exists an admissible control $\hat{u}_0(t)$ of the evader such that for any admissible controls $u_i(t)$ $(i \in J)$ of the pursuers no l-approach takes place for $t < \vartheta$, then the guaranteed pursuit time ϑ is called the *optimal pursuit time*.

Obviously, if $\sigma_i > 1$ for some $i \in J$, then for any initial state $(x_0^0,\ldots,x_k^0)$ the pursuit problem is solvable.

If $\sigma_i = 1$ for all $i \in J$, then it makes sense to consider the evasion problem.

The evasion problem. We say that the evasion problem is solvable for a given initial state if there is an S_1-valued function $u_0(x_0, x_1,\ldots,x_k)$, called an *evasion strategy*, such that for any admissible controls $u_1(t),\ldots,u_k(t)$ of the pursuers and for any solution $(x_0(t),\ldots,x_k(t))$(in the Carathéodory sense) of the system of equations $(x_i(0) = x_i^0, i = 0,\ldots,k)$

$$dx_0(t)/dt = u_0(x_0(t),\ldots,x_k(t)), \quad dx_i(t)/dt = u_i(t), \qquad i \in J,$$

no l-approach takes place for any $t \geqslant 0$.

Solvability conditions for the pursuit and evasion problems will be given in §4. Let us first consider some auxiliary problems.

2. For each $i \in J$ and each pair $z_i = (x_0, x_i)$ of vectors we define the set

$$X_i(z_i) = \{y \mid y \in R^n, F_i(y, z_i) \leqslant 0\}, \tag{5}$$

where

$$F_i(y, z_i) = \sigma_i|y - x_0| - |y - x_i| + l_i.$$

REMARK. For brevity we omit the index i in this and the next section.

Note that the set $X(z)$ has another representation:

$$X(z) = \{y \mid y \in R^n, \Phi(y, z) \leqslant 0\}, \tag{6}$$

where

$$\Phi(y, z) = (\sigma|y - x_0| + |y - x|)F(y, z).$$

Using the representations (5) and (6), we can easily prove the following properties of $X(z)$:

a) If $|x - x_0| \geqslant l$, then $X(z)$ is nonempty, and $x_0 \in X(z)$.

b) If $\sigma + l > 1$, then $X(z)$ is strictly convex, because the function $\Phi(y, z)$ is strictly convex in y when $\sigma + l > 1$.

c) $X(z)$ is bounded if $\sigma > 1$.

d) If $\sigma = 1$, then $X(z)$ contains the convex cone

$$\{y \mid y = x_0 + tu, |u| = 1, (u, e(z)) \geqslant (l/r(z)), t \geqslant 0\}$$

with vertex at x_0, where $e(z) = (x_0 - x)/|x_0 - x|$ and $r(z) = |x_0 - x|$.

e) $X(z)$ is axially symmetric in the sense that if the vector $x_0 + \xi e(z) + \eta f$, $\xi \in R^1$, $\eta \in R^1$, $f \in E(e(z))$, is contained in $X(z)$, then for any $f' \in E(e(z))$ the vector $x_0 + \xi e(z) + \eta f'$ is also contained in $X(z)$, where $E(e) = \{f \mid |f| = 1, (e, f) = 0\}$.

f) If the hyperplane $\Gamma^\alpha = \{y \mid (y - x_0, e(z)) = \alpha\}$ intersects $X(z)$, then $|y_\alpha - x_0|$ increases monotonically as α increases, where $y_\alpha \in \Gamma^\alpha \cap \partial X(z)$ and $\partial X(z) = \{y \mid F(y, z) = 0\}$.

Some properties of $X(z)$ are formulated in the following lemmas.

LEMMA 1. *The interior* int $X(z)$ *is nonempty if and only if*

$$|x - x_0| > l. \tag{7}$$

LEMMA 2. *Suppose that* $z^0 = (x_0^0, x^0)$, int $X(z^0) \neq \varnothing$ *and* $y \in X(z^0)$. *If the motion* $x_0(t)$ *corresponds to the initial state* $x_0(0) = x_0^0$ *and the control*

$$u_0(t) \equiv (y - x_0^0)/|y - x_0^0|,$$

while the motion $x(t)$ *corresponds to the initial state* $x(0) = x^0$ *and an arbitrary admissible control* $u(t)$, *then* (7) *holds for the vectors* $x_0(t)$ *and* $x(t)$ *for all* $t \in [0, \vartheta)$, *where* $\vartheta = |y - x_0^0|$.

The proofs of these lemmas are left to the reader.

3. Suppose that the motion of the vectors $x_0(t)$ and $x(t)$ in R^n described by (1) corresponds to the initial state $z^0 = (x_0^0, x^0)$ and the admissible controls $u_0(t) \in S_1$ and $u(t) \in S_\sigma$.

In R^n fix a hyperplane $\Gamma = \{y \mid (a, y) = 0\}$, $a \in R^n$, $|a| = 1$, which is assumed without loss of generality to pass through the origin, and let $\Gamma^- = \{y \mid (a, y) \leqslant 0\}$ and $\Gamma^+ = \{y \mid (a, y) > 0\}$.

We present two problems, and base the construction of optimal strategies for the pursuit of a single evader by several objects on solutions of them. First, define

$$K(z) + X(z) \cap \Gamma^-. \tag{8}$$

PROBLEM 1. *For a fixed initial state* $z^0 = (x_0^0, x^0)$ *such that* int $K(z) \neq \varnothing$ *and* $K(z)$ *is bounded, find a time* ϑ *and an* S_σ*-valued function* $u(\Gamma; z, u_0)$ *such that for any admissible control* $u_0(t)$ *and for each solution* (*in the Caretheódory sense*) *of the system of equations*

$$dx_0(t)/dt = u_0(t), \qquad dx(t)/dt = u(\Gamma; x_0(t), x(t), u_0(t)),$$
$$x_0(0) = x_0^0, \qquad x(0) = x^0, \tag{9}$$

there is a time $t \leqslant \vartheta$ at which either an l-approach takes place or

$$\operatorname{int} X(z(t)) \subset \Gamma^+ . \tag{10}$$

PROBLEM 2. *For a fixed initial state z^0 such that* $\operatorname{int} X(z^0) \subset \Gamma^+$, *find an S_σ-valued function $u(\Gamma; z, u_0)$ such that for any admissible control $u_0(t)$ and for each solution* (*in the Carathéodory sense*) *of the system* (9) *either an l-approach takes place or* (10) *holds for all $t \geqslant 0$.*

To solve Problem 1 we use the function

$$w(z) = \max\{|y - x_0| \mid y \in K(z)\}, \tag{11}$$

which turns out to be a Lipschitz function in a neighborhood of any point z for which $\operatorname{int} K(z) \neq \varnothing$ and $K(z)$ is bounded. The function $w(t) = w(z(t))$ is absolutely continuous for a motion $z(t)$ such that $\operatorname{int} K(z(t)) \neq \varnothing$, and the main technical difficulty in solving Problem 1 is to get an upper estimate for the derivative $dw(t)/dt$.

In order to determine a strategy $u^1(\Gamma; z, u_0)$ which solves Problem 1 we introduce some new concepts.

If $\operatorname{int} K(z) \neq \varnothing$ and $K(z)$ is bounded, then let

$$\hat{K}(z) = \{y \mid y \in K(z), |y - x_0| = w(z)\}.$$

If $y \in \hat{K}(z)$, then it follows from necessary conditions for an extremum [9] that there exist Lagrange multipliers $m = m(y)$ and $p = p(y)$, $m > 0$, such that

$$\psi_0(y) = m(\sigma\psi_0(y) - \psi(y)) + pa, \tag{12}$$

where $\psi_0(y) = (y - x_0)/|y - x_0|$ and $\psi(y) = (y - x)/|y - x|$.

Since at the same time $y \in \partial X(z)$, we have

$$w\psi_0(y) = (\sigma w + l)\psi(y) - re, \tag{13}$$

where $e = e(z)$, $r = r(z)$ and $w + w(z)$.

Comparing (12) and (13), we get that if $e(z) \neq a$ and the constraint Γ^- is essential (i.e., $\hat{K}(z) \subset \Gamma$) or if the constraint Γ^- is not essential, then for $y \in \hat{K}(z)$ the vectors $\psi_0(y)$ and $\psi(y)$ can be uniquely represented as linear combinations of the vectors $e(z)$ and a. Therefore, in the cases indicated the set $\hat{K}(z)$ consists of the single vector $y(z)$, $\hat{K}(z) = \{y(z)\}$.

In the case where $e(z) = a$ and $\hat{K}(z) \subset \Gamma$ we have

$$\hat{K}(z) = \{y \mid |y - x_0| = w, (y, a) = 0\}.$$

Here the Lagrange multipliers $m(y)$ in (12) are equal: $m(y) = m$ for all $y \in \hat{K}(z)$.

If $e(z) = a$ and the vectors u_0 and $e(z)$ are linearly independent, then the intersection of $\hat{K}(z)$ and the two-dimensional linear manifold passing through x_0 and spanned by u_0 and $e(z)$ consists of the two points $y_1 = y_1(z, u_0)$ and $y_2 = y_2(z, u_0)$. Let

$$\psi_0[\alpha] = \alpha\psi_0(y_1) + (1 - \alpha)\psi_0(y_2), \quad \psi[\alpha] = \alpha\psi(y_1) + (1 - \alpha)\psi(y_2), \tag{14}$$

where $\alpha \in [0, 1]$. The number $\alpha(z, u_0) \in [0, 1]$ is uniquely determined by the relation

$$W(\alpha(z, u_0), u_0) = \max_{\alpha \in [0,1]} W(\alpha, u_0) \tag{15}$$

$$(W(\alpha, u_0) = (m\sigma - 1)(\varphi_0[\alpha], u_0) - m\sigma|\psi[\alpha]|).$$

We define a function $u^1(\Gamma; z, u_0)$ by setting

$$u^1(\Gamma; z, u_0) = \sigma\psi(y(z)) \tag{16}$$

if $e(z) \neq a$,

$$u^1(\Gamma; z, u_0) = \sigma\psi[\alpha(z, u_0)]/|\psi[\alpha(z, u_0)]| \tag{17}$$

if $e(z) = a$ and the vectors u_0 and $e(z)$ are linearly independent, and

$$u^1(\Gamma; z, u_0) = \sigma e(z) \tag{18}$$

if $e(z) = a$ and the vectors u_0 and $e(z)$ are linearly dependent.

It can be shown that for any admissible control $u_0(t)$ and for a motion $x_0(t)$ corresponding to it the mapping $\mathcal{Q}$: $x(\cdot) \to \tilde{x}(\cdot)$ given by

$$\tilde{x}(t) = x^0 + \int_0^t u^1(\Gamma; x_0(\tau), x(\tau), u_0(\tau))\, d\tau, \qquad t \in [0, \vartheta],$$

is continuous.

From this, the Schauder fixed point theorem gives us the existence of a solution (in the Carathéodory sense) of the system (9) for the strategy $u^1(\Gamma; z, u_0)$ defined by (16)–(18).

THEOREM 1. *The strategy $u^1(\Gamma; z, u_0)$ determined by* (16)–(18) *solves Problem* 1, *and for any motion $z(t)$ of the system* (9) *such that* int $K(z(t)) \neq \varnothing$ *the relation*

$$dw(z(t))/dt \leqslant -1$$

holds for almost all t, so that at some time $t \leqslant \vartheta$, $\vartheta = w(z^0)$, either an l-approach takes place or the inclusion (10) *holds.*

Theorem 1 will be proved after the statement of a theorem on the solvability of Problem 2.

Let

$$w^1(z) = -\max\{-(a, y) \,|\, y \in X(z)\}. \tag{19}$$

If int $X(z) \subset \Gamma^+$, then, since $X(z)$ is strictly convex because of condition (3), there exists a unique vector $y_1(z)$ at which a maximum is attained on the right-hand side of (19), $w^1(z) > 0$.

Let

$$u^2(\Gamma; z) = \sigma\psi(y_1(z)). \tag{20}$$

It can be shown that the function $u^2(\Gamma; z)$ is continuous on the set $\{z \,|\, X(z) \subset \Gamma^+\}$; therefore, for any admissible control $u_0(t)$ there exists a solution (in the Carathéodory sense) of the system of equations (9) for the strategy $u^2(\Gamma; z)$.

THEOREM 2. *The strategy $u^2(\Gamma; z)$ defined by* (20) *solves Problem* 2, *and for any motion $z(t)$ either there is an l-approach at some time, or for almost all t*

$$dw^1(z(t))/dt \geqslant 0.$$

We give some general results that will be used in proving Theorems 1 and 2.

LEMMA 3. *Suppose that the multi-valued mapping $K(z) = \{y \,|\, \varphi(y, z) \leqslant 0\}$ is defined by means of a function $\varphi(y, z)$ that is convex in y and locally Lipschitz in y and* *z. Then in some neighborhood U of a point z^0 at which* int $K(z^0) \neq \varnothing$ *and $K(z^0)$ is bounded the multi-valued mapping $K(z)$ is Lipschitz*:

$$h(K(z^1), K(z^2)) \leqslant L|z^1 - z^2|, \qquad z^i \in U, \quad i = 1, 2,$$

where L is a constant, the quantity

$$h(A, B) = \max\left\{ \sup_{a \in A} \rho(a, B), \sup_{b \in B} \rho(b, A) \right\}$$

is the Hausdorff distance between the sets A and B, and

$$\rho(a, B) = \inf_{b \in B} |a - b|$$

is the distance from a to B.

Let the function $w(z)$ be defined as follows:

$$w(z) = \max\{f(y, z) | y \in K(z)\}, \tag{21}$$

where $f(y, z)$ satisfies a Lipschitz condition with respect to (y, z), and the multi-valued mapping $K(z)$ satisfies a Lipschitz condition with respect to z on the set U. It is not hard to show that this function $w(z)$ is a Lipschitz function on U.

Since, by Lemma 3, the multi-valued mapping $K(z)$ defined by (8) satisfies a Lipschitz condition in some neighborhood of z in which int $K(z) \neq \varnothing$ and $K(z)$ is bounded, the function $w(z)$ defined by (11) is a Lipschitz function in this neighborhood.

Analogous arguments give us also that $w^1(z)$ is a Lipschitz function in some neighborhood of z such that $X(z) \subset \Gamma^+$.

Let $w(z)$ be a function of the form (21), where

$$K(z) = \{y | \varphi(y, z) \leqslant 0, (a, y) \leqslant 0\},$$

and the functions $f(y, z)$ and $\varphi(y, z)$ are differentiable with respect to (y, z); $\nabla_y f(y, z)$, $\nabla_y \varphi \nabla_v(y, z)$ and $\nabla_z f(y, z)$, $\nabla_z \varphi(y, z)$ denote their gradients with respect to y and z. We set $\hat{K}(z) = \{y | f(y, z) = w(z)\}$, and let $m(y)$ be the Lagrange multiplier at a point $y \in \hat{K}(z)$,

$$\nabla_y f(y, z) = m(y) \nabla_y \varphi(y, z) + p(y) a. \tag{22}$$

We give an estimate for the upper derivative of $w(z)$ in the direction $v = (u_0, u)$ at the point $z = (x_0, x)$ (see [10]):

$$(w(z, v))^+ = \lim_{\lambda \searrow 0} \sup (w(z + \lambda v) - w(z))/\lambda.$$

LEMMA 4. *If* int $K(z) \neq \varnothing$ *and $K(z)$ is bounded, and for each $y \in \hat{K}(z)$ there is a vector d such that*

$$(a, d) = 0, \qquad (\nabla_y \varphi(y, z), d) < 0, \tag{23}$$

then the following estimate holds:

$$(w(z, v))^+ \leqslant \max_{y \in \hat{K}(z)} ((\nabla_z f(y, z), v) - m(y)(\nabla_z \varphi(y, z), v)). \tag{24}$$

PROOF. We choose a sequence $\lambda_i \searrow 0$, $i \to \infty$, such that

$$(w(z, v))^+ = \lim_{i \to \infty} (w(z_i) - w(z))/\lambda_i,$$

where $z_i = z + \lambda_i v$.

Let $y_i \in \hat{K}(z_i)$. By the conditions of the lemma, $K(z') \subset K$, $z' \in U$ (K a compact set) for some neighborhood U of z. Therefore, using the fact that the multi-valued mapping $\hat{K}(z)$ is upper semicontinuous with respect to inclusion [9], we can assume that $y_i \to y$ and $m_i \to m$ as $i \to \infty$, where $y \in \hat{K}(z)$ and $m_i = m(y_i)$ and $m = m(y)$ are the corresponding Lagrange multipliers in (12).

If the vector d satisfies (23), then we define the functions

$$F_i(\mu) = \varphi(y_i - \mu d, z) - \varphi(y_i, z), \qquad i = 1, 2, \dots.$$

Since

$$(d/d\mu)F_i(\mu) = \big(\nabla_y \varphi(y_i - \mu d, z), d\big) < 0,$$

for sufficiently large i and small μ, there is a sequence μ_i such that

$$F_i(\mu_i) = 0, \qquad \lim_{i\to\infty} \frac{\mu_i}{\lambda_i} = -\frac{(\nabla_z \varphi(y, z), v)}{\big(\nabla_y \varphi(y, z), d\big)}.$$

Because $\varphi(y_i - \mu_i d, z) = 0$ and $(y_i - \mu_i d, a) \leqslant 0$, this gives us that $y^i \in K(z)$, where $y^i = y_i - \mu_i d$.

By the definition of $w(z)$, we have that $f(y^i, z) \leqslant w(z)$. Therefore,

$$(w(z, v))^+ \leqslant (\nabla_z f(y, z), v) + \lim_{i\to\infty} \big(\nabla_y f(y, z), d\big)\mu_i/\lambda_i.$$

Using (28) and the definitions of the vector d and the sequence $\{\lambda_i\}$, we get (24). Lemma 4 is proved.

We show that, for the function $w(z)$ defined by (11), the conditions of Lemma 4 hold at a point z such that int $K(z) \neq \varnothing$, and $K(z)$ is bounded.

Suppose that the constraint Γ^- is essential (i.e., $\hat{K}(z) \subset \Gamma$); then it follows from (13)–(15) and property f) of $X(z)$ that if $y \in \hat{K}(z)$, then there exists a vector $d \in R^n$ such that

$$(d, \sigma\psi_0(y) - \psi(y)) < 0, \quad (d, \psi_0(y)) < 0, \quad (d, e) \leqslant 0, \quad (d, a) = 0. \tag{25}$$

If $\hat{K}(z) \not\subset \Gamma$, then analogous arguments give us the existence of a vector $d \in R^n$ such that

$$(d, \sigma\psi_0(y) - \psi(y)) < 0, \quad (d, \psi_0(y)) < 0, \quad (d, e) \leqslant 0. \tag{26}$$

The conditions of Lemma 4 follow from (25) and (26); therefore, the upper derivative of $w(z)$ at the point $z = (x_0, x)$ in the direction $v = (u_0, u)$ satisfies the estimate

$$(w(z, u_0, u))^+ \leqslant \max_{y\in\hat{K}(z)} \big((m\sigma - 1)(\psi_0(y), u_0) - m(\psi(y), u)\big), \tag{27}$$

where the Lagrange multiplier $m(y)$ in (12) is equal to m for all $y \in \hat{K}(z)$.

From (13)–(15), (25) and (26) it is not hard to get that $(d, \psi(y)) < 0$ and

$$m\sigma = 1 + (d, \psi(y))/(d, \sigma\psi_0(y) - \psi(y)) > 1. \tag{28}$$

If the constraint Γ^- is essential and $e(z) \neq a$ or if the constaint Γ^- is not essential, then $\hat{K}(z)$ consists of the single vector $y(z)$. Therefore, (27) and (28) give us immediately that

$$\max_{u_0\in S_1} \big(w(z, u_0, u^1(\Gamma; z, u_0))\big)^+ \leqslant -1, \tag{29}$$

where $u^1(\Gamma; z, u_0)$ is defined by (16).

Suppose that the constraint Γ^- is essential and $e(z) = a$; then

$$\hat{K}(z) = \{y \,|\, |y - x_0| = w, (y, e) = 0\}.$$

Let us obtain an upper estimate for the quantity

$$A = \max_{u_0 \in S_1} \min_{u \in S_\sigma(u_0)} \max_{y \in \hat{K}(z)} W(y, u_0, u),$$

where

$$W(y, u_0, u) = (m\sigma - 1)(\psi_0(y), u_0) - m(\psi(y), u),$$

$$S_\sigma(u_0) = \{u \,|\, |u| \leqslant \sigma, u = \xi e + \eta u_0, \xi \in R^1, \eta \in R^1\}.$$

From necessary conditions for an extremum [9] and the form of the set $\hat{K}(z)$ it follows that for $u \in S_\sigma(u_0)$

$$\max_{y \in \hat{K}(z)} W(y, u_0, u) = \max_{i=1,2} W(y_i(z, u_0), u_0, u).$$

Obviously,

$$\max_{i=1,2} W(y_i(z, u_0), u_0, u) = \max_{\alpha \in [0,1]} \tilde{W}(\alpha, u_0, u),$$

where

$$\tilde{W}(\alpha, u_0, u) = (m\sigma - 1)(\psi_0[\alpha], u_0) - m(\psi[\alpha], u).$$

Since the function $\tilde{W}(\alpha, u_0, u)$ is linear in α and u, a minimax theorem in [11] gives us that

$$\min_{u \in S_\sigma(u_0)} \max_{\alpha \in [0,1]} \tilde{W}(\alpha, u_0, u) = \max_{\alpha \in [0,1]} W(\alpha, u_0),$$

where $W(\alpha, u_0)$ is determined by (15). Hence

$$A = \max_{\alpha \in [0,1]} W(\alpha),$$

where $W(\alpha) = (m\sigma - 1)|\psi_0(\alpha)| - m\sigma|\psi[\alpha]|$.

The relations (13)–(15) give us that

$$w\psi_0[\alpha] = (\sigma w + l)\psi[\alpha] - re,$$

$$w^2|\psi_0[\alpha]|^2 = (\sigma w + l)^2|\psi[\alpha]|^2 + w^2 - (\sigma w + l)^2,$$

$$m = (\sigma w + l)/((\sigma^2 - 1)w^2 + \sigma l).$$

If we define the function of the argument $\tau = |\psi[\alpha]|$

$$\tilde{W}(\tau) = \left(\left((\sigma w + l)^2\tau^2 + w^2 - (\sigma w + l)^2\right)^{1/2} - \sigma(\sigma w + l)\right) / \left((\sigma^2 - 1)w^2 + \sigma l\right),$$

then it follows from the foregoing that ($\tau_0 = -(x, e)/(\sigma w + l)$)

$$\max_{\alpha \in [0,1]} W(\alpha) = \max_{\tau \in [\tau_0, 1]} \tilde{W}(\tau).$$

Simple computations show that $\tilde{W}(\tau)$ is strictly concave, $d^2\tilde{W}(\tau)/d\tau^2 < 0$, and $d\tilde{W}(\tau)/d\tau = 0$ at the point

$$\tau_m = \frac{\sigma^2}{(\sigma^2 - 1)}\left(1 - \frac{w^2}{(\sigma w + l)^2}\right).$$

Since $\tau_m \geqslant 1$ for $w > 0$, we have $A = \tilde{W}(1) = -1$.

From the definition (17), (19) of the function $u^1(\Gamma; z, u_0)$ and the value obtained for A it follows that (28) holds also if $\hat{K}(z) \not\subset \Gamma$ and $e(z) = a$.

It has thereby been shown that for any solution $z(t)$ of (9) corresponding to the strategy $u^1(\Gamma; z, u_0)$ we have

$$dw(z(t))/dt \leqslant -1$$

for almost all t in the case when an l-approach does not take place and the inclusion (10) does not hold. This implies that for the motion $z(t)$ either an l-approach takes place or (10) holds at some time $t \leqslant w(z^0)$.

Theorem 1 is proved.

To prove Theorem 2, for the function $-w^1(z)$ defined by (19) we give an estimate of the upper derivative at the point z in the direction (u_0, u).

It is not hard to verify the conditions of Lemma 4 for $-w^1(z)$ and to get that

$$\left(-w^1(z, u_0, u)\right)^+ \leqslant m((\sigma\psi_0(y_1(z)), u_0) - (\psi(y_1(z)), u)),$$

where $m > 0$ is the Lagrance multiplier in the extremal problem (19).

From this it follows that if for any solution $z(t)$ of (9) corresponding to the strategy $u^2(\Gamma; z)$ no l-approach takes place, then for almost all $t \geqslant 0$

$$dw^1(z(t))/dt \geqslant 0.$$

This means that (10) holds for all t.

REMARK 1. Theorems 1 and 2 give us that if int $K(z^0) \neq \varnothing$ and $K(z^0)$ is bounded at z^0, then the successive application of the strategies $u^1(\Gamma; z, u_0)$ and $u^2(\Gamma; z)$ ensures that for any motion $z(t)$ there is a time $\vartheta' \leqslant w(z^0)$ such that either an l-approach takes place before the time ϑ', or the inclusion (10) holds from the time ϑ' up to the time of an l-approach.

Let $u(\Gamma; z, u_0)$ denote the control procedure which consists of the successive application of the strategies $u^1(\Gamma; z, u_0)$ and $u^2(\Gamma; z)$ and ensures the indicated outcome.

REMARK 2. It is not hard to verify that the strategy $u^1(\Gamma; z, u_0)$ solves the problem of an optimal l-approach of the vectors $x_0(t)$ and $x(t)$ in the presence of the phase constraint $x_0(t) \in \Gamma^-$, $x(t) \in \Gamma^-$.

4. In this section we consider the original problem of pursuit of the vector $x_0(t)$ by vectors $x_1(t), \dots, x_k(t)$ whose motion is described by the equations (1).

Let us fix an initial state $x_0(0) = x_0^0, \dots, x_k(0) = x_k^0$, define a set $X_i = X_i(z_i^0)$ of the form (6) for each pair of vectors $z_i^0 = (x_0^0, x_i^0)$, and write $X = \bigcap_{i \in J} X_i$.

It is assumed that $|x_i^0 - x_0^0| > l$ for all $i \in J$.

Define the quantity $\vartheta = \vartheta(x_0^0, \dots, x_k^0)$ by

$$\vartheta = \sup\{|y - x_0^0| \mid y \in X\}. \tag{30}$$

LEMMA 5. *If $\vartheta = +\infty$, then the evasion problem is solvable.*

PROOF. It follows from the condition of the lemma that the set X and, hence, also the sets X_i are unbounded, i.e., $\sigma_i = 1$ for all $i \in J$.

Property d) of $X_i(z_i)$ implies that $U \neq \varnothing$, where

$$U = \bigcap_{i \in J} \{u \mid |u| = 1, (u, x_0^0 - x_i^0) \geqslant l_i\}.$$

Since $x_0 + tu_0 \in X$ for any $u_0 \in U$ and $t \geqslant 0$, it follows from Lemma 2 that a control $u_0(t) \equiv u_0$, $u_0 \in U$, ensures evasion of an l-approach at all times. This proves the lemma.

Suppose that $\vartheta < +\infty$. We determine conditions under which ϑ is an optimal pursuit time.

First, let $\hat{X}_i = \{y \mid y \in \partial X_i, |y - x_0^0| = \vartheta\}$. Then $X_i\ \hat{X}_i = \partial X_i \cap \Gamma_i$ by property e) of the set X_i, where the hyperplane Γ_i is defined as follows:

$$\Gamma_i = \{y \mid (y - y_i, e(z_i^0)) = 0\}, \qquad y_i \in \hat{X}_i. \tag{31}$$

We set $X_i^- = \{y \mid y \in X_i, (y - y_i, e(z_i^0)) \leqslant 0\}$ and $X_i^+ = X_i \backslash X_i^-$.

Let J' be a system of indices $i \in J$ such that

$$X = \bigcap_{i \in J'} X_i, \tag{32}$$

and which is minimal in the sense that the omission of any set from the intersection on the right-hand side of (32) leads to a violation of the equality (32).

Without loss of generality it can be assumed that $J' = \{1, \dots, k'\}$.

LEMMA 6. *If the quantity ϑ defined by* (30) *is finite*, (32) *holds, and*

$$\bigcap_{i \in J'} X_i^+ = \varnothing, \tag{33}$$

then ϑ is the optimal pursuit time.

PROOF. Let J' be a minimal system of indices satisfying (32) and (33), and for each $i \in J'$ define a hyperplane of the form (31).

A specification of Γ_i leads to a specification of a strategy $u_i^1(\Gamma_i;\, z_i, u_0)$ of the form (16)–(18) and a strategy $u_i^2(\Gamma_i;\, z_i)$ of the form (20), and these combine according to Remark 1 to form the control procedure $u_i(\Gamma_i;\, z_i, u_0)$.

We show that if for each $i \in J'$ the ith player, who controls the motion of the vector x_i, uses the control procedure $u_i(\Gamma_i;\, z_i, u_0)$, then an l-approach takes place no later than the time ϑ.

From Theorems 1 and 2 it follows that for any admissible control $u_0(t)$ the control procedure $u_i(\Gamma_i;\, z_i, u_0)$, $i \in J'$, ensures that either an l-approach takes place by the time ϑ ($\vartheta = w_i(z_i^0)$ for all $i \in J'$) or $X_i(z_i(\vartheta)) \subset \Gamma_i^+$.

The last inclusion means that $x_0(\vartheta) \in \bigcap_{i \in J'} X_i^+$, but this contradicts the condition of the lemma.

If $\hat{y} \in X$ and $|\hat{y} - x_0^0| = \vartheta$, then it follows from the definition of X that $\hat{y} \in X_i$ for each $i \in J$.

Therefore, by Lemma 2, the control $u_0(t) \equiv (\hat{y} - x_0^0)/|\hat{y} - x_0^0|$ ensures that the inequality $|x_i(t) - x_0(t)| > l_i$, $t \in [0, \vartheta)$, holds for any admissible control $u_i(t)$, $i \in J$, and the lemma is proved.

LEMMA 7. *If $\vartheta < +\infty$ and the system $J' = \{1, \dots, k'\}$ of indices is such that* (32) *holds and $k' \leqslant n$, then $\bigcap_{i \in J'} X_i^+ = \varnothing$.*

PROOF. If there exists a vector $\hat{y} \in \bigcap_{i \in J'} X_i^+$, then we define the vector $y(t) = \hat{y} + t\tilde{e}$, where the vector $\tilde{e} \neq 0$ is orthogonal to the vectors $e(z_1^0), \dots, e(z_{k'-1}^0)$, and $(\tilde{e}, e(z_{k'}^0)) \geqslant 0$. Since $k' \leqslant n$, $\tilde{e}$ exists.

Since $y(0) \in X$, $\lim_{t\to\infty}|y(t)| = \infty$, and $\vartheta < +\infty$, there is a t_* such that $y(t_*) \in \partial X$, where ∂X is the boundary of X. Moreover, $y(t_*) \in \partial X_j$ for some $j \in J'$, and $y(t_*) \in X_i^+$, $i \in J'$, because of property e) of the set $X_i(z_i)$.

Hence $|y(t_*) - x_0^0| > \vartheta$, which contradicts the definition of ϑ.

Consequently, $\bigcap_{i\in J'} X_i^+ = \varnothing$.

In conclusion we formulate the results of Lemmas 5–7 as the following theorem.

THEOREM 3. *Suppose that condition* (3) *holds for a differential game of pursuit by several objects with simple motions* (1), (2), *and that* $\vartheta = \vartheta(x_0^0,\dots,x_k^0)$ *is defined by* (30) *for the initial state* $(x_0^0,\dots,x_k^0)$. *Then the following assertions hold*:

1) *If* $\vartheta = +\infty$, *then the evasion problem is solvable for this initial state.*

2) *If* $\vartheta < +\infty$, *then the pursuit problem is solvable for this initial state.*

3) *If* $\vartheta < +\infty$ *and either there exists a minimal system* J' *of indices such that* (32) *and* (33) *hold or* $k < n$, *then* ϑ *is the optimal pursuit time for this initial state.*

BIBLIOGRAPHY

1. L. S. Pontryagin, *On the theory of differential games*, Uspekhi Mat. Nauk **21** (1966), no. 4(130), 219–274; English transl. in Russian Math. Surveys **21** (1966).

2. ______, *Linear differential games*. I, Dokl. Akad. Nauk SSSR **174** (1967), 1278–1280; English transl. in Soviet Math. Dokl. **8** (1967).

3. E. F. Mishchenko, M. S. Nikol'skiĭ and N. Satimov, *The problem of avoiding encounter in N-person differential games*, Trudy Mat. Inst. Steklov. **143** (1977), 105–128; English transl. in Proc. Steklov Inst. Math. **1980**, no. 1(143).

4. N. N. Krasovskiĭ and A. I. Subbotin, *Positional differential games*, "Nauka", Moscow, 1974; French transl., *Jeux différentiels*, "Mir", Moscow, 1977.

5. Rufus Isaacs, *Differential games*, Wiley, 1965.

6. L. A. Petrosyan, *Differential pursuit games*, Izdat. Leningrad. Univ., Leningrad., 1977. (Russian)

7. B. N. Pshenichnyĭ, *Simple pursuit by several objects*, Kibernetika (Kiev) **1976**, no. 3, 145–146; English transl. in Cybernetics **12** (1976).

8. F. L. Chernous'ko, *A problem of escape from several pursuers*, Prikl. Mat. Mekh. **40** (1976), 14–24; English transl. in J. Appl. Math. Mech. **40** (1976).

9. B. N. Pshenichnyĭ, *Necessary conditions for an extremum*, "Nauka", Moscow, 1969; English transl., Marcel Dekker, 1971.

10. Jacques Gauvin and Jon W. Tolle, *Differential stability in nonlinear programming*, SIAM J. Control Optimization **15** (1977), 294–311.

11. R. Tyrrell Rockafellar, *Convex analysis*, Princeton Univ. Press, Princeton, N. J., 1970.

12. Earl A. Coddington and Norman Levinson, *Theory of ordinary differential equations*, McGraw-Hill, 1955.

Translated by H. H. McFADEN

Труды Матем. Инст. Стеклов.
158(1981)

Proceedings of the Steklov
Institute of Mathematics
1983, Issue 4

ON THE LAURENT EXPANSION OF THE RIEMANN ZETA-FUNCTION

UDC 511.9

M. I. ISRAILOV

ABSTRACT. New representations for the coefficients γ_n in the Laurent expansion of the Riemann zeta-function about its pole are obtained and, with their help, known estimates for $|\gamma_n|$ are sharpened.

Bibliography: 13 titles.

1. The coefficients γ_n in the Laurent expansion of the Riemann zeta-function about its pole $s = 1$.

$$\zeta(s) = \frac{1}{s-1} + \sum_{n=0}^{\infty} \gamma_n (s-1)^n,$$

have been considered by Vacca, Glaisher and Hardy [1], Ramanujan [2], Kluyver [3], Wilton [4], Briggs and Chowla [5], [6], Lammel [7], Berndt [8], Lavrik [9] and other authors. The results obtained are in the form of different expressions for γ_n by means of series and estimates for the magnitude of the modulus of γ_n.

In this paper, we obtain further information on the quantities γ_n.

2. Set

$$b_{j,\nu} = \sum_{r_1=j}^{\nu} \frac{1}{r_1} \sum_{r_2=j-2}^{r_1-1} \frac{1}{r_2} \cdots \sum_{r_j=1}^{r_{j-1}-1} \frac{1}{r_j}. \tag{1}$$

We shall show that the following theorems hold.

THEOREM 1. *Let $f_n^{(\nu)}(x) = d^\nu(x^{-1}\log^n x)/dx^\nu$. Then, for all integral $n \geqslant 1$ and $k \geqslant 1$,*

$$\gamma_n = \sum_{r=[n/2]+1}^{k} \frac{B_{2r}}{2r} b_{n,2r-1} + \frac{(-1)^{k-1}}{n!\,(2k)!} \int_1^\infty B_{2k}^*(x) f_n^{(2k)}(x)\,dx. \tag{2}$$

But if the derivatives $f_n^{(2k_0)}(x)$ and $f_n^{(2k_0+2)}(x)$ have the same sign on the segment $N-1 \leqslant n \leqslant N$, then for $n \geqslant 1$

$$\gamma_n = \frac{(-1)^n}{n!}\left[\sum_{m=1}^{N} \frac{\log^n m}{m} - \frac{\log^{n+1} N}{n+1} - \frac{\log^n N}{2N} \right.$$
$$\left. - \sum_{r=1}^{k_0-1} \frac{B_{2r}}{2r} f_n^{(2r-1)}(N) - \theta \frac{B_{2k_0}}{(2k_0)!} f_n^{(2k_0-1)}(N)\right], \tag{3}$$

1980 *Mathematics Subject Classification.* Primary 10H05, 30B50; Secondary 65B15.

and $n = 0$

$$\gamma_0 = \sum_{m=1}^{N} \frac{1}{m} - \log N - \frac{1}{2N} + \sum_{r=1}^{k-1} \frac{B_{2r}}{2rN^{2r}} + \frac{\theta B_{2k}}{2kN^{2k}}, \tag{4}$$

where $0 < \theta < 1$, *the* B_j *and* $B_j^*(x)$ *denoting Bernoulli numbers and periodicized Bernoulli polynomials.*

THEOREM 2. *For* $n \geqslant 2k$ $(k = 1, 2, \dots)$

$$|\gamma_n| \leqslant C(k)(2k)^{-n}, \tag{5}$$

where

$$C(k) = \frac{|B_{2k}|}{2k}\left[1 + \sum_{j=1}^{2k} b_{j,2k}(2k)^j\right].$$

In particular,

$$C(1) = \tfrac{1}{2}, \quad C(2) = \tfrac{7}{12}, \quad C(3) = \tfrac{11}{3}, \quad C(4) \leqslant 73.3.$$

THEOREM 3. *For* $n \geqslant 0$

$$\gamma_{n+1} = \frac{1}{2(2^{n+1}-1)} \sum_{k=0}^{n} \gamma_k \gamma_{n-k} - \frac{(-1)^n}{2^{n+1}-1} \sum_{r=1}^{\infty} \frac{B_{2r}}{2r} \sum_{k=0}^{\min(n,2r-1)} \frac{(-1)^k 2^{n-k} b_{k,2r-1}}{(n-k)!} \zeta^{(n-k)}(2r+1), \tag{6}$$

where the series on the right is semiconvergent in the sense of [10], §13.7, $b_{0,\nu} = 1$, *and* $b_{k,\nu}$ *is defined by* (1) *for* $k \geqslant 1$.

In particular,

$$\gamma_1 = \frac{1}{2}\gamma^2 - \sum_{r=1}^{\infty} \frac{B_{2r}}{2r} \zeta(2r+1), \tag{7}$$

$$\gamma_2 = \frac{1}{3}\gamma\gamma_1 + \sum_{r=1}^{\infty} \frac{B_{2r}}{2r}\left[2\zeta'(2r+1) - \zeta(2r+1)\sum_{m=1}^{2r-1}\frac{1}{m}\right],$$

$$\gamma_3 = \frac{1}{14}(\gamma_1^2 + 2\gamma\gamma_2) - \frac{1}{7}\sum_{r=1}^{\infty}\frac{B_{2r}}{2r}\left[2\zeta''(2r+1) - 2\zeta'(2r+1)\sum_{m=1}^{2r-1}\frac{1}{m} + \zeta(2r+1)\sum_{m=2}^{2r-1}\frac{1}{m}\sum_{k=1}^{m-1}\frac{1}{k}\right],$$

where $\gamma = \gamma_0$ is Euler's constant.

On taking $k = 1$ in Theorem 1, we get the initial formula of [9], while the inequality of Theorem 2 improves known estimates for γ_n; in particular, for $k = 1$ we have the estimate of [9] and for $k = 2$ a more precise estimate than those of [7] and [8]. The equality (7) was obtained in a different way in [4]. With the help of (3) and (4) (and, as a check, with the help also of a certain modification of (6)), the values of γ_n are computed. This makes it possible to determine the numerical values of the coefficients in the principal term

$$P_k(\log x) = a_{k-1}^{(k)} \log^{k-1} x + \cdots + a_1^{(k)} \log x + a_0^{(k)} \tag{8}$$

of the divisor problem

$$\sum_{n \leqslant x} \tau_k(n) = xP_k(\log x) + \Delta_k(x),$$

as well as the values of the integrals

$$I_k = \int_1^\infty \frac{\Delta_k(x)}{x^2} dx. \tag{9}$$

Further, we compute the coefficients in the Laurent expansion of the logarithmic derivative of the zeta-function and obtain a sharper estimate of the constant A in the well-known inequality

$$\frac{\zeta'(\sigma)}{\zeta(\sigma)} < \frac{1}{\sigma - 1} + A.$$

3. We shall use the Bernoulli polynomials and numbers $B_n(x)$ and $B_n = B_n(0)$ given by the relation

$$\frac{te^{ix}}{e^t - 1} = \sum_{n=0}^\infty \frac{B_n(x)}{n!} t^n, \qquad |t| < 1.$$

Along with these, we introduce periodic functions $B_n^*(x)$ defined by the equalities

$$B_n^*(x) = B_n(x) \quad \text{for } 0 \leqslant x < 1 \qquad \text{and} \qquad B_n^*(x+1) = B_n^*(x).$$

The Fourier series of $B_{2k}^*(x)$ has the form

$$B_{2k}^*(x) = \frac{(-1)^{k-1} 2(2k)!}{(2\pi)^{2k}} \sum_{m=1}^\infty \frac{\cos 2\pi mx}{m^{2k}}.$$

In addition, we assume that all the derivatives of $f(x)$ are continuous for $x \geqslant a > 0$;

$$\int_a^\infty |f^{(2k)}(x)| dx < \infty \qquad (k \geqslant 1); \tag{10}$$

$$f^{(2k-1)}(x) \to 0 \quad \text{as } x \to \infty. \tag{11}$$

Setting

$$F(x) = \int_1^x f(t)\, dt, \qquad S_k(m) = \sum_{r=1}^k \frac{B_{2r}}{(2r)!} f^{(2r-1)}(m), \tag{12}$$

we have [10]

$$\sum_{a \leqslant m \leqslant N} f(m) = F(N) + \frac{1}{2} f(N) + S_k(N) + C_k + R_{k,N}, \tag{13}$$

where

$$C_k = \frac{1}{2} f(a) - F(a) - S_k(a) - \frac{1}{(2k)!} \int_a^\infty B_{2k}^*(x) f^{(2k)}(x)\, dx$$

$$R_{k,N} = \frac{1}{(2k)!} \int_N^\infty B_{2k}^*(x) f^{(2k)}(x)\, dx. \tag{14}$$

By virtue of (11), $R_{k,N} \to 0$ as $N \to \infty$. If conditions (11) and (12) are satisfied for all k, then C_k does not depend on k and N, but only on $f(x)$ and a. The analysis of the remainder given in [11] shows that if the derivatives $f^{(2k_0)}(x)$ and $f^{(2k_0+2)}(x)$

have the same sign on the interval $[a, N]$ then one can give another form of the Euler-Maclaurin formula [11]:

$$\sum_{a\leqslant m\leqslant N} f(m) = F(N) + \frac{1}{2}f(N) + S_{k_0-1}(N) + C_{k_0} + \frac{\theta B_{2k_0}}{(2k_0)!}f^{(2k_0-1)}(N), \tag{15}$$

$$C_{k_0} = \frac{1}{2}f(a) - F(a) - S_{k_0-1}(a) - \frac{\theta B_{2k_0}}{(2k_0)!}f^{(2k_0-1)}(a),$$

where $0 < \theta < 1$.

4. LEMMA. *Let n and ν be nonnegative integers and let $b_{0,\nu} = 1$. Then the νth derivative of the function $f_n(x) = x^{-1}\log^n x$ is determined by the formula*

$$f_n^{(\nu)}(x) = \frac{(-1)^\nu \nu!}{x^{\nu+1}} \sum_{j=0}^{\min(n,\nu)} \frac{(-1)^j n!}{(n-j)!} b_{j,\nu} \log^{n-j} x, \tag{16}$$

where the $b_{j,\nu}$ are computed by (1).

This lemma is proved by induction on ν.

PROOF OF THEOREM 1. Let $f(x) = f_n(x) = x^{-1}\log^n x$. Then, by virtue of (12), (14) and (16), for $k \geqslant 0$ the quantities $S_k(N)$, $R_{k,N}$ and $f_n^{(k)}(N)$ tend to zero as $N \to \infty$. Therefore, from (13) (from (15) as well) we have

$$C(n) = C_k(n) = \lim_{N\to\infty}\left(\sum_{a\leqslant m\leqslant N} \frac{\log^n m}{m} - \frac{\log^{n+1} N}{n+1}\right).$$

A comparison of this equality with the formula

$$\gamma_n = \frac{(-1)^n}{n!}\lim_{N\to\infty}\left(\sum_{m=1}^{N} \frac{\log^n m}{m} - \frac{\log^{n+1} N}{n+1}\right),$$

considered in [1]–[3], [5] and [7]–[9] shows that, for $a \geqslant 1$,

$$C(n) = (-1)^n n!\gamma_n - \sum_{1\leqslant m\leqslant a-1} \frac{\log^n m}{m}. \tag{17}$$

Hence, by (5), for any integral $k \geqslant 1$ and $a \geqslant 1$

$$\gamma_n = \frac{(-1)^n}{n!}\left[\sum_{1\leqslant m\leqslant a} \frac{\log^n m}{m} - \frac{\log^{n+1} a}{n+1} - \frac{\log^n a}{2a} - \sum_{r=1}^{k} \frac{B_{2r}}{(2r)!} f_n^{(2r-1)}(a) - \frac{1}{(2k)!}\int_a^\infty B_{2k}^*(x) f_n^{(2k)}(x)\,dx\right].$$

Setting $a = 1$ and taking into account that

$$f_n^{(2r-1)}(1) = \begin{cases} (-1)^{n+1}(2r-1)!n!b_{n,2r-1} & \text{for } n \leqslant 2r-1, \\ 0 & \text{for } n > 2r-1, \end{cases}$$

we get (2).

Similarly, taking $f(x) = x^{-1}\log^n x$ and $a = N - 1$ in (15), by virtue of (17) we find that

$$\sum_{m=1}^{N} \frac{\log^n m}{m} = \frac{\log^{n+1} N}{n+1} + \frac{\log^n N}{2N} + \sum_{r=1}^{k_0-1} \frac{B_{2r}}{(2r)!} f_n^{(2r-1)}(N)$$
$$+ (-1)^n n! \gamma_n + \frac{\theta B_{2k_0}}{(2k_0)!} f_n^{(2k_0-1)}(N),$$

whence (3) follows. For deriving (4), one has to take $f(x) = 1/x$ in (15). The theorem is thus proved.

PROOF OF THEOREM 2. Let $n \geqslant 2k$. Then from (2), (16) and (17) we have

$$\gamma_n = \frac{(-1)^{k-1}}{n!\,(2k)!} B_{2k}^*(x) f_n^{(2k)}(x)\, dx,$$

$$f_n^{(2k)}(x) = \frac{(2k)!}{x^{2k+1}} \left[\log^n x + \sum_{j=1}^{2k} \frac{(-1)^j n!}{(n-j)!} b_{j,2k} \log^{n-j} x \right].$$

Since

$$|B_{2k}^*(x)| \leqslant |B_{2k}|, \qquad \int_1^\infty \frac{\log^{n-j} x}{x^{2k+1}}\, dx = \frac{r(n-j)!}{(2k)^{n-j+1}},$$

therefore

$$|\gamma_n| \leqslant \frac{|B_{2k}|}{n!} \int_1^\infty x^{-2k-1} \left[\log^n x + \sum_{j=1}^{2k} \frac{n!}{(n-j)!} b_{j,2k} \log^{n-j} x \right] dx$$
$$= \frac{|B_{2k}|}{(2k)^{n+1}} \left[1 + \sum_{j=1}^{2k} b_{j,2k} (2k)^j \right],$$

which proves (5). Further, by direct calculation using (1), we find the values of $C(1), \dots, C(4)$ given above.

PROOF OF THEOREM 3. For $\operatorname{Re} s > 0$ one has the identity

$$\frac{1}{2}\zeta^2(s) - \frac{\zeta(2s-1)}{s-1} = \sum_{n=1}^{\infty} n^{-s} \left\{ \zeta(s) - \sum_{k=1}^{n-1} k^{-s} - \frac{n^{1-s}}{s-1} - \frac{n^{-s}}{2} \right\}.$$

Hence, taking into account the relations

$$\zeta(2s-1) = \frac{1}{2(s-1)} + \sum_{k=0}^{\infty} \gamma_k 2^k (s-1)^k,$$

$$a^{-s} = \sum_{k=0}^{\infty} \frac{(-1)^k \log^k a}{a \cdot k!} (s-1)^k \qquad (a > 1),$$

we conclude that

$$\sum_{n=0}^{\infty}\left[\frac{1}{2}\sum_{k=0}^{n}\gamma_k\gamma_{n-k}-(2^{n+1}-1)\gamma_{n+1}\right](s-1)^n$$
$$=\sum_{n=0}^{\infty}\left\{\sum_{m=1}^{\infty}\frac{(-1)^n}{m}\sum_{l=0}^{n}\frac{\log^{n-l}m}{(n-l)!l!}\right.$$
$$\left.\times\left[(-1)^l l!\gamma_l-\sum_{r=1}^{m-1}\frac{\log^l r}{r}+\frac{\log^{l+1}m}{l+1}-\frac{\log^l m}{2m}\right]\right\}(s-1)^n.$$

Consequently, for $n \geqslant 0$

$$\gamma_{n+1}=\frac{1}{2(2^{n+1}-1)}\sum_{l=0}^{n}\gamma_l\gamma_{n-l}-\frac{(-1)^n}{2^{n+1}-1}\sum_{m=1}^{\infty}\frac{1}{m}\sum_{l=0}^{n}\frac{\log^{n-l}m}{(n-l)!l!}$$
$$\times\left[(-1)^l l!\gamma_l-\sum_{r=1}^{m-1}\frac{\log^l r}{r}+\frac{\log^{l+1}m}{l+1}-\frac{\log^l m}{2m}\right]. \tag{18}$$

If in (3) and (4), we make k_0 and k tend to infinity, then for any $N = m \geqslant 1$ and $l \geqslant 0$ we shall have the semiconvergent series

$$(-1)^l l!\gamma_l=\sum_{r=1}^{m}\frac{\log^l r}{r}-\frac{\log^{l+1}m}{l+1}-\frac{\log^l m}{2m}-\sum_{r=1}^{\infty}\frac{B_{2r}}{(2r)!}f_l^{(2r-1)}(m).$$

On combining this with (18), we get

$$\gamma_{n+1}=\frac{1}{2(2^{n+1}-1)}\sum_{l=0}^{n}\gamma_l\gamma_{n-l}$$
$$+\frac{(-1)^n}{2^{n+1}-1}\sum_{m=1}^{\infty}\frac{1}{m}\sum_{l=0}^{n}\frac{\log^{n-l}m}{(n-l)!l!}\sum_{r=1}^{\infty}\frac{B_{2r}}{(2r)!}f_l^{(2r-1)}(m).$$

Finally, substituting here the expression for $f_l^{(2r-1)}(m)$ given by (16) and taking into account the equality

$$\sum_{l=j}^{n}\frac{1}{(n-l)!\,(l-j)!}=\frac{2^{n-j}}{(n-j)!},$$

we obtain the assertion in Theorem 3.

5. The coefficients $\bar{\gamma}_n$ in the Laurent expansion of the logarithmic derivative of the zeta-function

$$-\frac{\zeta'(s)}{\zeta(s)}=\frac{1}{s-1}+\sum_{n=0}^{\infty}\bar{\gamma}_n(s-1)^n$$

are expressed in terms of γ_n as follows:

$$\bar{\gamma}_n=\begin{cases}-\gamma_0 & \text{for } n=0,\\ -(n+1)\gamma_n-\sum_{k=0}^{n}\bar{\gamma}_k\gamma_{n-1-k} & \text{for } n\geqslant 1.\end{cases} \tag{19}$$

Given below are a part of the results of machine computation of γ_n, $\bar{\gamma}_n$ and

$$\varphi(s)=-\zeta'(s)/\zeta(s)-1/(s-1),$$

which have been computed with the help of (3), (4) and (19). All the signs given in the table are correct.

n	γ_n	$\bar{\gamma}_n$	s	$\varphi(s)$
0	0.577215665	−0.577215665	1.0	−0.577
1	0.072815846	0.187546233	1.1	−0.559
2	−0.004845182	−0.051688631	1.2	−0.541
3	−0.000342305	0.014751654	1.3	−0.525
4	0.000096889	−0.004524469	1.4	−0.510
5	−0.000006610	0.001446785	1.5	−0.495
6	−0.000000332	−0.000571535	1.6	−0.481
7	0.000000105	0.000155173	1.7	−0.468
8	−0.000000009	−0.000051341	1.8	−0.452
9		0.000017039	1.9	−0.442
10		−0.000005665	2.0	−0.430

From the last column of the table we see that, for $1 \leqslant \sigma \leqslant 2$,

$$-\zeta'(\sigma)/\zeta(\sigma) < 1/(\sigma - 1) - 0.42.$$

Further, by using the values of $\bar{\gamma}_n$ in the table, it is not difficult to show that, for $1 \leqslant \sigma \leqslant 2$,

$$-\zeta'(\sigma)/\zeta(\sigma) < 1/(\sigma - 1) - \gamma + 0.1875463(\sigma - 1),$$

which sharpens an estimate of Rosser and Schoenfeld [12]. In [13] were given, for $k = 1, \dots, 6$, expressions in terms of $\gamma_0, \gamma_1, \dots, \gamma_{k-2}$ for the coefficients $a_j^{(k)}$ in the principal term of the Dirichlet divisor problem (8) and for the integral (9).

By using the values of γ_n given above, we find that

$$\begin{array}{llllll}
a_1^{(2)} = 1, & a^{(2)} = 0.154; & & & & \\
a_2^{(3)} = 0.5, & a_1^{(3)} = 0.731, & a_0^{(3)} = 0.486; & & & \\
a_3^{(4)} = 0.167, & a_2^{(4)} = 0.654, & a_1^{(4)} = 1.931, & a_0^{(4)} = 0.273; & & \\
a_4^{(5)} = 0.042, & a_3^{(5)} = 0.314, & a_2^{(5)} = 0.905, & a_1^{(5)} = 0.930, & a_0^{(5)} = -5.186; & \\
a_5^{(6)} = 0.008, & a_4^{(6)} = 0.103, & a_3^{(6)} = 0.495, & a_2^{(6)} = 1.053, & a_1^{(6)} = 1.007, & a_0^{(6)} = 2.924; \\
I_1 = -0.423, & I_2 = 0.324, & I_3 = -0.056, & I_4 = 0.073, & I_5 = -0.720. &
\end{array}$$

We remark that, in the limits of our calculation, $\bar{\gamma}_n$ and I_n alternate in sign.

The author expresses his thanks to A. F. Lavrik for discussions, and to K. M. Endibaev for assistance in the numerical computations.

BIBLIOGRAPHY

1. G. H. Hardy, *Note on Dr. Vacca's series for* γ, Quart. J. Pure Appl. Math. **43** (1911/12), 215–216.

2. Srinivass Ramanujan, *Collected papers*, Cambridge Univ. Press, 1927.

3. J. C. Kluyver, *On certain series of Mr. Hardy*, Quart. J. Pure Appl. Math. **50** (1923/27), 185–192.

4. J. R. Wilton, *A note on the coefficients in the expansion of* $\zeta(s, x)$ *in powers of* $s - 1$, Quart. J. Pure Appl. Math. **50** (1923/27), 329–332.

5. W. E. Briggs and S. Chowla, *The power series coefficients of* $\zeta(s)$, Amer. Math. Monthly **62** (1955), 323–325.

6. William E. Briggs, *Some constants associated with the Riemann zeta-function*, Michigan Math. J. **3** (1955/56), 117–121.

7. Ernst Lammel, *Ein Beweis, dass die Riemannsche Zeta funktion* $\zeta(s)$ *in* $|s - 1| \leqslant 1$ *keine Nullstelle besitzt*, Univ. Nac. Tucumán Rev. Ser. A **16** (1966), 209–217.

8. Bruce C. Berndt, *On the Hurwitz zeta-function*, Rocky Mountain J. Math. **2** (1972), 151–157.

9. A. F. Lavrik, *The principal term in the divisor problem and the power series of the Riemann zeta-function in a neighborhood of its pole*, Trudy Mat. Inst. Steklov. **142** (1976), 165–173; English transl. in Proc. Steklov Inst. Math. **1979**, no. 3 (142).

10. G. H. Hardy, *Divergent series*, Clarenden Press, Oxford, 1949.

11. A. O. Gel'fond, *The calculus of finite differences*, 3rd ed., "Nauka", Moscow, 1967; English transl. Hindustan, Delhi, 1971.

12. J. Barkley Rosser and Lowell Schoenfeld, *Sharper bounds for the Chebyshev functions $\theta(x)$ and $\psi(x)$*, Math. Comp. **29** (1975), 243–269.

13. A. F. Lavrik, M. I. Israilov, and Zh. Ëdgorov, *Integrals containing the remainder of the divisor problem*, Acta Arith. **37** (1980), 381–389. (Russian)

Translated by V. N. SINGH

Труды Матем. Инст. Стеклов.
158(1981)

Proceedings of the Steklov
Institute of Mathematics
1983, Issue 4

ORTHOGONAL DECOMPOSITIONS OF SIMPLE LIE ALGEBRAS (TYPE A_n)

UDC 519.4

A. I. KOSTRIKIN, I. A. KOSTRIKIN AND V. A. UFNAROVSKIĬ

ABSTRACT. The notion of the decomposition $L = H \oplus H_1 \oplus \cdots \oplus H_r$ of a simple Lie algebra L over the complex numbers into a direct sum of Cartan subalgebras $H = H_0, H_i, 1 \leqslant i \leqslant r$, pairwise orthogonal with respect to the Killing form, is introduced in this paper. A stronger requirement of multiplicativity $[H_i, H_j] \subseteq H_k$ was earlier studied by Thompson for an algebra of type E_8 (J. Algebra **38** (1976), 525–530), but turns out to be too strong in the case of algebras of type A_n: for $n > 1$ it cannot occur. In this paper an orthogonal decomposition is constructed of a Lie algebra L of type A_n for any $n = p^m - 1$ (p a prime, m a natural number), and the automorphism group preserving the H_i is studied. The conjecture that algebras of type A_n have no orthogonal decompositions for $n \neq p^m - 1$ is advanced. On the contrary, if the orthogonality condition is removed, the assertion about a decomposition $L = H \oplus H_1 \oplus \cdots \oplus H_r$ holds for any n. Analogous results for simple Lie algebras of arbitrary types are announced.

Bibliography: 2 titles.

The theory of finite-dimensional simple Lie algebras over the complex number field **C**, which is intensively used in many branches of mathematics, has developed to such an extent that the absence of a ready answer to any particular question that arises is sometimes regarded as an irritating anomaly. Such cases, however, testify rather to the fundamental nature of the subject itself than to the gaps in its exposition. As experience shows, finite-dimensional simple Lie algebras over **C**, in a similar way to the classical linear groups, are an inexhaustible source of new problems, new approaches to old problems, and elegant interpretations of known facts.

The problem with which we are going to deal, and which, so far as we know, has not been stated before in full generality, consists in the study of the following question.

QUESTION 1. *Does a simple complex Lie algebra L admit a decomposition*

$$L = H \oplus H_1 \oplus \cdots \oplus H_r \tag{1}$$

into a direct sum of Cartan subalgebras H, where the H_i are pairwise orthogonal with respect to the Killing form $(* \mid *)$?

For brevity, we call (1) an *orthogonal decomposition* ($\mathcal{OD}$) of the algebra L.

1980 *Mathematics Subject Classification*. Primary 17B20, 17B40.

Thus the essential requirement is that $(H|H_i) = 0 = (H_i|H_j)$ for $i \neq j$. The restriction of the invariant symmetric bilinear form $(*|*)$ to H and the H_i $(1 \leqslant i \leqslant r)$, is, as we know, nondegenerate.

A weaker form of Question 1 is obviously

QUESTION 1′. *Does every simple complex Lie algebra L admit a decomposition (we shall call this property $\mathfrak{D}$) into a direct sum of Cartan subalgebras H and H_i?*

The answer to this is in the affirmative: the easy Theorem 1 in §1 states this for algebras of type A_n, the other algebras are considered in a different paper. An intrinsic explanation is given thereby of the fact, which at first sight seems accidental, that the dimension dim L of a simple Lie algebra L is divisible by its rank, i.e. the dimension of its Cartan subalgebra.

On the contrary, the answer to the main Question 1 apparently depends greatly on the type of the algebra L and on its rank. At any rate, it seems to us rather likely that the following holds:

CONJECTURE. *A Lie algebra L of type A_n admits $\mathfrak{O}\mathfrak{D}$ if and only if $n = p^m - 1$ (where p is an arbitrary prime and m an arbitrary natural number).*

In §1 we shall give some arguments in favor of this conjecture. Surprisingly enough, the technical difficulties arising are rather great, even in the simplest case of the algebra of type A_5, already celebrated by the perspicacious Winnie-the-Pooh in his humming (colloquial version):

> Let's take again this word we said "Afore"
> and on it let us ruminate.
> Why not go on a little bit more,
> with "A-five", and then "A-six", "A-seven", "A-eight"?(*)

In §2 it is shown that indeed A_6, A_7, and A_8, as, by the way, all algebras A_{p^m-1}, possess $\mathfrak{O}\mathfrak{D}$. The corresponding Theorem 3 is proved quite constructively, which makes it possible to compute the group $\mathrm{Aut}_{\mathfrak{O}\mathfrak{D}}(A_{p^m-1})$ of automorphisms of the algebra A_{p^m-1} that preserve the given $\mathfrak{O}\mathfrak{D}$ (Theorem 4 in §3). We did not aim, however, at a description of the orbits of all $\mathfrak{O}\mathfrak{D}$ under $\mathrm{Aut}(A_{p^m-1})$, since we intended to concentrate our attention at the moment on existence theorems. Elsewhere it has been shown that the number of orbits of $\mathfrak{O}\mathfrak{D}$ of a simple Lie algebra can be, a priori, greater than 1.

A stronger version of Question 1 is

QUESTION 2. *Does a simple complex Lie algebra L admit a decomposition into a direct sum* (1) *of Cartan subalgebras $H = H_0$, H_i, satisfying the conditions*

i) $[H_i, H_j] \subseteq H_k$, $0 \leqslant i, j, k \leqslant r$ *(multiplicativity), and*

ii) $(H_i|H_j) = 0$ *for* $i \neq j$ *(orthogonality)?*

(*) *Translator's note.* The verse given in Russian here bears no relation to anything I have been able to find in the Pooh books, and so I have rather freely translated it back into English so as to preserve at least some of its aptness. In the Russian, the word for "again" puns with the pronunciation of "A_5", and is contrasted with "A_6", "A_7", "A_8".

It is understood that for each pair of the subspaces H_i, H_j, there exists a third, H_k, which contains $[H_i, H_j]$. It can easily be shown that condition ii) is a consequence of condition i). However, we would prefer not to digress from the main stream of our exposition.

For brevity, a decomposition with properties i) and ii) will be called a *multiplicative orthogonal decomposition* ($\mathfrak{MOD}$). For algebras of type A_n, the notion of $\mathfrak{MOD}$ is not relevant since the answer to Question 2 for $n > 1$ turns out to be negative (Theorem 2). But for algebras of other types, Question 2 becomes attractive. In this connection we remark that an $\mathfrak{MOD}$ of the algebra E_8 led to a realization of the sporadic simple group of Thompson, having order $2^{15} \cdot 3^{10} \cdot 5^3 \cdot 7^2 \cdot 13 \cdot 19 \cdot 31$ (see [1]).

The authors have developed the construction of $\mathfrak{OD}$ and $\mathfrak{MOD}$ for practically all possible situations, and the aim of this paper is to describe a particular aspect of this construction in an extended form.

§1. $\mathfrak{D}$, $\mathfrak{OD}$ and $\mathfrak{MOD}$

Because Cartan subalgebras are conjugate, one of them, say H, can be assumed standard in (1). Replacing n by $n-1$ for convenience, we shall understand by L the Lie algebra of all $n \times n$ matrices $A = (a_{ij})$ with zero trace $\operatorname{Tr} A = a_{11} + a_{22} + \cdots + a_{nn}$. So

$$H = \{\operatorname{diag}(\lambda_1, \ldots, \lambda_n) \mid \lambda_i \in \mathbf{C}, \lambda_1 + \cdots + \lambda_n = 0\}.$$

The Cartan subalgebra H_1 is obtained from H by conjugating by some nonsingular matrix

$$X = (x_{ij})\colon H_1 = XHX^{-1}.$$

It is well known that

$$(A \mid B) = 2n \operatorname{Tr} AB, \quad \forall A, B \in L.$$

The orthogonality condition $(H \mid H_1) = 0$ thus means that

$$\operatorname{Tr} AB = 0, \quad \forall A \in H, B \in H_1.$$

If $B = (b_{ij})$, then $\operatorname{Tr}(E_{ii} - E_{jj})B = 0 \Rightarrow b_{ii} = b_{jj}$ $\forall i, j$. Since $b_{11} + b_{22} + \cdots + b_{nn} = 0$,

$$(H|B) = 0 \Leftrightarrow b_{ii} = 0; \quad i = 1, \ldots, n.$$

If $X^{-1} = (\hat{x}_{ij})$, then in the matrix

$$X(E_{jj} - E_{kk})X^{-1} = (b_{il})$$

the element $b_{ii} = x_{ij}\hat{x}_{ji} - x_{ik}\hat{x}_{ki}$ vanishes; that is, $x_{ij}\hat{x}_{ji} = x_{ik}\hat{x}_{ki}$, $k = 1, \ldots, n$. Summing over k, we obtain

$$n x_{ij} \hat{x}_{ji} = \sum_{k=1}^{n} x_{ik} \hat{x}_{ki} = 1.$$

Thus the entries x_{ij} of the matrix transforming H to H_1 in the orthogonal decomposition (1) are all nonzero, and moreover

$$\hat{x}_{ji} = \frac{1}{n x_{ij}}, \qquad \sum_{k=1}^{n} \frac{x_{ki}}{x_{kj}} = n\delta_{ij} = \sum_{k=1}^{n} \frac{x_{ik}}{x_{jk}}. \tag{2}$$

A matrix $X = (x_{ij})$ satisfying (2) will be called *suitable*. Clearly if X is suitable then so is X^{-1}, even though the Cartan subalgebra $X^{-1}HX$ is not necessarily in the decomposition (1).

We also remark that from a given Cartan subalgebra H_1 we cannot uniquely recover the suitable matrix X, but only up to multiplication on the right by an arbitrary monomial matrix. If $X_1, \ldots, X_r$ are matrices transforming H to $H_1, \ldots, H_r$ (in our case $r = n$), then from the conditions

$$X_s H X_s^{-1} = H_s, \qquad X_t H X_t^{-1} = H_t, \qquad s \neq t \quad (H_s | H_t) = 0,$$

we find that

$$\left(X_s H X_s^{-1} | X_t H X_t^{-1}\right) = 0,$$

and hence

$$\left(H | X_s^{-1} X_t H \left(X_s^{-1} X_t\right)^{-1}\right) = 0$$

(since $(A | B) = 0 \Leftrightarrow (CAC^{-1} | CBC^{-1}) = 0$). Thus, if X_s and X_t $(s \neq t)$ are suitable then the matrix $X_s^{-1} X_t$ will also be suitable. But if the matrices $X = (x_{ij})$, $Y = (y_{ij})$ and $X^{-1}y$ are suitable, then along with the relations of type (2), the following relations must be satisfied:

$$\left(\sum_{k=1}^{n} \frac{x_{ki}}{y_{kj}}\right)\left(\sum_{k=1}^{n} \frac{y_{kj}}{x_{ki}}\right) = n;$$
$$\sum_{s=1}^{n}\left(\sum_{k=1}^{n} \frac{y_{ks}}{x_{ki}}\right)\left(\sum_{k=1}^{n} \frac{x_{kj}}{y_{ks}}\right) = \delta_{ij} n^2, \qquad 1 \leqslant i, j \leqslant n. \tag{3}$$

Using the nondegeneracy of the restriction of the Killing form to any Cartan subalgebra, we can easily show that our arguments are reversible; that is, the following statement holds.

PROPOSITION 1. *The algebra A_{n-1} possesses an $\mathfrak{OD}$ if and only if there exist n suitable matrices $X_1, \ldots, X_n$ of dimension n such that the matrices $X_i^{-1} X_j$ are also suitable for any $1 \leqslant i, j \leqslant n$, $i \neq j$.*

Unfortunately, when solving the systems of nonlinear equations of type (2) and (3) for $n \geqslant 6$ there arise considerable difficulties. On the contrary, the construction of $\mathfrak{D}$ in the algebra A_{n-1} is quite easy, as is shown by

THEOREM 1. *Any Lie algebra L of type A_{n-1} is decomposable into a direct sum of Cartan subalgebras (has property $\mathfrak{D}$).*

PROOF. Let us consider the monomial matrices

$$R_{k+1} = \operatorname{diag}(\lambda_k, \lambda_{k+1}, \ldots, \lambda_{k+n-1}) \cdot P; \qquad k = 0, 1, \ldots, n-1,$$

where

$$P = \begin{Vmatrix} 0 & 0 & \ldots & 0 & 0 & 1 \\ 1 & 0 & \ldots & 0 & 0 & 0 \\ 0 & 1 & \ldots & 0 & 0 & 0 \\ \cdots & \cdots & \cdots & \cdots & \cdots & \cdots \\ 0 & 0 & \ldots & 1 & 0 & 0 \\ 0 & 0 & \ldots & 0 & 1 & 0 \end{Vmatrix} \tag{4}$$

is the matrix corresponding to a cyclic permutation of $(1, 2, \dots, n-1, n)$, the indices j of the parameters λ_j being taken modulo n, with $\lambda_j \neq 0$, $j = 0, 1, \dots, n-1$. It is clear that the linear span

$$H_s = \left\langle R_s, R_s^2, \dots, R_s^{n-1} \right\rangle, \qquad 1 \leqslant s \leqslant n,$$

is a Cartan subalgebra in L. We assert that for an appropriate choice of the parameters λ_j (and in fact for almost any choice) we have the required decomposition into a direct sum of subspaces

$$L = H \oplus H_1 \oplus \cdots \oplus H_n$$

(recall that H is the subalgebra of diagonal matrices). It suffices to verify the linear independence of the matrices

$$R_{k+1}^m = \operatorname{diag}\left(\Lambda_{k-m}^{(m)}, \Lambda_{k-n+1}^{(m)}, \dots, \Lambda_{k-m+n-1}^{(m)} \right) P^m, \qquad 0 \leqslant k \leqslant n-1,$$

for a fixed m. Here

$$\Lambda_{k+s}^{(m)} = \lambda_{k+s+1} \cdot \lambda_{k+s+2} \cdot \cdots \cdot \lambda_{k+s+m-1} \cdot \lambda_{k+s+m}, \qquad 0 \leqslant s \leqslant n-1.$$

The linear independence of these matrices is equivalent to the nonvanishing of the circulant

$$\begin{vmatrix} \Lambda_0^{(m)} & \Lambda_1^{(m)} & \dots & \Lambda_{n-1}^{(m)} \\ \Lambda_1^{(m)} & \Lambda_2^{(m)} & \dots & \Lambda_0^{(m)} \\ \dots & \dots & \dots & \dots \\ \Lambda_{n-1}^{(m)} & \Lambda_0^{(m)} & \dots & \Lambda_{n-2}^{(m)} \end{vmatrix} = \prod_{k=0}^{n-1} \left(\Lambda_0^{(m)} + \varepsilon^k \Lambda_1^{(m)} + \cdots + \varepsilon^{(n-1)k} \Lambda_{n-1}^{(m)} \right),$$

where ε is a primitive nth root of unity. Putting

$$\lambda_0 = \lambda^{-1}, \qquad \lambda_1 = \lambda, \qquad \lambda_2 = \lambda_3 = \cdots = \lambda_{n-1} = 1,$$

we have

$$\Lambda_0^{(m)} = \lambda, \qquad \Lambda_{n-m}^{(m)} = \lambda^{-1}, \qquad \Lambda_t^{(m)} = 1, \qquad t \neq 0, n-m.$$

If λ is sufficiently large, say $\lambda = n+1$, then

$$\left|\Lambda_0^{(m)}\right| = n + 1 > \left|\varepsilon^k \Lambda_1^{(m)} + \cdots + \varepsilon^{(n-1)k} \Lambda_{n-1}^{(m)}\right|$$

and the determinant is nonzero. Theorem 1 is proved.

The above proof shows that property $\mathfrak{D}$ is almost trivial. On the other hand, for the algebras under consideration property $\mathfrak{MOD}$ turns out to be too strong.

THEOREM 2. *For $n > 1$ a Lie algebra of type A_n does not possess an $\mathfrak{MOD}$.*

PROOF. We shall start from an $\mathfrak{OD}$

$$L = H \oplus H_1 \oplus \cdots \oplus H_r$$

of the Lie algebra L of type A_{n-1}. Let $X = (x_{ij})$ be a suitable matrix corresponding to H_1 (see the beginning of this section), and $X^{-1} = (\hat{x}_{ij})$. We consider the matrix

$$\mathfrak{D}_{s,t} = \begin{Vmatrix} d_1 & & & \\ & d_2 & & \\ & & \ddots & \\ & & & d_n \end{Vmatrix}, \qquad d_j = x_{sj}^{-1} x_{tj}, \ s \neq t.$$

Since $d_1 + \cdots + d_n = \Sigma_{j=1}^{n} x_{sj}^{-1} x_{tj} = n\delta_{s,t}$ (see (2)), $\mathfrak{D}_{s,t} \in H$. Therefore,

$$Y_{s,t} = (y_{lk}) = X\mathfrak{D}_{s,t}X^{-1} \in H_1,$$

and moreover, as is shown by (2),

$$y_{sk} = \sum_{j=1}^{n} x_{sj} d_j \hat{x}_{jk} = \sum_{j=1}^{n} x_{sj} x_{sj}^{-1} x_{ij} (n x_{kj})^{-1} = \delta_{kt},$$

$$y_{lt} = \sum_{j=1}^{n} x_{lj} x_{sj}^{-1} x_{tj} (n x_{tj})^{-1} = \delta_{ls}.$$

We see that the matrices $Y_{s,1}, \ldots, Y_{s,s-1}, Y_{s,s+1}, \ldots, Y_{s,n}$, which each have exactly one 1 in row number s (at the intersection with the tth column; $t = 1, \ldots, s-1, s+1, \ldots, n$) form a basis for the Cartan subalgebra H_1. Hence H_1 does not contain a nonzero matrix with zero sth row. Applying these considerations in particular to $s = 1$ and $s = n$, we conclude that in any linear combination of the basis matrices

$$Y_{1,2}, Y_{1,3}, \ldots, Y_{1,n}$$

the nth row is nonzero. The same conclusion remains valid with respect to any other Cartan subalgebra H_k, $k = 1, \ldots, n$. But the matrix

$$[(E_{22} - E_{33}), [(E_{11} - E_{22}), Y_{1,2}]] = \begin{Vmatrix} 0 & -2 & 0 & 0 & \ldots & 0 \\ -2y_{21} & 0 & -2y_{23} & -y_{24} & \ldots & -y_{2n} \\ y_{31} & 0 & 0 & 0 & \ldots & 0 \\ 0 & 0 & 0 & 0 & \ldots & 0 \\ \cdots & \cdots & \cdots & \cdots & \cdots & \cdots \\ 0 & 0 & 0 & 0 & \ldots & 0 \end{Vmatrix},$$

which belongs by the definition of an $\mathfrak{MOD}$ to some Cartan subalgebra H_k, contains a zero row for $n > 3$. Thus the general case of algebras of type A_n, $n > 2$, is exhausted.

In the case of an algebra L of type A_2, additional information is required about the orthogonal decomposition

$$L = H \oplus H_1 \oplus H_2 \oplus H_3.$$

We already know that $H_1 = \langle A, B \rangle$, where

$$A = \begin{Vmatrix} 0 & 1 & 0 \\ u & 0 & a \\ x & 0 & 0 \end{Vmatrix}, \qquad B = \begin{Vmatrix} 0 & 0 & 1 \\ y & 0 & 0 \\ v & b & 0 \end{Vmatrix},$$

and the linear independence of corresponding rows (as well as columns) of the matrices A and B leads to the condition $axby \neq 0$. Since $AB - BA = [A, B] = 0$, we have $u = 0 = v$ and $x = ab = y$; that is,

$$A = \begin{Vmatrix} 0 & 1 & 0 \\ 0 & 0 & a \\ ab & 0 & 0 \end{Vmatrix}, \qquad B = \begin{Vmatrix} 0 & 0 & 1 \\ ab & 0 & 0 \\ 0 & b & 0 \end{Vmatrix}, \qquad ab \neq 0.$$

This is the general form of the basis matrices of the Cartan subalgebra H_1 orthogonal to H_0. Now without any computations we take

$$H_2 = \langle C, D\rangle; \quad C = \begin{Vmatrix} 0 & 1 & 0 \\ 0 & 0 & c \\ cd & 0 & 0 \end{Vmatrix}, \quad D = \begin{Vmatrix} 0 & 0 & 1 \\ cd & 0 & 0 \\ 0 & d & 0 \end{Vmatrix}, \quad cd \neq 0;$$

$$H_3 = \langle F, G\rangle; \quad F = \begin{Vmatrix} 0 & 1 & 0 \\ 0 & 0 & f \\ fg & 0 & 0 \end{Vmatrix}, \quad G = \begin{Vmatrix} 0 & 0 & 1 \\ fg & 0 & 0 \\ 0 & g & 0 \end{Vmatrix}, \quad fg \neq 0.$$

The orthogonality condition $(H_1 \mid H_2) = 0$ is expressed by the relations $\operatorname{Tr} AD = 0 = \operatorname{Tr} BC$, from which it follows that

$$ad = bc, \qquad ab + bc + cd = 0.$$

Analogously,

$$(H_1|H_3) = 0 \Rightarrow ag = bf, \qquad ab + bf + fg = 0;$$
$$(H_2|H_3) = 0 \Rightarrow cg = df, \qquad cd + df + fg = 0.$$

Thus

$$c = \varepsilon a, \qquad d = \varepsilon b; \qquad f = \varepsilon^2 a, \qquad g = \varepsilon^2 b$$

(ε is a primitive cube root of unity), and the entire orthogonal decomposition is determined by two independent parameters $a, b \neq 0$.

We have obtained more than is required in order to use the $\mathcal{MOD}$ condition. From $[H_1, H_2] \subseteq H_k$ it follows that one of the matrices

$$[A, C] = \begin{Vmatrix} 0 & 0 & c - a \\ ac(d - b) & 0 & 0 \\ 0 & ab - cd & 0 \end{Vmatrix},$$

$$[A, D] = \begin{Vmatrix} cd - ab & 0 & 0 \\ 0 & -d(c - a) & 0 \\ 0 & 0 & a(b - d) \end{Vmatrix}$$

must be nonzero; but this is impossible since for example $c - a = (\varepsilon - 1)a \neq 0$.

In the case of a Lie algebra L of type A_1 the orthogonal decomposition containing the diagonal subalgebra can be uniquely written in the form

$$L = \left\langle \begin{Vmatrix} 1 & 0 \\ 0 & -1 \end{Vmatrix} \right\rangle \oplus \left\langle \begin{Vmatrix} 0 & 1 \\ a & 0 \end{Vmatrix} \right\rangle \oplus \left\langle \begin{Vmatrix} 0 & 1 \\ -a & 0 \end{Vmatrix} \right\rangle, \qquad a \neq 0.$$

Clearly it possesses the $\mathcal{MOD}$ property. The theorem is proved.

REMARK. The automorphism induced by conjugation by the matrix $\begin{Vmatrix} 1 & 0 \\ 0 & \beta \end{Vmatrix}$, $\beta^2 = a^{-1}$, transforms the decomposition just given into the standard one, with parameter $a = 1$. Similarly the parameters $a, b \neq 0$ determining an $\mathcal{OD}$ of the Lie algebra of type A_2 can be made equal to 1 using the conjugating matrix $\operatorname{diag}(\rho^{-1}a, 1, \rho)$, where $\rho^3 = ab^{-1}$. It is quite plausible that in general for Lie algebras of type A_{p^n-1} the $\mathcal{OD}$ is unique up to conjugacy.

The description of the $\mathcal{OD}$ of Lie algebras of type A_1 and A_2 which emerged in the proof of Theorem 2, and all the decompositions constructed in §2, suggest the monomiality of all $\mathcal{OD}$ containing the standard subalgebra H, in the sense that every subalgebra H_i admits a basis of nonsingular monomial matrices. In connection with this observation the following proposition serves apparently as a serious argument in favor of the absence of an $\mathcal{OD}$ for the algebra of type A_5.

PROPOSITION 2. *A Lie algebra L of type A_5 does not admit a monomial $\mathcal{OD}$.*

OUTLINE OF THE PROOF. It would be rather tedious to go into all the details, which often require direct item-by-item examination of a large number of particular cases. So we isolate the most important features, characteristic, to some extent, of algebras of type A_{n-1} for any n.

1. Suppose the subalgebra H_i has a monomial basis

$$M_{i,1},\dots,M_{i,n-1}.$$

To each matrix $M_{i,j}$ there corresponds a permutation $\pi_{i,j}\in S_n$. It is shown that the permutations $\{E,\pi_{i,1},\dots,\pi_{i,n-1}\}$ form an Abelian group $S(H_i)$, each element of which is a product of disjoint cycles of equal length. For $n=p_1\cdots p_r$ (the p_i are distinct primes), in particular for $n=6$ and 91, the group $S(H_i)$ is cyclic.

2. For $n=6$ we prove (unfortunately by case-by-case examination and direct computations) that $(H_i\,|\,H_j)\neq 0$ if the groups $S(H_i)$ and $S(H_j)$ are distinct.

3. If all the groups $S(H_i)$ coincide and are cyclic (see step 1), then the orthogonality condition $(H_i\,|\,H_j)=0$, $i\neq j$, leads to the following system:

$$\sum_{s=0}^{n-1} x_s x_{1+s}\cdots x_{k+s}=0,\qquad 0\leqslant k\leqslant n-2 \tag{$*$}$$

(the indices of the variables being taken modulo n: $0,1,\dots,n-1$).

Let $M_{1,1}=\operatorname{diag}(a_0,\dots,a_{n-1})P$, $M_{2,1}=\operatorname{diag}(b_0,\dots,b_{n-1})P$ and $M_{3,1}=\operatorname{diag}(c_0,\dots,c_{n-1})P$, where without loss of generality P is the permutation matrix of form (4), and all the a_i, b_i and c_i are nonzero. The last condition actually means that the matrices of the subalgebra H_i have no zero rows (see the proof of Theorem 2).

The conditions $(H_1\,|\,H_2)=(H_2\,|\,H_3)=(H_3\,|\,H_1)=0$ lead to three systems of type $(*)$ with the variables $x_i=a_i/b_i$, $y_i=b_i/c_i$ and $z_i=c_i/a_i$, $0\leqslant i\leqslant n-1$, so that $x_iy_iz_i=1$. Hence the existence of three pairwise orthogonal subalgebras requires the existence of three solutions $(x_0,\dots,x_{n-1})$, $(x'_0,\dots,x'_{n-1})$, $(x''_0,\dots,x''_{n-1})$ of the system $(*)$, satisfying the condition $x_ix'_ix''_i=1$, $0\leqslant i\leqslant n-1$.

4. For $n=6$ it turns out that $x_4=\alpha x_1$, $x_5=\beta x_2$ and $x_6=\gamma x_3$, where (α,β,γ) is one of 19 triples obtained by permuting the components in

$$(-1,-1,-1),\ \left(1,-2+\sqrt{3},-2+\sqrt{3}\right),\ \left(1,-2-\sqrt{3},-2-\sqrt{3}\right),$$
$$\left(-1,-1-\sqrt{3}-\sqrt{\sqrt{12}+3},-1-\sqrt{3}+\sqrt{\sqrt{12}+3}\right),$$
$$\left(-1,-1+\sqrt{3}-i\sqrt{\sqrt{12}-3},-1+\sqrt{3}+i\sqrt{\sqrt{12}-3}\right).$$

Analogously: $x'_4=\alpha'x'_1$, $x'_5=\beta'x'_2$, $x'_6=\gamma'x'_3$; $x''_4=\alpha''x''_1$, $x''_5=\beta''x''_2$, $x''_6=\gamma''x''_3$, where (α',β',γ') and $(\alpha'',\beta'',\gamma'')$ belong to the same set of triples. From the condition $x_ix'_ix''_i=1$ it follows that $\alpha\alpha'\alpha''=1$, $\beta\beta'\beta''=1$ and $\gamma\gamma'\gamma''=1$. However, a direct verification shows that no three triples from the above family of 19 can satisfy this condition. Hence there do not even exist three subalgebras H_1, H_2, H_3 (in addition to the standard H) which are pairwise orthogonal.

§2. Orthogonal decompositions of Lie algebras of type A_{p^n-1}

We start with the simplest case, $n = 1$. In an algebra L of type A_{p-1} we introduce the matrices

$$D = \begin{Vmatrix} 1 & & & & \\ & \varepsilon & & & \\ & & \varepsilon^2 & & \\ & & & \ddots & \\ & & & & \varepsilon^{p-1} \end{Vmatrix}, \qquad P = \begin{Vmatrix} 0 & 0\dots 0 & 1 \\ 1 & 0\dots 0 & 0 \\ 0 & 1\dots 0 & 0 \\ \cdots & \cdots & \cdots \\ 0 & 0\dots 1 & 0 \end{Vmatrix},$$

which was used to some extent in §1 (see (4)). Here ε is a primitive pth root of unity. It is obvious that

$$H = \langle D, D^2, \dots, D^{p-1} \rangle; \qquad H_0 = \langle P, P^2, \dots, P^{p-1} \rangle$$

are Cartan subalgebras in L, where $(H \mid H_0) = 0$. Here and below the standard Cartan subalgebra will be denoted by H. It can be verified directly that

$$P^b D^a = \varepsilon^{-ab} D^a P^b$$

and hence

$$[J_{a,b}, J_{a',b'}] = \varepsilon^{-ab'}\{\varepsilon^{(ab'-a'b)} - 1\} J_{a+a',b+b'}, \tag{5}$$

where

$$J_{a,b} = D^a P^b.$$

From the commutation rule (5) and the obvious relation

$$\operatorname{Tr} J_{a,b} = 0, \quad \forall (a, b) \neq (0,0),$$

it follows that the subspaces

$$H = \langle J_{a,0}; 1 \leqslant a \leqslant p-1 \rangle, \quad H_k = \langle J_{kb,b}; 1 \leqslant b \leqslant p-1 \rangle, \quad 0 \leqslant k \leqslant p-1,$$

which are commutative subalgebras in L, are pairwise orthogonal. Since on the other hand the restriction of the form $(* \mid *)$ is nondegenerate, i.e. $(J_{kb,b} \mid J_{k(p-b),p-b}) \neq 0$, all H_k are Cartan subalgebras. So we have arrived at an $\mathfrak{O}\mathfrak{D}$:

$$L = H \oplus H_0 \oplus H_1 \oplus \cdots \oplus H_{p-1}.$$

All these considerations are based on the fact that the number p is prime. Replacing p by a composite number m and taking $\varepsilon^m = 1$, we do not obtain even an ordinary decomposition (property $\mathfrak{D}$), since the subspaces defined above in general have nontrivial intersections.

Now let n be an arbitrary natural number, and let

$$u = (a_1, b_1; a_2, b_2; \dots; a_n, b_n)$$

be an element of a $2n$-dimensional vector space $V = Z_p^{2n}$ (actually it is convenient to take $a_k, b_k = 0, 1, \dots, p-1$). Later it will be necessary and rather convenient to interpret V as the additive group $\mathbf{F}_{q^2}^+$ of the finite field $\mathbf{F}_{q^2}$, $q = p^n$. Let

$$F_q = Z_p(\theta) = \langle \sigma_1, \dots, \sigma_n \rangle;$$
$$F_{q^2} = F_q(\tau) = \langle \sigma_1, \tau_1; \dots; \sigma_n, \tau_n \rangle,$$

where

$$\begin{aligned}
&\theta^n + t_1\theta^{n-1} + t_2\theta^{n-2} + \cdots + t_{n-1}\theta + t_n = 0, \\
&\sigma_1 = 1, \qquad \tau_1 = \tau\theta^{n-1}, \\
&\sigma_2 = \theta + t_1, \qquad \tau_2 = \tau\theta^{n-2}, \\
&\sigma_3 = \theta^2 + t_1\theta + t_2, \qquad \tau_3 = \tau\theta^{n-3}, \\
&\cdots\cdots\cdots\cdots\cdots\cdots\cdots\cdots \\
&\sigma_n = \theta^{n-1} + t_1\theta^{n-2} + \cdots + t_{n-1}, \qquad \tau_n = \tau \cdot 1.
\end{aligned} \tag{6}$$

In the basis (6) we set

$$u = (a_1, b_1; \ldots; a_n, b_n) = \sum_i a_i\sigma_i + \sum_i b_i\tau_i.$$

That our choice of basis (6) is natural will follow from further considerations.

As usual, for the tensor product

$$A \otimes B = (a_{ij}B)$$

of matrices $A = (a_{ij})$ and B we have

$$(A \otimes B)(A' \otimes B') = AA' \otimes BB', \qquad \operatorname{Tr} A \otimes B = (\operatorname{Tr} A)(\operatorname{Tr} B).$$

The linearly independent matrices

$$J_u = J_{a_1, b_1} \otimes \cdots \otimes J_{a_n, b_n}$$

with $u \neq 0$, constitute a basis for the Lie algebra L of type A_{p^n-1} ($\operatorname{Tr} J_u = 0$ for $u \neq 0$ and $J_0 = E$). Operations with these are actually performed by the same rules as operations with the matrices $J_{a,b}$ (see (5)):

$$\begin{gathered}
J_u^k = \varepsilon^{k(k-1)B(u,u)/2} J_{ku}, \qquad J_u J_{u'} = \varepsilon^{B(u,u')} J_{u+u'}, \\
[J_u, J_{u'}] = \varepsilon^{B(u',u)}\{\varepsilon^{\langle u|u'\rangle} - 1\} J_{u+u'}.
\end{gathered} \tag{7}$$

Here

$$\begin{gathered}
u = (a_1, b_1; \ldots; a_n, b_n), \quad u' = (a'_1, b'_1; \ldots; a'_n, b'_n); \\
B(u, u') = -\sum_{i=1}^n b_i a'_i, \\
\langle u|u'\rangle = B(u, u') - B(u', u) = \sum_{i=1}^n \begin{vmatrix} a_i & b_i \\ a'_i & b'_i \end{vmatrix}
\end{gathered} \tag{8}$$

is a nondegenerate skew-symmetric bilinear form (over Z_p) on $\mathbf{F}_{q^2}^+$.

We note, although we do not require it yet, that the matrices of form $\varepsilon^s J_u$ form an extraspecial group Γ of order p^{2n+1}. The commutator operation in Γ is expressed by the formula

$$(\varepsilon^s J_u)(\varepsilon^t J_v)(\varepsilon^s J_u)^{-1}(\varepsilon^t J_v)^{-1} = \varepsilon^{\langle u|v\rangle} J_0.$$

In our basis (6) the form $\langle u | u'\rangle$ is given by the relations

$$\langle \sigma_i|\sigma_j\rangle = 0 = \langle \tau_i|\tau_j\rangle, \qquad \langle \sigma_i|\tau_j\rangle = \delta_{ij} = -\langle \tau_j|\sigma_i\rangle, \tag{8'}$$

which show that $V_\theta = \langle \theta_1, \ldots, \theta_n\rangle$ and $V_\tau = \langle \tau_1, \ldots, \tau_n\rangle$ are maximal isotropic subspaces in $V = V_\theta + V_\tau = \mathbf{F}_{q^2}^+$. Every vector of $V_\theta = \mathbf{F}_q^+$ has the form

$$u = (a_1, 0; \ldots; a_n, 0).$$

According to (7) and (8), the space

$$H = \left\langle J_u | u \in \mathbf{F}_q^* \right\rangle$$

is a commutative subalgebra in L of dimension $q-1$ ($\mathbf{F}_q^* = \mathbf{F}_q \backslash \{0\}$). Since $\operatorname{Tr}(J_u J_{-u}) = q\varepsilon^{B(u,-u')} \neq 0$, the restriction of the form $(\,*\,|\,*\,)$ to H is nondegenerate, and so H is a Cartan subalgebra. As can easily be seen, H consists of diagonal matrices.

Noting now that the elements

$$1, \qquad \tau + \alpha, \qquad \alpha \in \mathbf{F}_q,$$

form a complete system of coset representatives of the multiplicative group $\mathbf{F}_{q^2}^*$ with respect to its subgroup $\mathbf{F}_q^*$, we consider, in addition to H, the system of $(q-1)$-dimensional subspaces

$$H_\alpha = \left\langle J_{(\tau+\alpha)\beta} | \beta \in \mathbf{F}_q^* \right\rangle, \qquad \alpha \in \mathbf{F}_q.$$

THEOREM 3. *A Lie algebra L of type A_{p^n-1} possesses an* $\mathcal{OD}$ *of the form*

$$L = H \oplus \sum_{\alpha \in \mathbf{F}_q} H_\alpha,$$

where H and H_α are the subspaces defined above.

PROOF. All the subspaces H and H_α are pairwise orthogonal, L is their direct sum, and the restriction of the Killing form $(\,*\,|\,*\,)$ to H_α is nondegenerate. Almost everything in this assertion is trivial, since $(J_u | J_{u'}) \neq 0$ only for $u + u' = 0$, but for every $J_u \in H_\alpha$ the element J_{-u} lies in the same subspace H_α.

Strangely enough, a proof is required only for the commutativity of the subspaces H_α (for H it was discussed above), which was obvious for $n = 1$. The formulas (7) reduce this problem to the proof of the relation $\langle (\tau+\alpha)\beta | (\tau+\alpha)\gamma \rangle = 0$ for all $\alpha \in \mathbf{F}_q$ and $\beta, \gamma \in \mathbf{F}_q^*$. According to (8′), we have

$$\left\langle (\tau+\alpha)\beta | (\tau+\alpha)\gamma \right\rangle = \left\langle \tau\beta | \alpha\gamma \right\rangle + \left\langle \alpha\beta | \tau\gamma \right\rangle = \left\langle \tau\beta | \alpha\gamma \right\rangle - \left\langle \tau\gamma | \alpha\beta \right\rangle.$$

Thus the required relation follows from the following three points.

1. According to (6) and (8),

$$\left\langle \tau\sigma_k | \sigma_l \right\rangle = \sum_{i=0}^{k-1} t_i \left\langle \tau\theta^{k-1-i} | \sigma_l \right\rangle = \sum_{i=0}^{k-1} t_i \left\langle \tau_{n-k+1+i} | \sigma_l \right\rangle$$
$$= \begin{cases} -t_{k+l-n-1}, & \text{if } k + l \geqslant n+1, \\ 0, & \text{if } k + l < n+1. \end{cases}$$

Since k and l play symmetric roles here, $\langle \tau\sigma_k | \sigma_l \rangle = \langle \tau\sigma_l | \sigma_k \rangle$, and by linearity with respect to each argument we obtain

$$\left\langle \tau\nu | \mu \right\rangle = \left\langle \tau\mu | \nu \right\rangle, \quad \forall \nu, \mu \in \mathbf{F}_q.$$

This is just what is required in the particular case $\alpha = 1$.

2. The following formula holds:

$$\left\langle \tau\nu | \theta\mu \right\rangle = \left\langle \tau\theta\nu | \mu \right\rangle, \quad \forall \nu, \mu \in \mathbf{F}_q,$$

or, equivalently,

$$\left\langle \tau\sigma_k | \theta\sigma_l \right\rangle = \left\langle \tau\theta\sigma_k | \sigma_l \right\rangle, \quad 1 \leqslant k, l \leqslant n.$$

Indeed

$$\theta\sigma_l = \begin{cases} \sigma_{l+1} - t_l\sigma_1, & \text{if } l < n \\ -t_n\sigma_1, & \text{if } l = n. \end{cases}$$

For $k = n$ and $l = n$, according to point 1, we have

$$\langle \tau\sigma_n | \theta\sigma_n \rangle = \langle \tau\theta\sigma_n | \sigma_n \rangle.$$

For $k < n$ and $l = n$:

$$\langle \tau\sigma_k | \theta\sigma_n \rangle = \langle \tau\theta\sigma_n | \sigma_k \rangle = -t_n \langle \tau_n | \sigma_k \rangle = 0.$$

Analogously,

$$\langle \tau\theta\sigma_k | \sigma_n \rangle = \sum_{i=0}^{k-1} t_i \langle \tau\theta^{k-i} | \sigma_n \rangle = 0.$$

Finally for $k < n$ and $l < n$, we have

$$\langle \tau\sigma_k | \theta\sigma_l \rangle = \langle \tau\sigma_k | \sigma_{l+1} \rangle - t_l \langle \tau\sigma_k | \sigma_1 \rangle = \begin{cases} -t_{k+l-n} & \text{for } k + l \geqslant n, \\ 0 & \text{for } k + l < n. \end{cases}$$

Analogously,

$$\langle \tau\theta\sigma_k | \sigma_l \rangle = \langle \tau\sigma_{k+1} | \sigma_l \rangle - t_k \langle \tau\sigma_1 | \sigma_l \rangle = \begin{cases} -t_{k+l-n}, & k + l \geqslant n, \\ 0, & k + l < n. \end{cases}$$

3. From point 2, by induction on s, we obtain

$$\langle \tau\nu | \theta^s\mu \rangle = \langle \tau\theta\nu | \theta^{s-1}\mu \rangle = \langle \tau\theta^s\nu | \mu \rangle,$$

whence, by the bilinearity of the form, we have

$$\langle \tau\nu | \rho\mu \rangle = \langle \tau\rho\nu | \mu \rangle. \tag{9}$$

Here ν, μ, ρ are completely arbitrary elements of $\mathbf{F}_q$, so

$$\langle \tau\beta | \alpha\gamma \rangle = \langle \tau\beta\gamma | \alpha \rangle = \langle \tau\gamma | \alpha\beta \rangle,$$

as required.

REMARK. The monomial basis $\{J_v\}$ of a Lie algebra of type A_{p^n-1} has been encountered earlier on various occasions (see, for example, [2]). It seems, however, that on all occasions only such situations were considered where v runs through elements of a $2n$-dimensional vector space V equipped with a nondegenerate skew-symmetric form $\langle * | * \rangle$ but without any additional structure. However, the proof of Theorem 3 relied significantly on the realization of V in the form $\mathbf{F}_{q^2}^+$, and the identification of V with $\mathbf{F}_{q^2}^+$ by the choice of the basis (6) was coordinated with the form $\langle * | * \rangle$ in such a way that the multiplicative property (9) holds, ensuring the commutativity of H_α.

§3. The group of automorphisms preserving an $\mathcal{OD}$

It is well known that the automorphism group $\mathrm{Aut}(L)$ of a Lie algebra L of type A_{p^n-1} has a subgroup $\mathrm{Int}(L)$ of inner automorphisms of L of index 2, all elements of which are induced by conjugation $X \mapsto FXF^{-1}$ by arbitrary nonsingular matrices F. So $\mathrm{Int}(L) \cong \mathrm{PGL}(p^n, C)$, and $\mathrm{Aut}(L) = (\mathrm{Int}(L))\langle T \rangle$ is a semidirect product, where T: $X \mapsto -{}^tX$ is an outer automorphism of order 2.

We should like to study the group $\mathrm{Aut}_{\mathcal{OD}}(L)$ of those automorphisms from $\mathrm{Aut}(L)$ which preserve an $\mathcal{OD}$; that is, $\Phi(H_i) = H_j\ \forall\Phi \in \mathrm{Aut}_{\mathcal{OD}}(L)$ (by H_i and H_j we understand any subalgebras of the decomposition). Since the main difficulty is connected with the study of $\mathrm{Int}_{\mathcal{OD}}(L) = \mathrm{Int}(L) \cap \mathrm{Aut}_{\mathcal{OD}}(L)$, we note at once that $T \in \mathrm{Aut}_{\mathcal{OD}}(L)$. Indeed for every $u = (a_1, b_1; \ldots : a_n, b_n)$ we put $\bar{u} = (a_1, -1_1; \ldots ; a_n, -b_n)$. Then

$$T(J_u) = -\varepsilon^{-B(u,u)} J_{\bar{u}},$$

whence

$$T(H) = H, \qquad T(H_\alpha) = H_{-\alpha}, \quad \forall\alpha \in \mathbf{F}_q.$$

We now turn to the subgroup $\mathrm{Int}_{\mathcal{OD}}(L)$. As already remarked, an $\mathcal{OD}$ for the algebra L consists of monomial matrices J_u. Hence it is natural to concentrate first of all on the subgroup $G(J)$ of inner automorphisms of the algebra L of the form Φ: $J_u \mapsto \lambda J_v = FJ_uF^{-1}$. According to (7), $J_u^p = E = J_v^p$, so that $\lambda^p E = (FJ_uF^{-1})^p = FEF^{-1} = E$, whence $\lambda^p = 1$.[1] So

$$\Phi(J_u) = FJ_uF^{-1} = \varepsilon^{f(u)} J_{\varphi(u)} \quad \text{and} \quad \Phi = \Phi_{\varphi, f},$$

where again ε is a primitive pth root of unity, while the functional f and the operator φ (over Z_p) remain to be studied. Using the obvious consideration that Φ is an automorphism of the associative algebra $M_n(\mathbf{C})$ (as distinct from T!), and applying (7) several times, we obtain

$$\begin{aligned}\varepsilon^{f(u+v)} J_{\varphi(u+v)} &= \Phi(J_{u+v}) = \varphi^{-B(u,v)}\Phi(J_uJ_v) = \varepsilon^{-B(u,v)}\Phi(J_u)\Phi(J_v) \\ &= \varepsilon^{-B(u,v)+f(u)+f(v)} J_{\varphi(u)} J_{\varphi(v)} = \varepsilon^{-B(u,v)+B(\varphi(u),\varphi(v))+f(u)+f(v)} J_{\varphi(u)+\varphi(v)}.\end{aligned}$$

Analogously,

$$\varepsilon^{f(ku)} J_{\varphi(ku)} = \varepsilon - (k(k-1)/2)^{\{B(u,u)-B(\varphi(u),\varphi(u))\}+kf(u)} J_{k\varphi(u)}.$$

Thus we have

$$\begin{aligned}&\varphi(u+v) = \varphi(u) + \varphi(v), \qquad \varphi(ku) = k\varphi(u); \\ &f(u+v) = f(u) + f(v) - B(u,v) + B(\varphi(u), \varphi(v)), \\ &\quad f(ku) = \frac{k(k-1)}{2}\{-B(u,u) + B(\varphi(u), \varphi(u))\} + kf(u).\end{aligned} \tag{10}$$

The first two conditions in (10) mean that φ: $V \to V$ is a (nonsingular) linear operator over Z_p. In their turn, the last two conditions make it possible to express $f(u)$ uniquely in terms of $f(\sigma_i)$ and $f(\tau_i)$, $i = 1, \ldots, n$ (see (6)). Moreover,

$$\begin{aligned}&f(u) + f(v) - B(u,v) + B(\varphi(u), \varphi(v)) \\ &\qquad = f(u+v) = f(v+u) = f(v) + f(u) - B(v,u) + B(\varphi(v), \varphi(u)),\end{aligned}$$

whence

$$B(\varphi(u), \varphi(v)) - B(\varphi(v), \varphi(u)) = B(u,v) - B(v,u),$$

---

[1]. Here and below, $p > 2$. For $p = 2$ we have $J_u^2 = (-1)^{B(u,u)}E$ and $\lambda^4 = 1$. The changes in the subsequent considerations are insignificant and do not affect the statement of the final result.

or, recalling (8),

$$\langle \varphi(u)|\varphi(v)\rangle = \langle u|v\rangle;$$

that is, $\varphi \in \mathrm{Sp}(2n, Z_p)$, which is the result we should have expected.

We note that the automorphisms $\Phi_{1,l}$ (in this case the functional l is linear) form an elementary abelian normal subgroup $\mathfrak{E}_{2n}$ of order p^{2n}, since

$$\Phi_{\varphi,f}\Phi_{1,l}\Phi_{\varphi,f}^{-1}(J_u) = \Phi_{\varphi,f}\Phi_{1,l}\left\{\varepsilon^{-f(\varphi^{-1}(u))}J_{\varphi^{-1}(u)}\right\}$$

$$= \Phi_{\varphi,f}\left\{\varepsilon^{-f(\varphi^{-1}(u))+l(\varphi^{-1}(u))}J_{\varphi^{-1}(u)}\right\} = \Phi_{1,l\circ\varphi^{-1}}(J_u)$$

and so

$$\Phi_{\varphi,f}\Phi_{1,l}\Phi_{\varphi,f}^{-1} = \Phi_{1,l\circ\varphi^{-1}}.$$

We remark for the future that $\mathfrak{E}_{2n} \subset \mathrm{Int}_{\mathcal{O}\mathcal{D}}(L)$.

We arrive at the following conclusion.

The group $G(J)$ of automorphisms of L preserving the basis $\{J_u\}$ is isomorphic to the semidirect product

$$(\mathfrak{E}_{2n}\mathrm{Sp}(2n, p))\langle T\rangle.$$

Now let us deal with the group

$$G(J, \mathcal{O}\mathcal{D}) = \mathrm{Int}_{\mathcal{O}\mathcal{D}}(L) \cap G(J).$$

We isolate in it the subgroup $G_H(J, \mathcal{O}\mathcal{D})$ of those automorphisms leaving invariant the subalgebra H of diagonal matrices. Thus, let $\Phi_{\varphi,f} \in G_H(J, \mathcal{O}\mathcal{D})$. Then $\varphi(\sigma_i) = \sum_{k=1}^n a_{ki}\sigma_k$. We note that $\mathfrak{E}_{2n} \lhd G_H(J, \mathcal{O}\mathcal{D})$, and so we ignore f and study only φ, identified with Φ. The Cartan subalgebra H is transformed under $\Phi_{\varphi,f}$ into some subalgebra H_α. This implies that

$$\varphi(\tau_j) = (\tau + \alpha)\beta_j, \qquad \beta_j = \sum_{s=1}^{n} a'_{sj}\theta^{n-s} \in \mathbf{F}_q^*.$$

This form is convenient for us, since $\tau\theta^{n-s} = \tau_s$ (see (6)). Since φ is symplectic, we have

$$\delta_{ij} = \langle \sigma_i|\tau_j\rangle = \langle \varphi(\sigma_i)|\varphi(\tau_j)\rangle = \left\langle \sum_k a_{ki}\sigma_k \middle| \sum_s a'_{sj}\tau_s + \alpha\beta_j\right\rangle = \sum_{k=1}^{n} a_{ki}a'_{kj}.$$

Thus the matrices $A = (a_{ij})$ and $A' = (a'_{ij})$ are connected by the relation $A \cdot {}^tA' = E$. Conversely, these conditions ensure that φ is symplectic, since

$$\langle \varphi(\sigma_i)|\varphi(\sigma_j)\rangle = 0 = \langle \sigma_i|\sigma_j\rangle;$$

$$\langle \varphi(\tau_i)|\varphi(\tau_j)\rangle = \langle (\tau + \alpha)\beta_i|(\tau + \alpha)\beta_j\rangle = \langle \tau\beta_i|\alpha\beta_j\rangle - \langle \tau\beta_j|\alpha\beta_i\rangle = 0 = \langle \tau_i|\tau_j\rangle.$$

The penultimate equality was proved at the end of §2 (before the Remark).

Further, the preservation of the $\mathcal{O}\mathcal{D}$ is expressed by the conditions

$$\Phi(H_\nu) = H_{h(\nu)}, \quad \forall \nu \in \mathbf{F}_q,$$

i.e.

$$\varphi((\tau + \nu)\theta^{n-j}) = (\tau + h(\nu))\lambda_j$$

for some $\lambda_j \in \mathbf{F}_q$ $(j = 1,\ldots,n)$. Putting

$$\varphi(\nu\theta^{n-j}) = M_j(\nu)\lambda_j,$$

we have

$$\varphi((\tau+\nu)\theta^{n-j}) = \varphi(\tau_j + \nu\theta^{n-j}) = \varphi(\tau_j) + \varphi(\nu\theta^{n-j}) = (\tau+\alpha)\beta_j + M_j(\nu)\lambda_j,$$

whence

$$(\tau + h(\nu))\lambda_j = (\tau+\alpha)\beta_j + M_j(\nu)\lambda_j.$$

Comparing the coefficients from $\mathbf{F}_q$ for τ on both sides, we conclude that $\beta_j = \lambda_j$, and hence

$$h(\nu) = \alpha + M_j(\nu).$$

Hence $M_j(\nu) = M(\nu)$ is independent of j, and we have

$$\varphi(\nu\theta^{n-j}) = M(\nu)\lambda_j.$$

From this we obtain that M is an operator on $\mathbf{F}_q$ linear over Z_p. So the automorphism φ is determined by the elements

$$M(\theta^k),\quad 0 \leqslant k \leqslant n-1; \qquad \lambda_j,\quad 1 \leqslant j \leqslant n,$$

which, as we shall soon see, are not independent.

By definition,

$$\varphi(\theta^s \cdot \theta^{n-j}) = M(\theta^s)\lambda_j. \tag{11}$$

For $j = s+1$ we have

$$M(1)\lambda_1 = M(\theta)\lambda_2 = \cdots = M(\theta^k)\lambda_{k+1} = \cdots = M(\theta^{n-1})\lambda_n,$$

i.e.

$$M(\theta^k) = M(1)\lambda_1/\lambda_{k+1}.$$

Analogously for $j = s+2$ we have

$$M(1)\lambda_2 = \cdots = M(\theta^k)\lambda_{k+2} = \cdots$$

and

$$M(\theta^k) = M(1)\lambda_2/\lambda_{k+2}.$$

Comparing the two expressions for $M(\theta^k)$, we get the recurrence relation

$$\lambda_{k+2} = \lambda_2\lambda_{k+1}/\lambda_1, \qquad 0 \leqslant k \leqslant n-2,$$

whence

$$\lambda_{k+2} = (\lambda_2/\lambda_1)^{k+1}\lambda_1.$$

Applying to the relation

$$\theta^n + \sum_{i=1}^{n} t_i\theta^{n-i} = 0$$

the automorphism φ, we obtain, using the equality $M(\theta^n)\lambda_n = M(\theta)\lambda_1$,

$$M(\theta)\lambda_1 + \sum_{i=1}^{n} t_i M(\theta^{n-i})\lambda_n = 0,$$

whence, after substituting the expressions for $M(\theta^k)$ and λ_k, we have

$$\left(\frac{\lambda_1}{\lambda_2}\right)^n + \sum_{i=1}^{n} t_i \left(\frac{\lambda_1}{\lambda_2}\right)^{n-i} = 0.$$

Without going into details, we note that all the other consequences of (11) for arbitrary s and j will follow from the ones already obtained.

Thus,

$$\lambda_1/\lambda_2 = \theta_m = \theta^{p^m}, \qquad m = 0, 1, \ldots, n-1,$$

and so

$$\lambda_j = \theta_m^{1-j}\lambda_1; \qquad 1 \leqslant j \leqslant n,$$
$$M(\theta^k) = M(1)\theta_m^k; \qquad 0 \leqslant k \leqslant n-1.$$

The automorphism φ is completely determined by giving the two elements $\alpha \in \mathbf{F}_q$ and $\lambda = \lambda_1 \in \mathbf{F}_q^*$, and by the choice of the number $m = 0, 1, \ldots, n-1$. As for the element $\mu = M(1)$, the relation $A \cdot {}^t A' = E$ mentioned above allows us to recover it uniquely, by the formula $\mu\lambda = \varphi(\theta^{n-1})$.

We see that any automorphism φ of the group $G_H(J, \mathcal{OD})$ is determined by the independent parameters $\lambda \in \mathbf{F}_q^*$, $\alpha \in \mathbf{F}_q$ and $m = 0, 1, \ldots, n-1$, and acts on the basis elements as follows:

$$\varphi(\theta^{n-j}) = \mu\lambda\theta_m^{1-j}, \quad \theta_m = \theta^{p^m}; \qquad \varphi(\tau_j) = (\tau + \alpha)\theta_m^{1-j}\lambda.$$

There are altogether $nq(q-1)$, $q = p^n$, such automorphisms.

We note that the group $G_H(J, \mathcal{OD})$ described above acts transitively on the set $\{H_\nu \mid \nu \in \mathbf{F}_q\}$.

Now we show that the whole of the group $G(J, \mathcal{OD})$ acts (doubly) transitively on the set of all the subspaces, for which it suffices to produce an automorphism transforming H into H_0. As such an automorphism, we can take the mapping

$$\Phi_0 : \begin{cases} J_\alpha \mapsto J_{\tau\alpha}; \\ J_{\tau\beta} \mapsto J_{-\beta}; \\ J_u = J_{(\tau+\alpha)\beta} \mapsto \varepsilon^{-B(u,u)} J_{(\tau - 1/\alpha)\alpha\beta}, \qquad \alpha \neq 0. \end{cases}$$

We note that for $n = 1$ the automorphism Φ_0 is induced by conjugation by a suitable (in the sense of §1) matrix

$$F = (f_{ij})/\sqrt{p}, \qquad f_{ij} = \varepsilon^{-(i-1)(j-1)}.$$

Moreover, it is not hard to show that $\Phi_0^4 = E$, and Φ_0^2 is the unique (for $p > 2$) nontrivial automorphism of $G(J, \mathcal{OD})$ leaving invariant all the subspaces H and H_α, $\alpha \in \mathbf{F}_q$.

Now we immediately obtain

$$|G(J, \mathcal{OD})| = (q+1)|G_H(J, \mathcal{OD})| = (q+1)nq(q-1) = nq(q^2-1).$$

It is time to prove that

$$\mathrm{Int}_{\mathcal{OD}}(L) = G(J, \mathcal{OD});$$

that is, that every automorphism $\Phi \in \mathrm{Int}_{\mathcal{OD}}(L)$ preserves the basis $\{J_u\}$ of the algebra L. Indeed, let

$$\Phi(J_u) = \sum_{J_\nu \in H_\nu} \lambda_\nu J_\nu.$$

Our problem is to verify that exactly one coefficient λ_v is nonzero. Using the double transitivity of $G(J, \mathcal{OD})$ we can suppose without loss of generality that $J_u \neq H$, and $H_\nu \neq H$. Let $\Phi(H) = H_\mu$. We consider an automorphism $\Psi \in G(J, \mathcal{OD})$ such that $\Psi(H_\mu) = H$. Then $(\Psi\Phi)(H) = H$, and hence $(\Psi\Phi)(X) = FXF^{-1}\ \forall X \in L$, where F is some monomial matrix. In particular, $(\Psi\Phi)(J_u) = FJ_uF^{-1}$ is also a monomial matrix (not in H). On the other hand,

$$(\Psi\Phi)(J_u) = \sum_{J_\nu \in H_\nu} \lambda_\nu \Psi(J_\nu),$$

and all the matrices $\Psi(J_v)$ are monomial and lie in the same Cartan subalgebra (distinct from H). It remains to note that for distinct v the nonzero entries of such matrices are situated in distinct places. Hence only one coefficient λ_v can be nonzero.

Thus we have proved

THEOREM 4. *The group of automorphisms* $\mathrm{Aut}_{\mathcal{OD}}(L)$ *is a semidirect product*

$$\{\mathcal{E}_{2n}\mathrm{Int}_{\mathcal{OD}}(L)\}\langle T\rangle,$$

where $\mathcal{E}_{2n}$ *is an elementary abelian normal subgroup of order* $q^2 = p^{2n}$, $\langle T\rangle$ *is a cyclic group of order* 2, *and* $\mathrm{Int}_{\mathcal{OD}}(L)$ *is a group of order* $nq(q^2-1)$ *which was explicitly described above.*

We have no space here to dwell on the identification of $\mathrm{Int}_{\mathcal{OD}}(L)$ with the group

$$\Gamma L(2, q) = SL(2, q)\langle\sigma\rangle$$

which is a semidirect product of SL(2, q) by the cyclic group $\langle\sigma\rangle$ of automorphisms of the field $\mathbf{F}_q$.

BIBLIOGRAPHY

1. J. G. Thompson, *A conjugacy theorem for* E_8, J. Algebra **38** (1976), 525–530.
2. Iulian Popovici, *Graduations spéciales simples*, Bull. Soc. Roy. Sci. Liège **39** (1970), 218–228.

Translated by L. QUEEN

Труды Матем. Инст. Стеклов.
158(1981)

Proceedings of the Steklov
Institute of Mathematics
1983, Issue 4

ON THE POSSIBILITY OF EVADING ENCOUNTER IN DIFFERENTIAL GAMES IN A CRITICAL CASE

UDC 517.92

E. F. MISHCHENKO AND N. SATIMOV

ABSTRACT. In this article a sufficient condition for evading encounter is given for quasilinear differential games not satisfying the rotatability condition.
Bibliography: 13 titles.

A. Let us consider a quasilinear differential game described by the equation

$$\dot{z} = Cz + f(u, v), \tag{1}$$

where $z \in R^n$ (n-dimensional Euclidean space), C is a constant matrix, u and v are control parameters with $u \in P$ the pursuit parameter and $v \in Q$ the evasion parameter, P and Q are nonempty compact subsets of R_p and R^q, respectively, and $f(u, v)$ is a function defined and continuous on $P \times Q$. A nonempty terminal set M is designated in R^n, and it is assumed to be a subspace in what follows. The game is regarded as over if the phase point z falls in M.

We say that it is possible to evade encounter with M from a point $z_0 \in R^n \setminus M$ if for any measurable function $u(t)$, $0 \leqslant t < \infty$, $u(t) \in P$, it is possible to construct a measurable function $v(t)$, $0 \leqslant t < \infty$, $v(t) \in Q$, such that the solution $z(t), 0 \leqslant t < \infty$, of the equation $\dot{z} = Cz + f(u(t), v(t))$, $z(0) = z_0$, does not fall in M for any value of t, $0 < t < \infty$: $z(t) \notin M$ for all $t > 0$. Moreover, to find the value $v(t)$ of the parameter v at each time $t \geqslant 0$ it is permitted to use only the values $u(t)$ and $z(s)$ for $s \in [t - \theta, t]$, where θ is a certain positive number.

If it is possible to evade encounter with M from each point $z_0 \in R^n \setminus M$, then we say that escape is possible in the game (1).

Many references deal with the problem of escape in nonlinear and quasilinear games (see, for example, [1]–[13]). In most of them the so-called rotatability condition is assumed [2].

In this article we single out the class of differentiability games which do not satisfy the rotatability condition but in which escape is possible. Analogues of known examples are considered as illustrations.

B. We introduce some notation and state our assumptions and our main result.

Let L be the orthogonal complement of the subspace M in R^n, W_1 and W_2 one-dimensional subspaces of L which are arbitrary so far, and π_1 and π_2 the orthogonal projections of R^n onto W_1 and W_2, respectively. Positive numbers depending on the game (1) but depending neither on the initial point z_0 nor on the control functions $u(t)$ and $v(t)$, and only such numbers, will be called *constant*.

1980 *Mathematics Subject Classification*. Primary 90D26; Secondary 49A45.

ASSUMPTION 1. There exist a positive integer k and one-dimensional subspaces W_1 and W_2 such that:

a) each set $\pi_j C^i f(P, Q)$ is a singleton for all i and j, $i = 0, 1, \ldots, k - 2$, $j = 1, 2$, where $k_1 = k$ and $k_2 = n + 1$; and, moreover,

b) the set $R = \bigcap_{u \in P} \pi_1 C^{k-1} f(u, Q)$ contains an interior (with respect to W_1) point.

When Assumption 1 holds, the Hamilton-Cayley theorem gives us that each set $\pi_2 C^i f(P, Q)$, $i \geqslant n$, is a singleton. Let $a_i = \pi_2 C^i f(P, Q)$, $i = 0, 1, \ldots$. We consider the function

$$g(t, z) = \pi_2 e^{tC} z + \sum_{i=0}^{\infty} a_i \frac{t^{i+1}}{(t+1)!}, \qquad 0 \leqslant t \leqslant 1.$$

ASSUMPTION 2. $g(t, z) \not\equiv 0$ for all $z \in R^n \setminus M$.

THEOREM. *If Assumptions* 1 *and* 2 *are satisfied, then escape is possible in the game* (1). *Moreover, the process of evading encounter with M from a point $z_0 \in R^n \setminus M$ can be managed in such a way that the following inequality holds for the distance $\xi(t)$ from $z(t)$ to M*:

$$\xi(t) > \delta(z(i\theta_0))(t - i\theta_0)^k, \tag{2}$$

where $t \in [i\theta_0, (i + 1)\theta_0]$, $i = 0, 1, \ldots,$ θ_0 is a constant, and $\delta(\cdot)\colon R^n \setminus M \to (0, \infty)$.

C. PROOF OF THE THEOREM. 1. By Assumption 1, R has an interior point. Let l be such a point. Obviously, there exists a constant c_1 such that the segment $l + \Gamma$, where $\Gamma = \{\omega\colon |\omega| \leqslant c_1\} \subset W_1$, is entirely contained in R. We consider the following equation in $v \in Q$ (cf. [4]):

$$\pi_1 C^{k-1} f(u, v) = l + \omega. \tag{3}$$

According to Assumption 1, this equation is solvable. Denote by $v_\omega(u)$ the solution of (3) which is least in the lexicographical sense. Let $u(t)$, $0 \leqslant t \leqslant 1$, be an arbitrary measurable function with $u(t) \in P$. The function $v_\omega(t) = v_\omega(u(t))$, $0 \leqslant t \leqslant 1$, is measurable [4]. Let $z(t)$, $0 \leqslant t \leqslant 1$, be a solution of the equation

$$\dot{z} = Cz + f(u(t)), v_\omega(u(t)), \qquad z(0) = z_0, 0 \leqslant t \leqslant 1, z_0 \in R^n \setminus M.$$

For the solution $z(t)$, $0 \leqslant t \leqslant 1$, we have (see (3), and also [4])

$$\pi_1 z(t) = h(t, z_0) + \frac{t^k}{k!}[\omega + f(t)], \qquad \pi_2 z(t) = g(t, z_0), \tag{4}$$

where

$$h(t, z_0) = \pi_1 e^{tC} z_0 + \sum_{i=0}^{k-1} b_i \frac{t^{i+1}}{(i+1)!}, \qquad b_i = \pi_1 C^i f(P, Q),$$

$$i = 0, 1, \ldots, k - 2, b_{k-1} = l,$$

$$t^k f(t) = \int_0^t (t - s)^k \left[\pi_1 C^k + \frac{t - s}{k + 1} \pi_1 C^{k+1} + \cdots \right] f(u(s), v_\omega(s))\, ds.$$

Clearly, $|f(t)| \leqslant c_2 t$ for all $t \in [0, 1]$, where c_2 is a constant.

2. We claim that there exist a point $\omega_0 \in \Gamma$ and a constant θ_0 such that the functions

$$h(t, z_0) + t^k[\omega_0 + f(t)]/k! \quad \text{and} \quad g(t, z_0)$$

do not vanish simultaneously on $(0, \theta_0)$. (It is assumed that a coordinate has been introduced on the one-dimensional subspace W_1, and the points of W_1 will be taken with respect to this coordinate; for convenience the points of W_1 will be identified with numbers.)

Indeed, the analytic function $g(t, z_0)$, $0 \leqslant t \leqslant 1$, has finitely many zeros (Assumption 2). Let $\tau_1, \ldots, \tau_m, \tau_{m+1}, \ldots, \tau_r$ be all the positive zeros of $g(t, z_0)$, $0 \leqslant t \leqslant 1$, and suppose that $h(\tau_i, z_0) = 0$ for $i = 1, \ldots, m$ and $h(\tau_i, z_0) \neq 0$ for $i = m+1, \ldots, r$ (the cases where $h(\tau_i, z_0) = 0$ for all $i = 1, \ldots, r$ or $h(\tau_i, z_0) \neq 0$ for all $i = 1, \ldots, r$ are not excluded). Clearly, there exists a number $\omega, |\omega| \leqslant c_1$, such that $h(\tau_i, z_0) + \tau_i^k\omega/k! \neq 0$ for all $i = m+1, \ldots, r$. Next, it is obvious that there is a constant c_3 such that $h(\tau_i, z_0) + \tau^k\omega/k! \neq 0$ for all i and ω, $i = m+1, \ldots, r$, $c_3/2 \leqslant |\omega| \leqslant c_3$. Let $\omega_0 = 3c_3/4$ and

$$\theta_0 = \min\{1, \theta, \omega_0/4c_2\}. \tag{5}$$

With these ω_0 and θ_0 we have $c_3/2 \leqslant \omega_0 + f(t) \leqslant c_3$ for all $t \in (0, \theta_0]$. Hence

$$h(\tau_i, z_0) + \tau_i^k[\omega_0 + f(\tau_i)]/k! \neq 0$$

for all $i = m+1, \ldots, r$. For $i = 1, \ldots, m$

$$h(\tau_i, z_0) + \tau_i^k[\omega_0 + f(\tau_i)]/k! = \tau^k[\omega_0 + f(\tau_i)]/k! \geqslant \tau_i^k c_3/2k! > 0.$$

Hence, the function $h(t, z_0) + t^k[\omega_0 + f(t)]/k!$, $0 < t \leqslant \theta_0$, does not vanish at the points $\tau_1, \ldots, \tau_r$, i.e., the functions $h(t, z_0) + t^k[\omega_0 + f(t)]/k!$ and $g(t, z_0)$ do not have common zeros on $(0, \theta_0]$. The assertion is proved.

3. In this subsection we describe a rule for controlling the parameter v which permits the phase point to evade encounter with M.

Let z_0 be an arbitrary point of $R^n \setminus M$. By the previous subsection, there exist a point $\omega_0 \in \Gamma$ and a constant θ_0 such that if v is chosen according to (3) with $\omega = \omega_0$, then the vector-valued function $(\pi_1 z(t), \pi_2 z(t))$ does not vanish on $(0, \theta_0]$, i.e., $z(t) \notin M$ for all $t \in [0, \theta_0]$ ($z(0) = z_0 \notin M$). For $t = \theta_0$ we repeat the procedure for controlling the parameter v, this time with $z_0 = z(\theta_0)$, and so on.

The estimate (2) can be derived in the usual way [1, 2].

The proof of the theorem is completed.

D. In this section we consider examples.

1. A differential game is described by the equations

$$\dot{z}_1 = v - u, \qquad \dot{z}_2 = 0,$$

where $z_1, z_2 \in R^1$, $P = \{u\colon -1 \leqslant u \leqslant 0\}$, $Q = \{v\colon 1 \leqslant v \leqslant 3\}$ and $M = \{0\}$.

It is not hard to show that Assumption 2 of our theorem is not satisfied for this example (Assumption 1 is automatically satisfied). Obviously, it is not possible to evade encounter with the set M from the points of the set $\{(z_1, 0)\colon z_1 < 0\}$ (for if we set $u(t) = -1, 0 \leqslant t < \infty$, then from each point $(z_1^0, 0)$, $z_1^0 < 0$, the phase point falls in the terminal set M by a time not exceeding $|z_1^0|/2$).

2. The problem of "the boy and the crocodile" [1]. In reduced coordinates this game is described by the equations

$$\dot{z}_1 = z_2 + v, \qquad \dot{z}_2 = -u,$$

where $z_1, z_2 \in R^2$ and

$$P = \{u = (u_1, u_2): |u_1| \leqslant 1, u_2 = u_2^0 \neq 0\},$$

$$Q = \{v = (v_1, v_2): |v_1| \leqslant 1, v_2 = v_2^0\}, \qquad M = \{z_1 = 0\}.$$

We show that Assumptions 1 and 2 are satisfied for this example. To prove this it suffices to set $W_1 = \{z_{12} = 0,\ z_2 = 0\}$ and $W_2 = \{z_{11} = 0,\ z_2 = 0\}$, where $(z_{11}, z_{12}) = z_1$.

Elementary computations show that $\pi_1 P = \{0\}$ and $\pi_1 Q = I$, where I is the unit interval of W_1, $\pi_2 C^i P$ and $\pi_2 C^i Q$ are singletons for all $i \geqslant 0$, and $g(t, z) = z_{12} + z_{22}t + v_2^0 t - u_2^0 t^2/2$. Consequently, $g(t, z) \not\equiv$ for all z (since $u_2^0 \neq 0$).

3. *Control example* [1]. In reduced coordinates this game is described by the equations

$$\dot z_1 = z_2 + z_3, \qquad \dot z_2 = -\alpha z_2 - \rho u, \qquad \dot z_3 = -\beta z_3 + \sigma u,$$

where $z_1, z_2, z_3 \in R^2$, α, β, ρ and σ are nonnegative numbers, and

$$P = \{u = (u_1, u_2): |u_1| \leqslant 1, u_2 = u_2^0\},$$

$$Q = \{v = (v_1, v_2): |v_1| \leqslant 1, v_2 = v_2^0\}, \qquad M = \{z_1 = 0\}.$$

It is not hard to show that our theorem holds for this example if at least one of the following two conditions holds:

1. $\sigma > \rho$, $\alpha \neq \beta$, $\sigma\alpha v_2^0 \neq \rho\beta u_2^0$;
2. $\sigma > \rho$, $\alpha = \beta$, $\sigma v_2^0 \neq \rho u_2^0$.

REMARK 1. The escape conditions in [1]–[13] are not satisfied for Examples 2 and 3.

BIBLIOGRAPHY

1. L. S. Pontryagin and E. F. Mishchenko, *The problem of evasion in linear differential games*, Differentsial'nye Uravneniya **7** (1971), 436–445; English transl. in Differential Equations **7** (1971).

2. L. S. Pontryagin, *A linear differential evasion game*, Trudy Mat. Inst. Steklov **112** (1971), 30–63; English transl. in Proc. Steklov Inst. Math. **112** (1971).

3. N. N. Krasovskiĭ and A. I. Subbotin, *Positional differential games*, "Nauka", Moscow, 1974; French transl., *Jeux différentiels*, "Mir", Moscow, 1977.

4. E. F. Mishchenko and N. Satimov, *The evasion problem in differential games with nonlinear controls*, Differentsial'nye Uravneniya **9** (1973), 1792–1797; English transl. in Differential Equations **9** (1973).

5. R. V. Gamkrelidze and G. K. Kharatishvili, *A differential game of evasion with nonlinear control*, SIAM J. Control **12** (1974), 332–349.

6. M. S. Nikol'skiĭ, *On a linear escape problem*, Dokl. Akad. Nauk SSSR **218** (1974), 1024–1027; English transl. in Soviet Math. Dokl. **15** (1974).

7. N. Satimov, *On the escape problem in differential games with nonlinear controls*, Dokl. Akad. Nauk SSSR **216** (1974), 744–747; English transl. in Soviet Math. Dokl. **15** (1974).

8. ______, *On a way to avoid encounter in differential games*, Mat. Sb. **99** (**141**) (1976), 380–393; English transl. in Math. USSR Sb. **28** (1976).

9. B. N. Pshenichnyĭ, *The evasion problem*, Kibernetika (Kiev) **1975**, no. 4, 120–127; English transl. in Cybernetics **11** (1975).

10. P. B. Gusyatnikov, *On the l-evasion of encounter in a linear differential game*, Prikl. Mat. Mekh. **38** (1974), 417–421; English transl. in J. Appl. Math. Mech. **38** (1974).

11. E. F. Mishchenko and H. Satimov, *The problem of evading encounter in n-person differential games*, Dokl. Akad. Nauk SSSR **224** (1975), 285–288; English transl. in Soviet Math. Dokl. **16** (1975).

12. P. B. Gusyatnikov, *l-evasion in linear differential games*, Differentsial'nye Uravneniya **12** (1976), 446–455; English transl. in Differential Equations **12** (1976).

13. Rufus Isaacs, *Differential games*, Wiley, 1965.

Translated by H. H. McFADEN

Труды Матем. Инст. Стеклов.
158(1981)

Proceedings of the Steklov
Institute of Mathematics
1983, Issue 4

AN ESTIMATE OF THE MODULUS OF COMPLETE RATIONAL TRIGONOMETRIC SUMS OF DEGREES THREE AND FOUR

UDC 511

V. I. NECHAEV AND V. L. TOPUNOV

ABSTRACT. A sharp upper bound is obtained for the modulus of complete rational trigonometric sums of degrees 3 and 4.

Tables: 4. Bibliography: 6 titles.

1. Notation. **Z** denotes the ring of integers, and **Z**$[x]$ the ring of polynomials in one variable over **Z**; q, n and s are natural numbers, and p is a prime. Let $f(x) \in \mathbf{Z}[x]$; $M_n(q)$ denotes the set of all polynomials $f = a_0 + a_1x + \cdots + a_nx^n \in \mathbf{Z}[x]$ such that $(a_1, \ldots, a_n, q) = 1$; and

$$S(q, f) = \sum_{x=1}^{q} e^{2\pi i f(x)/q}; \qquad \rho_n(q, f) = |S(q, f)|\, q^{-1+1/n};$$

$$c(n, q) = \sup_{f \in M_n(q)} \rho_n(q, f); \qquad c(n) = \sup_q c(n, q).$$

Gauss [1] showed that $c(2) = \sqrt{2}$. Weil's inequality [2]

$$f \in M_n(p) \Rightarrow |S(p, f)| \leqslant (n-1)\sqrt{p}$$

allowed the estimate [3] for $n \geqslant 3$

$$\ln C(n) < 5n^2/\ln n.$$

Stechkin [4] showed that $\ln c(n) \leqslant n + O(n/\ln n)$. Chen [5] established that $\ln c(n) < 6.1n$, and that, specifically, $\ln c(3) < 18.3$ and $\ln c(4) < 22$.

In this paper we study trigonometric sums of third and fourth degree and determine sharp estimates for $c(3)$ and $c(4)$. The solution of this problem was reduced (by Nechaev) to the item-by-item examination of a finite set of values $\rho_n(q, f)$, which was implemented (by Topunov) on the ES-1022 and BÈSM-6 computers.

2. LEMMA 1 ([6], Chapter 3, problem 12, d). *Assume that the natural numbers $q_1, \ldots, q_k$ are pairwise relatively prime. Then*

$$c(n, q_1 \cdot \ldots \cdot q_k) = c(n, q_1) \cdot \ldots \cdot c(n, q_k).$$

1980 *Mathematics Subject Classification.* Primary 10G10.

LEMMA 2 ([3], Lemma 5). *Let* $f = a_0 + a_1x + \cdots + a_nx^n \in M_n(p)$, *and assume that* t *is the greatest integer such that* $p^t \mid (a_1, 2a_2, \ldots, na_n)$. *If the congruence*

$$p^{-t}f'(x) \equiv 0 \pmod{P} \tag{1}$$

has no solution, then $S(p^s, f) = 0$ *for* $s \geqslant 2t + 2$.

If the congruence (1) has solutions and $\lambda_1, \ldots, \lambda_r$ are all its distinct roots, and if m_ν $(1 \leqslant \nu \leqslant r)$ is the largest integer such that

$$f_\nu = p^{-m_\nu}(f(px + \lambda_\nu) - f(\lambda_\nu)) \in Z[x],$$

then for $s \geqslant 2t + 2$

$$S(p^s, f) = \sum_{\nu=1}^{r} e^{2\pi i f(\lambda_\nu)/p^s} S(p^{s-1}, P^{m_\nu - 1} f_\nu). \tag{2}$$

Moreover

$$S(p^{s-1}, p^{m_\nu - 1} f_\nu) = \begin{cases} p^{s-1}, & \text{if } s < m_\nu; \\ p^{m_\nu - 1} S(p^{s-m_\nu}, f_\nu), & \text{if } s \geqslant m_\nu. \end{cases} \tag{2}$$

One verifies easily that if $n = p = 3$, $t \geqslant 1$ and $3 \mid a_2$, then (2) holds also for $s = 2t + 1$.

LEMMA 3 (see [3], Lemmas 8 and 9). *Assume* $n \geqslant 3$. *Then*

$$c(n, p^s) \leqslant \begin{cases} \max(1, c(n, p)), & \text{if } p \geqslant (n-1)^{n/(n-2)}; \\ 1, & \text{if } p \geqslant (n-1)^{2n/(n-2)}. \end{cases}$$

From Lemma 3 follows

LEMMA 4.

$$c(3, p^s) \leqslant \begin{cases} 1, & \text{if } p \geqslant 67; \\ \max(1, c(3, p)), & \text{if } 11 \leqslant p \leqslant 61. \end{cases}$$

$$c(4, P^s) \leqslant \begin{cases} 1, & \text{if } p \geqslant 83; \\ \max(1, c(4, p)), & \text{if } 11 \leqslant p \leqslant 79. \end{cases}$$

3. We augment the natural numbers by a symbol ∞ such that $k < \infty$ for all natural numbers k. For every integer z we shall write $p^\zeta \| z$ only in the following cases:

(1) if $z \neq 0$ and ζ is the largest integer such that $p^\zeta \mid z$; or

(2) if $z = 0$ and $\zeta = \infty$.

Let $n \geqslant 3$ and $f = a_0 + a_1x + \cdots + a_nx^n \in M_n(p)$, and assume that the symbols $\alpha, \beta, \gamma, \delta$ are determined by the conditions

$$p^\alpha \| a_1, \quad p^\beta \| 2a_2, \quad p^\gamma \| 3a_3, \quad p^\delta \| 4a_4.$$

We denote by $M_n^{(i)}(p)$ $(i = 1, 2, 3, 4)$ the subset of $M_n(p)$ as defined for various values of i and p in Table 1.

TABLE 1

i	$p=2$	$p=3$	$p>3$
1	$\alpha<\beta, \alpha<\gamma, \alpha<\delta$ or $\alpha=\beta=\gamma=1$	$\alpha<\beta, \alpha<\gamma, \alpha<\delta$	$\alpha<\beta, \alpha<\gamma, \alpha<\delta$
2	$\alpha\geqslant\beta, \beta<\gamma, \beta<\delta$ or $\alpha>\beta=\gamma=1$	$\alpha\geqslant\beta, \beta<\gamma, \beta\leqslant\delta$	$\alpha\geqslant\beta, \beta<\gamma, \beta<\delta$
3	$\alpha\geqslant\gamma, \beta>\gamma, \delta>\gamma+1$ or $\alpha\geqslant 1, \beta\geqslant 2, \gamma=1, \delta=2$	$\alpha\geqslant\gamma, \beta\geqslant\gamma, \gamma<\delta$	$\alpha\geqslant\gamma, \beta\geqslant\gamma, \gamma<\delta$
4	$\alpha\geqslant\delta, \beta\geqslant\delta, \gamma\geqslant\delta$	$\alpha\geqslant\delta, \beta\geqslant\delta, \gamma>\delta$	$\alpha\geqslant\delta, \beta\geqslant\delta, \gamma\geqslant\delta$

Evidently

$$M_n(p) = M_n^{(1)}(p) \cup M_n^{(2)}(p) \cup M_n^{(3)}(p) \cup M_n^{(4)}(p).$$

One easily verifies the following assertion.

LEMMA 5. *Let k and l be integers, $(k,p)=1$, $i\in\{1,2,3,4\}$ and $f\in M_n^{(i)}(p)$; then*

$$\varphi(x) = f(kx+l) - f(l) \in M_n^{(i)}(p)$$

and

$$S(p^s, f) = S(p^s, \varphi).$$

Denote by $D_n^{(i)}(p)$ $(i=2,3,4)$ the sets of polynomials with integral coefficients whose form is indicated in Table 2 for various values of i and p $(a, b, \alpha, \beta, \gamma, \delta, \nu \in \mathbf{Z}$; $(a,p)=(b,p)=1$; $\alpha\geqslant 0, \beta\geqslant 0, \gamma\geqslant 0, \delta\geqslant 0, \nu\geqslant 0)$.

TABLE 2

i	$p=2$	$p=3$	$p>3$
2	$ax^2+2^\gamma ax^3+2^\nu bx^4$ $\delta=\nu+2$	$ax^2+3^\nu ax^3+3^\delta bx^4$ $\gamma=\nu+1$	$ax^2+p^\gamma ax^3+p^\delta bx^4$
3	$2^\alpha ax+bx^3+2^\nu bx^4$ or $2^\alpha ax+2bx^3+2^\nu bx^4$ $\delta=\nu+2$	$3^\alpha ax+bx^3+3^\delta bx^4$	$p^\alpha ax+bx^3+p^\delta bx^4$
4	$2^\alpha ax+2^\nu ax^2+bx^4$ $\beta=\nu+1, \delta=2$	$3^\alpha ax+3^\beta ax^2+bx^4$	$p^\alpha ax+p^\beta ax^2+bx^4$

It follows from Lemma 5 that if $n=4$ and s is arbitrary, then for every polynomial f from $M_n^{(i)}(p)$ $(i=2,3,4)$ one can find a polynomial φ in $D_4^{(i)}(p)$ such that

$$|s(p^s, f)| = |s(p^s, \varphi)|.$$

This remark allows us to reduce the number of item-by-item examinations and thereby cut the machine time required for the computations.

From Lemma 2 and known properties of trigonometric sums one easily obtains

LEMMA 6. *Let* $f \in M_4^{(1)}(p)$. *Then*

$$|S(p^s, f)| \leqslant \begin{cases} 0, & \text{if } p = 2 \text{ and } s \geqslant 4, \text{ or if } p = 3 \text{ and } s \geqslant 2, \\ & \text{or if } p > 3; \\ 8, & \text{if } p = 2 \text{ and } s \leqslant 3; \\ 4\sqrt{2+\sqrt{2}}, & \text{if } p = 2, s \leqslant 3 \text{ and } \delta > 2; \\ 3, & \text{if } p = 3 \text{ and } s = 1. \end{cases}$$

Moreover,

$$\rho_3(8, 2x + x^2 + 6x^3) = \sqrt{2+\sqrt{2}} = 1.847759\dots.$$

If we apply Lemma 2 to a polynomial from $D_4^{(i)}(p)$ $(i = 2, 3, 4)$ and make use of the resulting recursive relation, we can verify after the necessary computations that the following three lemmas hold.

LEMMA 7. *Let* $f \in M_4^{(2)}(p)$. *Then*

$$|S(p^s, f)| \leqslant \begin{cases} p^{s/2}, & \text{if } p > 2; \\ 2 \cdot 2^{s/2}, & \text{if } p = 2 \text{ and } \delta = 2; \\ \sqrt{2} \cdot 2^{s/2}, & \text{if } p = 2 \text{ and } \delta > 2. \end{cases}$$

LEMMA 8. *Let* $f \in M_4^{(3)}(p)$. *Then*

$$|S(p^s, f)| \leqslant \begin{cases} p^{2s/3}, & \text{if } p \geqslant 5 \text{ and } s \geqslant 3, \\ & \text{or if } p \geqslant 11 \text{ and } s \geqslant 2; \\ 2\cos(\pi/27) \cdot 3^{2s/3}, & \text{if } p = 3; \\ \sqrt{2+\sqrt{2}} \cdot 2^{2s/3}, & \text{if } p = 2. \end{cases}$$

Moreover,

$$\rho_3(27.24x + 16x^3) = 2\cos(\pi/27) = 1.986476\dots.$$

LEMMA 9. *Let* $f \in M_4^{(4)}(p)$. *Then*

$$|S(p^s, f)| \leqslant \begin{cases} p^{3s/4}, & \text{if } p \geqslant 5 \text{ and } s \geqslant 3, \\ & \text{or if } p \geqslant 11 \text{ and } s \geqslant 2; \\ c(4.9) \cdot 3^{3s/4}, & \text{if } p = 3; \\ c(4.128) \cdot 2^{3s/4}, & \text{if } p = 2. \end{cases}$$

Moreover,

$$c(4.9) = \rho_4(9.5x^2 + 5x^4) = 1.639530\dots,$$
$$c(4.128) = \rho_4(128.64x + 82x^3 + 3x^4) = 2.263192\dots.$$

4. It follows from Lemmas 1, 4 and 6–9 that in order to obtain the exact values of $c(3)$ and $c(4)$ it suffices to find the following values:

(1) $c(3, p)$ for all primes p from 5 to 61;

(2) $c(4, p)$ for all primes p from 5 to 79;

(3) $c(3, p^2)$ for $p = 5$ and $p = 7$; and

(4) $c(4, p^2)$ for $p = 5$ and $p = 7$.

These values were obtained using a computer. The results of the computations are summarized in Tables 3 and 4. These tables do not contain those values p for which $c(3, p)$ and $c(r, p)$ are less than 1.

TABLE 3

p	$s = s_3(p)$	$\varphi_p(x) = a_1x + a_2x^2 + a_3x^3$			$\rho_s(p^s, \varphi_p(x))$	p	$s = s_3(p)$	$\varphi_p(x) = a_1x + a_2x^2 + a_3^3$			$\rho_2(p^s, \varphi_p(x))$
		a_1	a_2	a_3				a_1	a_2	a_3	
2	3	2	1	6	1.847759...	23	1	1	0	8	1.070062...
3	3	2	0	16	1.986476...	29	1	1	0	18	1.084647...
5	1	1	0	4	1.237350...	31	1	1	0	3	1.051803...
7	1	0	0	1	1.295584...	37	1	0	0	21	1.083795...
11	1	1	0	8	1.190056...	41	1	1	0	32	1.049026...
13	1	0	0	6	1.257653...	43	1	0	0	26	1.019238...
17	1	1	0	15	1.122756...	47	1	1	0	34	1.016841...
19	1	0	0	3	1.196716...						

TABLE 4

p	$s = s_4(p)$	$\psi_p(x) = a_1x + a_2x^2 + a_3x^3 + a_4x^4$				$\rho_4(p^s, \psi_p(x))$	p	$s = s_4(p)$	$\psi_p(x) = a_1x + a_2x^2 + a_3x^3 + a_4x^4$				$\rho_4(p^s, \psi_p(x))$
		a_1	a_2	a_3	a_4				a_1	a_2	a_3	a_4	
2	7	64	82	0	3	2.263192...	23	1	0	2	0	4	1.153157...
3	2	0	5	0	5	1.639530...	29	1	0	2	0	11	1.147475...
5	2	25	23	0	1	1.332241...	31	1	0	6	0	24	1.147375...
7	1	7	5	0	1	1.276621...	37	1	0	19	0	36	1.054115...
11	1	0	2	0	1	1.284991...	41	1	0	0	0	6	1.141010...
13	1	0	11	0	11	1.201907...	43	1	0	1	0	26	1.009466...
17	1	0	0	0	10	1.268828...	59	1	0	30	0	47	1.0335007...
19	1	0	10	0	1	1.188147...							

5. From the lemmas and tables we obtain the following theorems.

THEOREM 1. *Let $s = s_3(p)$. Then*

(1)
$$c(3) = \prod_{2\leqslant p\leqslant 47} c(3, p^s);$$

(2) *if $2 \leqslant p \leqslant 47$, then*

$$c(3, p^s) = \rho_3\big(p^s, \varphi_p(x)\big);$$

(3) $\ln c(3) = 2.834\ldots.$

THEOREM 2. *Let* $s = s_4(p)$. *Then*

(1)
$$c(4) = \prod_{2 \leqslant p \leqslant 43} c(4, p^s) \cdot c(4.59);$$

(2) *if* $2 \leqslant p \leqslant 43$ *or* $p = 59$, *then*
$$c(4, p^s) = \rho_4(p^s, \psi_p(x));$$

(3) $\ln c(4) = 3.33\ldots$.

BIBLIOGRAPHY

1. Carl Friedrich Gauss, *Summatio quarundam serierum singularium*, Comment. Soc. Reg. Sci. Gottingen. Recentiores **1** (1808/11); reprinted in his *Werke*, Band II, Königl. Ges. Wiss. Göttingen, 1863, 1876, pp. 9–45.

2. A. Weil, *On some exponential sums*, Proc. Nat. Acad. Sci. U.S.A. **34** (1948), 204–207.

3. V. I. Nechaev, *An estimate of the complete rational trigonometric sum*, Mat. Zametki **17** (1975), 839–849; English transl. in Math. Notes **17** (1975).

4. S. B. Stechkin, *Estimate of a complete rational trigonometric sum*, Trudy Mat. Inst. Steklov **143** (1977), 188–207; English transl. in Proc. Steklov Inst. Math. 1980, no. 1(143).

5. Jing Run Chen [Chen Jing-run], *On Professor Hua's estimate of exponential sums*, Sci. Sinica **20** (1977), 711–719.

6. I. M. Vinogradov, *Foundations of the theory of numbers*, 8th ed., "Nauka", Moscow, 1972; English transl. of 5th ed., Dover, New York, 1954; of 6th ed., Pergamon Press, 1955.

Translated by P. HAFNER

Труды Матем. Инст. Стеклов.
158(1981)

Proceedings of the Steklov
Institute of Mathematics
1983, Issue 4

A MARTINGALE APPROACH IN PROBLEMS ON FIRST CROSSING TIME OF NONLINEAR BOUNDARIES

UDC 519.21

A. A. NOVIKOV

ABSTRACT. The problem of computing the distribution of the first crossing time $\tau = \inf\{t \geqslant 0,\ S_t > f(t)\}$ is considered, where $f(t)$ is a nonlinear boundary with $f(0) > 0$, and S_t is a process with independent increments and $\mathbf{E}S_t = 0$ for t in $[0, \infty)$ or $\{0, 1, 2, \dots\}$. All the moments of τ for stable processes of order α without positive jumps and for a boundary $f(t) = at^{1/\alpha} + c$ are found with the help of a technique based on martingale theory. The asymptotic behavior of the probability $\mathbf{P}\{\tau > T\}$ as $T \to \infty$ is found for certain other cases. In particular, Theorems 2 and 3 imply that if S_t is a standard Wiener process and $f(t)$ is a monotone convex or concave boundary, then

$$\int_1^\infty |f(t)|\, t^{-3/2}\, dt < \infty \Leftrightarrow \mathbf{P}\{\tau > T\} \sim T^{-1/2}\mathbf{E}\rho_\tau\sqrt{\frac{2}{\pi}}, \qquad T \to \infty,$$

where $\mathbf{E}\rho_\tau$ is a finite positive constant.

Bibliography: 28 titles.

§1. Introduction and main results(¹)

1. Let S_t, $t \geqslant 0$, be a process with independent homogeneous increments with $S_0 = 0$ and $\mathbf{E}S_t = 0$, $t \in R^+$ or $t \in Z^+ = \{0, 1, 2, \dots\}$, defined on a probability space $(\Omega, \mathfrak{F}, \mathbf{P})$.

In this article we consider the behavior of the tail of the distribution of the stopping time

$$\tau = \inf\{t: S_t > f(t)\} \qquad (f(0) > 0 \text{ for } t \in R^+,\ f(1) \geqslant 0 \text{ for } t \in Z^+)$$

under certain assumptions about the nonlinear (nonrandom) boundary $f(t)$ and the distribution of the jump S_1.

2. In the case of linear boundaries there are many results on the distribution of τ and on other boundary functionals connected with it (see, for example, the monographs of Borovkov [1], Korolyuk [2], Skorokhod [3] and Takács [4]). Considerably less work has been done in the nonlinear case. There are some general results in this direction in, for example, the articles of Borovkov [5], Rotar′ [6], Portnoy [7] and Lai and Wijsman [8] (there are interesting applications to the theory of sequential tests in the last article). Direct probabilistic methods are the main tools in the references

1980 *Mathematics Subject Classification.* Primary 60G40, 60G42, 60G44, 60J50; Secondary 33A40, 33A70, 41A60, 60E07, 60J30.

(¹) Our main results were reported at the International Working Conference on Stochastic Differential Systems held in Visegrád, Hungary in September 1980 (see [28]).

cited. The use of the traditional methods for studying linear problems (the method of factorization identities [1], [9], the potential method [2], the differential equation method [3]) in nonlinear boundary problems involves considerable difficulties. However, we note in this connection the articles [10] and [11] of Mogul′skiĭ and Pecherskiĭ, in which the authors succeeded in obtaining a number of important results for the case of sums of independent random variables (i.e., for the case of discrete time in our interpretation) with the help of a modification of the method of factorization identities.

In this article we use an approach based on a martingale technique to investigate the distribution of τ. This technique is based, in particular, on the simple (but fundamental for martingale theory) fact that for any uniformly integrable martingale Z_t and any stopping time τ (s.t. in what follows) the following identity holds:

$$\mathbf{E}Z_\tau = \mathbf{E}Z_0.$$

For a suitable choice of the martingale Z_t we can deduce from this relation useful information about the behavior of the s.t. τ under consideration. Of course, this device has also been used previously (see, for example, Shepp [12], Robbins and Siegmund [13] and Novikov [14]). It is by this device (with the use of an appropriate portion of analytic techniques) that in this paper all the moments of the s.t. τ have been found for a boundary of the form $f(t) = at^{1/\alpha} + c$ in the case of a stable process S_t of order α, $1 < \alpha \leqslant 2$, without positive jumps (see Theorem 1). The same device is used in an essential way in deriving the asymptotic behavior of the probability $\mathbf{P}\{\tau > t\}$ as $T \to \infty$ under certain general assumptions about $f(t)$ and S_t (see Theorem 2).

Another important element of the martingale technique used intensively in this article is the replacement of the original probability measure $\mathbf{P}$ by an auxiliary measure $\tilde{\mathbf{P}}$. This new measure can be introduced by means of a certain nonnegative uniformly integrable martingale Z_t according to the formula $\tilde{\mathbf{P}}\{A\} = \mathbf{E}I\{A\}Z_\infty$ ($I\{A\}$ is the indicator function of A, and $\mathbf{E}Z_\infty = 1$), with the Z_t selected in such a way that the nonlinear boundary $f(t)$ is transformed (in a certain sense) into a constant level. This enables us to get lower and upper estimates for $\mathbf{P}\{\tau > t\}$ (see §4) that can be used later to obtain conditions for the boundary functional $\mathbf{E}S_\tau$ in the asymptotic expression for $\mathbf{P}\{\tau > t\}$ in Theorem 2 to be finite and positive (see Theorem 3).

3. Main results. 1). The first result is one of those very few cases in the theory of nonlinear boundary problems in which it is possible to compute the moments of τ in explicit form.

We introduce the function

$$H(\nu, \alpha, x) = \frac{1}{\Gamma(-\nu)} \int_0^\infty y^{-\alpha\nu - 1} \exp\left(xy - \frac{1}{\alpha} y^\alpha\right) dy, \tag{1}$$

defined initially for $\operatorname{Re}\nu < 0$ ($1 < \alpha \leqslant 2$, $|x| < \infty$, and $\Gamma(z)$ is the Euler gamma function). The function $H(\nu, \alpha, x)$ has an analytic continuation to the right half-plane with respect to the parameter ν (this continuation will be denoted by the same symbol), and for $\alpha = 2$ it does not have singular points at all, while for $1 < \alpha < 2$ it has only poles on the real axis as singular points (the first pole is at $\nu = 1$; see the Appendix).

Let $\nu_\alpha(x)$ be the smallest positive zero of $H(\nu, \alpha, x)$. It can be shown (see the Appendix) that $\nu_\alpha(x)$ is monotone and continuous (in the variable x), and that $0 < \nu_\alpha(x) < 1/\alpha$ for $x > 0$, $\nu_\alpha(0) = 1/\alpha$, and $\nu_\alpha(x) > 1/\alpha$ for $x < 0$.

THEOREM 1. *Suppose that* $\mathbf{E}\exp(\lambda S_t) = \exp((1/\alpha)\lambda^\alpha t)$ *for* $\lambda \geqslant 0$, $1 < \alpha \leqslant 2$, $t \in R^+$, *and that*

$$f(t) = a(t+b)^{1/\alpha} + c \qquad (b \geqslant 0, ab^{1/\alpha} + c > 0).$$

If $\nu < \nu_\alpha(a)$, *then* $\mathbf{E}\tau^\nu < \infty$ *and*

$$\mathbf{E}(\tau+b)^\nu = b^\nu \frac{H(\nu, \alpha, -cb^{-1/\alpha})}{H(\nu, \alpha, a)} \qquad (b > 0), \tag{2}$$

$$\mathbf{E}\tau^\nu = \frac{c^{\alpha\nu}}{H(\nu, \alpha, a)} \qquad (b = 0, c > 0). \tag{3}$$

If $\nu \geqslant \nu_\alpha(a)$, *then* $\mathbf{E}\tau^\nu = \infty$.

We remark that the result of the theorem for $\alpha = 2$ (i.e., in the case of a Wiener process) was obtain by the author earlier in [14].

Theorem 1 will be proved in §2, and the properties of $H(\nu, \alpha, x)$ needed in the proof will be derived in the Appendix.

2). The result of Theorem 1 is not typical, in the sense that there are hardly any other classes of processes with independent increments and classes of nonlinear boundaries $f(t)$ for which it is possible to obtain explicit formulas for the moments of τ. However, it turns out to be possible to get the asymptotic behavior of $\mathbf{P}\{\tau > t\}$ under sufficiently broad assumptions about S_t and $f(t)$. Theorem 2 will give the form of this asymptotic behavior for boundaries that grow no more rapidly than $t^{1/\alpha}$, $1 < \alpha \leqslant 2$, where α is a parameter characterizing the rate of decrease of the tail of a negative jump of S_t. It will be convenient for us to use the notation $f(t) \ll g(t)$ to indicate the existence of constants $C > 0$ and $t_0 \geqslant 0$ such that $f(t) \leqslant Cg(t)$ for all $t \geqslant t_0$.(2) Moreover, it will be convenient to regard a boundary $f(t)$ defined for $t \in Z^+$ to be defined by continuity on R^+ with the help of linear interpolation.

THEOREM 2. *Suppose that the positive jumps of* S_t *are bounded by a constant* (*say*, L) *for* $t \in R^+$ *or* Z^+, *and that*

$$\ln \mathbf{E}\exp(\lambda S_1) \equiv \varphi(\lambda) \sim d\lambda^\alpha, \qquad \lambda \downarrow 0, \tag{4}$$

where $d > 0$ *and* $1 < \alpha \leqslant 2$. *Suppose, moreover, that* $\sup_{t \geqslant 0} f(t) < \infty$ *or that* $f(t)$ *is a continuous function such that*

$$0 \ll f(t) \ll t^{1/\alpha}.$$

Then $\mathbf{E}S_\tau$ *exists and is nonnegative, and*

$$\mathbf{P}\{\tau > t\}T^{1/\alpha} \to \mathbf{E}S_\tau\big(d^{1/\alpha}\Gamma(1 - 1/\alpha)\big)^{-1}, \qquad T \to \infty. \tag{5}$$

This is proved in §3. We remark that the assumption (4) holds when either $\mathbf{E}S_1^2 < \infty$ (the case $\alpha = 2$) or the tail of a negative jump varies in a certain regular way (see Remark 1 at the end of §3).

(2) The symbol $\ll$ was introduced by I. M. Vinogradov.

3). Theorem 2 gives the exact asymptotic behavior of $\mathbf{P}\{\tau > t\}$ only under the additional condition that $0 < \mathbf{E}S_\tau < \infty$. It can easily be shown (see Remark 2 in §3) that if the conditions of Theorem 2 holds, then for a nondecreasing boundary $f(t)$

$$\int_1^\infty \frac{f(t)}{t^{1/\alpha+1}}\,dt = \infty \Rightarrow \mathbf{E}S_\tau = \infty \tag{6}$$

and that for a continuous nonincreasing boundary $f(t)$

$$\int_1^\infty \frac{|f(t)|}{t^{1/\alpha+1}}\,dt = \infty \Rightarrow \mathbf{E}S_\tau = 0. \tag{7}$$

On the other hand, it is completely obvious that $\mathbf{E}S_\tau > 0$ when $f(t) > 0$, and $\mathbf{E}|S_\tau| < \infty$ when $\sup_{t\geqslant 0} f(t) < \infty$. The problem of finding necessary and sufficient conditions for the inequalities $0 < \mathbf{E}S_\tau < \infty$ turned out to be very nontrivial. In this article it is solved only under the additional condition of Cramér on a negative jump of S_t:

A_-: *There exists a* $\lambda < 0$ *such that* $\mathbf{E}\exp(\lambda S_1) \equiv \exp(\varphi(\lambda)) < \infty$.

Let $\lambda_- = \inf\{\lambda < 0\colon \mathbf{E}\exp(\lambda S_1) < \infty\}$.

THEOREM 3. *Suppose that condition A_- holds, and that the positive jumps of S_t are bounded by a constant L for $t \in R^+$ or Z^+.*

1°) *If $f(t)$ is a nondecreasing concave function, then* $\mathbf{E}S_\tau > 0$ *and*

$$\int_1^\infty \frac{f(t)}{t^{3/2}}\,dt < \infty \Leftrightarrow \mathbf{E}S_\tau < \infty.$$

2°) *If $f(t)$ is a nonincreasing convex function and there exists a number $t_0 > 0$ such that $f(t_0) > 0$ and*[3] *$f'(t_0) > \varphi'(\lambda_-)$, then* $\mathbf{E}|S_\rho| < \infty$, *and*

$$\int_1^\infty \frac{|f(t)|}{t^{3/2}}\,dt < \infty \Leftrightarrow \mathbf{E}S_\tau > 0.$$

This theorem is proved in §4 after the derivation of estimates of the form $T^{-1/2} \ll \mathbf{P}\{\tau > T\} \ll T^{-1/2}$ (see Theorem 5). In the case of a Wiener process such estimates were also obtained independently by Uchiyama [15] and Gärtner [16] in studying nonlinear parabolic equations by probability methods.

4). To complete the picture we formulate another result about the asymptotic behavior of $\mathbf{P}\{\tau > T\}$ for the case where condition A_- holds and the boundary $f(t)$ decreases sufficiently rapidly (but not more rapidly than a linear function).

THEOREM 4. *Suppose that condition A_- holds for $t \in R^+$ or Z^+, and that the boundary is continuous, nonincreasing, convex and such that there is a number $t_0 > 0$ for which $f(t_0) > 0$ and $f'(t_0) > \varphi'(\lambda_-)$. Let*

$$\Lambda_T(f) = \int_{t_0}^T [f'(t)g(t) - \varphi(g(t))]\,dt,$$

where $g(t)$ is a solution of the equation $\varphi'(g(t)) = f'(t)$, $t \geqslant t_0$. If

$$\ln T = o(\Lambda_T(f)), \qquad T \to \infty,$$

[3] Let $f'(t) = f(t+1) - f(t)$ in the case $t \in Z^+$.

then

$$\mathbf{P}\{\tau > T\} = \exp\{-\Lambda_T(f)(1+o(1))\}, \qquad T \to \infty.$$

In the case of discrete time this result was proved by the author in [17] with the help of a change-of-measure method. Its proof in the case of continuous time is completely analogous to that in the case of discrete time and is therefore omitted.

5. A remark on the notation. As already mentioned, the technique used in this article is based on martingale theory. The required facts from this theory will be given as needed. All martingales here are considered with respect to the natural σ-algebras $\mathfrak{F}_t = \sigma(S_u, u \leqslant t)$, but the probability measures may be different: the basic probability measure $\mathbf{P}$ is always used in §§2 and 3, while either $\mathbf{P}$ or the new measure $\tilde{\mathbf{P}}$ is used in §4 (this will be stated explicitly).

In the process of the proofs all constants whose values are not essential are denoted by the letter C (with or without indices).

§2. The case of stable processes of order α, $1 < \alpha \leqslant 2$, and the boundary $f(t) = a(t+b)^{1/\alpha} + c$

1. In this section we consider stable processes with continuous time such that

$$\mathbf{E}\exp(\lambda S_t) = \exp(d\lambda^\alpha t), \qquad d > 0, \lambda \geqslant 0, 1 < \alpha \leqslant 2.$$

As is known (see, for example, [3]), such processes are characterized by the absence of positive jumps. For convenience of notation we shall set $d = 1/\alpha$; the general case can be reduced to this by a simple recalculation. We remark that for this choice of the parameter α the process S_t is a standard Wiener process in the case $\alpha = 2$.

2. The function $H(\nu, \alpha, x)$. For $\alpha = 2$ the function $H(\nu, \alpha, x)$ is connected with the so-called parabolic cylinder function $D_\nu(z)$ by the relation

$$H(\nu, 2, x) = D_{2\nu}(-x)\exp(\tfrac{1}{4}x^2)$$

(see [18], formula 3.462.1). The result of Theorem 1 in the author's paper [14] is formulated precisely in terms of $H(\nu, 2, x)$.

Throughout the rest of this section it is assumed that $1 < \alpha < 2$. For these values of α the function $H(\nu, \alpha, x)$ defined by (1) does not reduce to any other special function (according to the widely known reference books of Gradshteĭn and Ryzhik [18], Abramowicz and Stegun [19] and Erdélyi et al. [20]). However, it is actually easy to get information about the behavior of $H(\nu, \alpha, x)$ by following familiar procedures for investigating special functions (see, for example, Olver [21]). The properties of $H(\nu, \alpha, z)$ needed to prove Theorem 1 are derived in the Appendix.

3. LEMMA 1. *Suppose that* $\mathbf{E}\exp(\lambda S_t) = \exp((1/\alpha)\lambda^\alpha t)$ *for* $\lambda \geqslant 0$. *Then for all* $b > 0$, *any* c, *and* $\operatorname{Re}\nu < 1$ *the process*

$$\Psi_t(\nu) = (t+b)^\nu H\big(\nu, \alpha, (S_t - c)/(t+b)^{1/\alpha}\big)$$

is a martingale.

PROOF. First of all we observe that the process

$$Z_t(\lambda) = \exp\{\lambda(S_t - c) - \lambda^\alpha(t+b)/\alpha\}, \qquad \lambda \geqslant 0,$$

is a martingale. This follows from the assumption that the process S_t has independent increments. In the domain $\operatorname{Re}\nu < 0$ the martingale $Z_t(\lambda)$ and the process $\Psi_t(\nu)$ are connected by the following relation:

$$\Psi_t(\lambda) = \frac{1}{\Gamma(-\alpha\nu)} \int_0^\infty Z_t(\lambda)\lambda^{-\alpha_\nu - 1}\, d\lambda \qquad (\operatorname{Re}\nu < 0). \tag{9}$$

It is easy to see this if we make the change of variable $\lambda(t+b)^{1/\alpha} = y$ in the integral and use the notation in (1). The martingale property of the process $\Psi_t(\nu)$ for $\operatorname{Re}\nu < 0$ follows from the martingale property of $Z_t(\lambda)$ by Fubini's theorem.

To prove Lemma 1 in the domain $\operatorname{Re}\nu < 1$ it is necessary to integrate by parts in (9). Here the second derivative of $Z_t(\lambda)$ for $\lambda > 0$ appears, and it is conveniently written in the form

$$d^2 Z_t(\lambda)/d\lambda^2 = \left[(S_t - c) - \lambda^{\alpha-1}(t+b)\right]^2 Z_t(\lambda) - (\alpha - 1)\lambda^{\alpha-2}(t+b)Z_t(\lambda). \tag{10}$$

Since $Z_t(\lambda)$ is a martingale for $\lambda > 0$, so is $d^2 Z_t(\lambda)/d\lambda^2$ (the possibility of differentiating under the expectation sign is easy to verify). Using the martingale property of these processes, we get the following estimate from (10) with the help of the inequality $|a+b| \leqslant |a| + |b|$:

$$\mathbf{E}\left|\frac{d^2}{d\lambda^2} Z_t(\lambda)\right| \leqslant 2(\alpha-1)\lambda^{\alpha-2}(t+b)\exp\left(-\lambda c - \frac{\lambda^\alpha}{\alpha} b\right) + \frac{d^2}{d\lambda^2} Z_0(\lambda). \tag{11}$$

Let us now consider the process

$$\Phi_t(\nu) = \frac{1}{\Gamma(2-\alpha\nu)} \int_0^\infty \lambda^{-\alpha\nu+1} \frac{d^2}{d\lambda^2} Z_t(\lambda)\, d\lambda \qquad (\operatorname{Re}\nu < 1),$$

which is obtained by integrating by parts twice in (9) and, hence, coincides with $\Psi_t(\nu)$ for $\operatorname{Re}\nu < 0$. From the martingale property of the process $d^2 Z_t(\lambda)/d\lambda^2$, $\lambda > 0$, and (11) it follows by Fubini's theorem that $\Phi_t(\nu)$ is a martingale for $\operatorname{Re}\nu < 1$. Moreover, (11) ensures that $\Phi_t(\nu)$ is analytic as a function of ν in the domain $\operatorname{Re}\nu < 1$. But since

$$\Phi_t(\nu) + \Psi_t(\nu) = (t+b)^\nu H\left(\nu, \alpha, (S_t - c)(t+b)^{-1/\alpha}\right)$$

in the domain $\operatorname{Re}\nu < 0$, the uniqueness of the analytic continuation implies that this equality holds also in the domain $\operatorname{Re}\nu < 1$, and the lemma is proved.

4. Proof of Theorem 1. For the proof we assume that $b > 0$. The case $b = 0$ is easily obtained by passing to the limit.

a) Let us first show that

$$\mathbf{E}\tau^\nu < \infty \quad \text{for } \nu < \nu_\alpha(a). \tag{12}$$

By the martingale stopping theorem, Lemma 1 implies that for $\operatorname{Re}\nu < 1$

$$\mathbf{E}(t\wedge\tau + b)^\nu H\left(\nu, \alpha, \frac{S_{t\wedge\tau} - c}{(t\wedge\tau + b)^{1/\alpha}}\right) = b^\nu H(\nu, \alpha, -cb^{-1/\alpha}). \tag{13}$$

Since S_t does not have positive jumps, the inequality

$$(S_{t\wedge\tau} - c)(t\wedge\tau + b)^{-1/\alpha} \leqslant a$$

always holds for the s.t. $t \wedge \tau$[4] under consideration, and the equality sign can be used in this inequality on the set $\{\tau \leqslant t\}$. We consider the case of real $\nu < \nu_\alpha(a)$ in (13). The function $H(\nu, \alpha, x)$ is positive for $\nu < \nu_\alpha(a)$ and $x \leqslant a$ (recall that $\nu_\alpha(x) \geqslant \nu_\alpha(a)$ when $x \leqslant a$); therefore, (13) implies that

$$\mathbf{E}I\{\tau < t\}(\tau + b)^\nu \leqslant b^\nu \frac{H(\nu, \alpha, -cb^{-1/\alpha})}{H(\nu, \alpha, a)} < \infty$$

for all $t \geqslant 0$. The assertion (12) follows from this.

b) We now show that (2) holds when $\operatorname{Re}\nu < \nu_\alpha(a)$. To do this, note first that in the domain $\operatorname{Re}\nu < 0$ and under the constraint $x \leqslant a$ the function $H(\nu, \alpha, x)$ is majorized by a constant (for fixed ν):

$$|H(\nu, \alpha, x)| \leqslant \frac{1}{|\Gamma(-\alpha\nu)|} \int_0^\infty y^{-\alpha \operatorname{Re}\nu - 1} \exp\left(ay - \frac{1}{\alpha} y^\alpha\right) dy$$

(see (1)). Therefore, we can pass to the limit as $t \to \infty$ under the expectation sign on the left-hand side of (13) for fixed ν in the domain $\operatorname{Re}\nu < 0$, and the result is

$$\mathbf{E}(\tau + b)^\nu H(\nu, \alpha, a) = b^\nu H(\nu, \alpha, -cb^{-1/\alpha}) \qquad (\operatorname{Re}\nu < 0).$$

For real $\nu < 0$ the function $H(\nu, \alpha, a)$ is positive; therefore, the desired relation (2) certainly holds for these ν. We now observe that, by (12), the function $\mathbf{E}(\tau + b)^\nu$ is analytic in the domain $\operatorname{Re}\nu < \nu_\alpha(a)$. This means that the right-hand side of (2) (which so far has been proved for $\operatorname{Re}\nu < 0$) also has an analytic continuation from the negative axis to the whole domain $\operatorname{Re}\nu < \nu_\alpha(a)$, and it must coincide with $\mathbf{E}(\tau + b)^\nu$.

Assume that $H(\nu, \alpha, a)$ has a complex zero $\nu_\alpha^*(a)$ in the domain $\operatorname{Re}\nu < \nu_\alpha(a)$. But if here $H(\nu_\alpha^*(a), \alpha, -cb^{-1/\alpha}) \neq 0$, then, by the uniqueness of the analytic continuation, the function $\mathbf{E}(\tau + b)^\nu$ must have a pole at $\nu = \nu_\alpha^*(a)$, and this cannot be. The only remaining possibility is that $H(\nu_\alpha^*(a), \alpha, -cb^{-1/\alpha}) = 0$. But the parameters b and c are connected with the parameter a only by the restriction $-cb^{-1/\alpha} \leqslant a$. Therefore, we have actually established (under the assumption that $\nu_\alpha^*(a)$ exists) that $H(\nu_\alpha^*(a), \alpha, x) = 0$ for all $x \leqslant a$. The latter is impossible, since the derivatives of $H(\nu, \alpha, x)$ with respect to x are nonzero (see (A.2)).

Thus, we have proved that $H(\nu, \alpha, x)$ does not have zeros in the domain $\operatorname{Re}\nu < \nu_\alpha(x)$ for all x.

It now remains to observe that, since the right-hand side of (2) is an analytic function in the domain $\operatorname{Re}\nu < \nu_\alpha(a)$, the equality (2) there follows from the uniqueness of the analytic continuation, and, hence, all the assertions of Theorem 1 are valid in the case $b > 0$ under consideration.

c) If $b = 0$, then assertion 3) of Theorem 1 follows from what has been proved, by the theorem on monotone passage to the limit under the expectation sign and the formula (see (A.3))

$$\lim_{b \to 0} b^\nu H(\nu, \alpha, -cb^{-1/\alpha}) = c^{\alpha\nu} \qquad (c > 0, \operatorname{Re}\nu < 1).$$

Theorem 1 is proved.

---

[4] $t \wedge \tau = \min(t, \tau)$.

5. Some remarks. 1. The behavior of $\nu_\alpha(a)$ as $a \to -\infty$ in the case $\alpha = 2$ differs essentially from that in the case $1 < \alpha < 2$:

$$\lim_{a\to-\infty} \nu_2(a) = \infty, \qquad \lim_{a\to-\infty} \nu_\alpha(a) = 1 \quad (1 < \alpha < 2).$$

The first of these relations follows, for example, from the estimates $\frac{1}{8}a^2 + \frac{1}{2} + \frac{1}{2}a \leqslant \nu_2(a) \leqslant \frac{1}{8}a^2 + \frac{1}{2}$ $(a < 0)$ found by the author in [22]. The second relation is proved in the Appendix (see (A.5)).

2. It would be interesting to determine how the probability $\mathbf{P}\{\tau > T\}$ behaves as $T \to \infty$. To all appearances,

$$\mathbf{P}\{\tau > T\} \sim CT^{-\nu_\alpha(a)}, \qquad T \to \infty, \tag{14}$$

where C is a positive constant. This result would follow immediately from Theorem 1 with the help of a Tauberian theorem if we could show that the function $\mathbf{P}\{\tau > T\}T^{\nu_\alpha(a)}$ is monotone for sufficiently large T and that the zero $\nu_\alpha(a)$ has multiplicity 1. Another path is to use the formula for the inverse Laplace transform, but this will require some more information about the complex zeros of $H(\nu, \alpha, a)$ for $\nu > \nu_\alpha(a)$. The only case in which we were able to prove (14) by this path is the case $\alpha = 2$. In this case the function $D_\nu(z)$ does not have complex zeros at all (for real z), and the asymptotic behavior (14) is easy to get with the help of the inverse Laplace transform.

We remark that (14) for $\alpha = 2$ was recently obtained by another method by Uchiyama [15], and still earlier for $\alpha = 2$ and $a > 0$ by Sato [23]. The authors of these papars were apparently not acquainted with [14]. where this asymptotic behavior had in fact already been found.

3. The stopping time τ in Theorem 1 was considered by Mackevičius [24] in connection with a certain optimal stopping problem for a process S_t. He showed that τ is finite and that it is optimal for a certain value of the parameter a.

§3. The asymptotic behavior of $\mathbf{P}\{\tau > T\}$ in the case $|f(t)| \ll t^{1/\alpha}$

1. Proof of Theorem 2. a) First of all we show that $\mathbf{E}S_\tau$ exists and is nonnegative. Since S_t is a martingale, $t \in R^+$ or Z^+, it follows that $\mathbf{E}S_{\tau\wedge\tau} = 0$ and, hence,

$$\mathbf{E}S_\tau I\{\tau \leqslant t\} = -\mathbf{E}S_t I\{\tau > t\} \geqslant -f(t)\mathbf{P}\{\tau > t\}. \tag{15}$$

We remark that $\mathbf{P}\{\tau < \infty\} = 1$ (for example, by the law of the iterated logarithm), and consider first the case $\sup_{t\geqslant 0} f(t) < \infty$. Then (15) implies that([5])

$$\lim_{t\to\infty} \mathbf{E}S_\tau^- \{\tau \leqslant t\} \leqslant \lim_{t\to\infty} \mathbf{E}S_\tau^+ \{\tau \leqslant t\}.$$

But $S_\tau^+ \leqslant \sup_{t\geqslant 0} f(t) + L$ by the boundedness of the jumps, and, consequently, $\mathbf{E}S_\tau^- \leqslant \mathbf{E}S_\tau^+ < \infty$ by the monotone convergence theorem. Hence, $\mathbf{E}|S_\tau| < \infty$ and $\mathbf{E}S_\tau \geqslant 0$.

Suppose now that $f(t)$ is a continuous function such that $0 \ll f(t)$, i.e., there exists a $t_0 \geqslant 0$ such that $f(t) \geqslant 0$ for $t \geqslant t_0$. Assume also that $f(t_0) = 0$. (If there is no such number t_0, then $f(t) > 0$ for all $t \geqslant 0$ by continuity, and the assertion to be proved is

---

([5]) $x^+ = \max(x, 0)$ and $x^- = \max(-x, 0)$.

obvious). Since $S_\tau^- \leqslant -\inf_{t\geqslant 0} f(t) < \infty$ in this case, $\mathbf{E}S_\tau$ is defined. Moreover,

$$\mathbf{E}S_\tau = \mathbf{E}S_\tau I\{\tau \leqslant t_0\} + \mathbf{E}S_\tau I\{\tau < t_0\}.$$

The second term is nonnegative, because $S_\tau \geqslant f(\tau) \geqslant 0$ for $\tau \geqslant t_0$. The first term is also nonnegative because of (15). (In the discrete case it is necessary to consider the integer part $[t_0]$ instead of t_0.)

b) We prove the identity

$$\mathbf{E}\exp(\lambda S_\tau - \varphi(\lambda)\tau) = 1 \qquad (\lambda \geqslant 0). \tag{16}$$

To do this let us consider the martingale $Z_t = \exp(\lambda S_t - \varphi(\lambda)t)$. By the martingale stopping theorem, the process $Z_{t\wedge\tau}$ is also a martingale. The boundedness of the jumps and the conditions imposed on the functions $f(t)$ and $\varphi(\lambda)$ give us that

$$\ln Z_{t\wedge\tau} \leqslant \lambda(f(t\wedge\tau) + L) - \varphi(\lambda)(t\wedge\tau) \leqslant C(\lambda),$$

where $C(\lambda)$ is a bounded constant (recall that $f(t) \ll t^{1/\alpha}$). Therefore, $Z_{t\wedge\tau}$ is a uniformly integrable martingale, and thus (16) holds.

c) Let us prove the theorem under the additional assumption that $\mathbf{E}S_\tau < \infty$.

On the basis of (16),

$$\begin{aligned}1 - \mathbf{E}\exp(-\varphi(\lambda)\tau) &= \lambda\mathbf{E}S_\tau\exp(-\varphi(\lambda)\tau)\\ &\quad+\mathbf{E}\big(e^{\lambda S_\tau} - \lambda S_\tau - 1\big)\exp(-\varphi(\lambda)\tau).\end{aligned} \tag{17}$$

We show later that

$$\Delta(\lambda) \equiv \mathbf{E}\big(e^{\lambda S_\tau} - \lambda S_\tau - 1\big)\exp(-\varphi(\lambda)\tau) = o(\lambda), \qquad \lambda\downarrow 0. \tag{18}$$

This implies the assertion of the theorem at once. Indeed, writing $\varphi(\lambda) = \mu$, we get from (4), (17) and (18) that

$$1 - \mathbf{E}\exp(-\mu\tau) \sim (\mu/d)^{1/\alpha}\mathbf{E}S_\tau, \qquad \mu\downarrow 0,$$

and on the basis of a Tauberian theorem (see [25]) we get the theorem at once. (Note that $\mathbf{P}\{\tau > T\} \sim \mathbf{P}\{\tau \geqslant T\}$ as $T \to \infty$.)

Let us now prove (18). We fix ε, $0 < \varepsilon < 1$, and apply the estimates

$$0 \leqslant e^x - x - 1 \leqslant I\{x < -\varepsilon\}\,|x| + I\{|x| \leqslant \varepsilon\}2|x|\varepsilon e^\varepsilon + I\{x > \varepsilon\}xe^x$$

to the second term in (17). Then

$$\begin{aligned}0 \leqslant \Delta(\lambda) &\leqslant \lambda\mathbf{E}I\{S_\tau < -\varepsilon/\lambda\}\,|S_\tau| + 2\lambda\varepsilon e^\varepsilon\mathbf{E}|S_\tau|\\ &\quad+\lambda\mathbf{E}I\{S_\tau > \varepsilon/\lambda\}S_\tau\exp(\lambda S_\tau - \varphi(\lambda)\tau).\end{aligned} \tag{19}$$

In the last term we have $\lambda S_\tau - \varphi(\lambda)\tau \leqslant C$ for all sufficiently small λ. Indeed, since $\varphi(\lambda) \geqslant d(1-\varepsilon)\lambda^\alpha$ for all $0 \leqslant \lambda \leqslant \lambda_0(\varepsilon)$ and $S_\tau \leqslant f(\tau) + L \leqslant C_1\tau^{1/\alpha} + C_2$, the needed uniform boundedness follows from the fact that

$$\sup_{x\geqslant 0}\big(C_1x^{1/\alpha} - C_2x\big) \leqslant C.$$

It now follows from (19) that

$$\limsup_{\lambda\to 0}(1/\lambda)\Delta(\lambda) \leqslant 2\varepsilon e^\varepsilon\mathbf{E}|S_\tau|.$$

The desired relation (18) now follows from the fact that ε is arbitrary.

d) *The case* $\mathbf{E}S_\tau = \infty$. Since the jumps are bounded, this implies that $\mathbf{E}f(\tau) = \infty$. We consider the "truncated" boundary $f_A(t) = f(t) \wedge A$, $A < \infty$, and let τ_A be the crossing time for this boundary. Then $\mathbf{E}S_{\tau_A} < \infty$, $\tau_A \to \tau$ $\mathbf{P}$-a.s. as $A \to \infty$, $\mathbf{P}\{\tau > T\} \geqslant \mathbf{P}\{\tau_A > T\}$, and

$$\liminf_{A\to\infty} \mathbf{E}S_{\tau_A} \geqslant \liminf_{A\to\infty} \mathbf{E}f(\tau_A) \wedge A = \mathbf{E}f(\tau) = \infty. \tag{20}$$

By the fact proved in c),

$$\lim_{T\to\infty} \mathbf{P}\{\tau > T\}T^{1/\alpha} \geqslant \lim_{T\to\infty} \mathbf{P}\{\tau_A > T\}T^{1/\alpha} = \mathbf{E}S_{\tau_A}\left(d^{1/\alpha}\Gamma(1-1/\alpha)\right)^{-1},$$

and, hence, by (20),

$$\lim_{T\to\infty} \mathbf{P}\{\tau > T\}T^{1/\alpha} = \infty.$$

This completes the proof of Theorem 2.

2. Remarks. 1) The condition (4) holds only if either

$$\mathbf{P}\{S_1 < x\} \sim \frac{d|x|^{-\alpha}}{|\Gamma(a-a)|}, \qquad x \to -\infty,\ 1 < \alpha < 2,$$

or $\mathbf{E}|S_1|^2 < \infty$ (the case $\alpha = 2$; here $\mathbf{E}S_1^2 = 2d$).

It is not hard to see this by using Tauberian theorems in the case $1 < \alpha < 2$ (the case $\alpha = 2$ is trivial) for the Laplace transform of the jump S_1 (see [25]; of course, the boundedness of the positive jumps is also assumed).

It is easy to see that Theorem 2 generalizes to the case where

$$\mathbf{P}\{S_1 < x\} = \frac{dL(-x)}{|\Gamma(1-\alpha)|}|x|^{-\alpha}, \qquad x \to -\infty,\ 1 < \alpha < 2,$$

where $L(x)$ is a slowly varying function as $x \to \infty$. Here the factor $t^{1/\alpha}$ in (5) is replaced by the factor $\gamma(T)$, where $\gamma(T)$ is the function inverse to $T^\alpha/L(T)$.

2) The constant $\mathbf{E}S_\tau$ is certainly equal to ∞ if $f(t)$ is a nondecreasing boundary such that $\int_1^\infty f(t)t^{-1/\alpha-1}\,dt = \infty$ (i.e., (6) is valid). Indeed, in this case $\mathbf{E}S_\tau > 0$ and, by Theorem 2,

$$\lim_{T\to\infty} \mathbf{P}\{\tau > T\}T^{1/\alpha} > 0. \tag{21}$$

But, since the jumps across the boundary are bounded,

$$\mathbf{E}S_\tau = \infty \Leftrightarrow \mathbf{E}f(\tau) = \infty. \tag{22}$$

For any nondecreasing function

$$\mathbf{E}f(t) = \infty \Leftrightarrow \int_1^\infty \mathbf{P}(\tau > t)\,df(t) = \infty,$$

$$\int_1^\infty f(t)t^{-1/\alpha-1}\,dt = \infty \Leftrightarrow \int_1^\infty t^{-1/\alpha}\,df(t) = \infty; \tag{23}$$

therefore, it now follows from (21) and (22) that $\mathbf{E}S_\tau = \infty$.

In a similar way it can be shown that if $f(t)$ is a nonincreasing boundary and $\int_1^\infty |f(t)|t^{-1/\alpha-1}\,dt = \infty$, then the constant $\mathbf{E}S_\tau$ is zero, (i.e., the implication (7) holds). Indeed, $\sup_{t\geqslant 0} f(t) < \infty$ in this case; therefore, $\mathbf{E}|S_\tau| < \infty$. This implies that

$$\int_1^\infty \mathbf{P}\{\tau > t\}d|f(t)| < \infty.$$

But since (see (23)) $\int_1^\infty t^{-1/\alpha} d|f(t)| = \infty$, it follows that necessarily

$$\lim_{T\to\infty} \mathbf{P}\{\tau > T\} T^{1/\alpha} = 0,$$

and, by Theorem 2, $\mathbf{E}S_\tau = 0$.

3) For stable processes of order α, $1 < \alpha \leqslant 2$, without positive jumps and for continuous boundaries $f(t)$ such that $f(t) \ll t^{1/\alpha - \delta}$ for some $\delta > 0$, we have the estimate $\mathbf{E}S_\tau < \infty$.

Indeed, the estimate $f(t) \leqslant C_\varepsilon + \varepsilon t^{1/\alpha}$, $t \geqslant 0$, is valid for any $\varepsilon > 0$. Let τ_ε be the time at which the stable process reaches the boundary $C_\varepsilon + \varepsilon t^{1/\alpha}$. Since $\tau \leqslant \tau_\varepsilon$, Theorem 1 implies that

$$\mathbf{E}\tau^\nu \leqslant \mathbf{E}\tau_\varepsilon^\nu < \infty \quad \text{for } \nu < \nu_\alpha(\varepsilon), \tag{24}$$

where $\nu_\alpha(\varepsilon) = 1/\alpha - c_\alpha \varepsilon + o(\varepsilon)$ as $\varepsilon \to 0$ (see the Appendix, (A.7)). Because $f(t) \leqslant C_1 + C_2 t^{1/\alpha - \delta}$ for all t and the boundary $f(t)$ is attained without being jumped across, it follows that

$$\mathbf{E}S_\tau = \mathbf{E}f(\tau) \leqslant C_1 + C_2 \mathbf{E}\tau_\varepsilon^{1/\alpha - \delta}.$$

It is clear that by choosing a small ε we can ensure that $1/\alpha - \delta < \nu_\alpha(\varepsilon)$ for any given $\delta > 0$. Therefore, we get the estimate $\mathbf{E}S_\tau < \infty$ by virtue of (24).

4) CONJECTURES. In all likelihood Theorem 2 remains true if the assumption that the jumps are bounded is replaced by the right-sided Cramér condition:

A_+ : *There is a* $\lambda > 0$ *such that* $\mathbf{E}\exp(\lambda S_1) < \infty$.

If $f(t)$ is required to be monotone and concave or convex, it is apparently necessary that

$$\int_1^\infty \frac{|f(t)|}{t^{1/\alpha + 1}} dt < \infty \Leftrightarrow 0 < \mathbf{E}S_\tau < \infty.$$

So far, such an assertion has been proved (see Theorem 3) only under the additional condition A_- and the assumption that the positive jumps are bounded (instead of condition A_+).

§4. Estimates for $\mathbf{P}\{\tau > T\}$ in the case $\int_0^\infty |f'(t)|^2 \, dt < \infty$ and a proof of Theorem 3

1. Preliminary remarks. In this section we derive upper and lower estimates for the probability $\mathbf{P}\{\tau > T\}$ which lead directly to the assertion of Theorem 3. The derivation of these estimates will be carried out only for continuous time. The case of discrete time is treated in exactly the same way, except that it is necessary to replace integrals everywhere by the corresponding sums and derivatives by the corresponding increments.

2. THEOREM 5. *Suppose that condition* A_- *holds, and that the positive jumps are bounded by a constant* L *for all* $t \in R^+$ *or* Z^+. *Suppose, moreover, that* $f(t)$ *is absolutely continuous and*

$$\int_0^\infty (f'(t))^2 \, dt < \infty.$$

2°) *If* $f(t)$ *is a nonincreasing and convex function and there exists a number* $t_0 > 0$ *such that* $f(t_0)$ *and* $f'(t_0) > \varphi'(\lambda_-)$, *then*

$$\mathbf{P}\{\tau > T\} \gg T^{-1/2} \exp\left\{-C\int_1^T |f(t)|\, t^{-2/3}\, dt\right\}. \tag{26}$$

3. Proof. We precede the direct derivation of (25) and (26) by some auxiliary statements.

1) *Properties of* $\varphi(\lambda)$. From condition A_- it follows immediately that for all $\lambda > \lambda_-$ the function $\varphi(\lambda)$ is infinitely differentiable, and

$$\mathbf{E}S_1 \exp(\lambda S_1 - \varphi(\lambda)) = \varphi'(\lambda), \quad \mathbf{E}(S_1 - \varphi'(\lambda))^2 \exp(\lambda S_1 - \varphi(\lambda)) = \varphi''(\lambda).$$

Since $\mathbf{E}S_1 = 0$, the first relation gives us that $\varphi'(0) = 0$, while the second gives us that $0 < \varphi''(\lambda) < \infty$ for $\lambda > \lambda_-$ (we are, of course, considering nonsingular distributions). Therefore, $\varphi(\lambda)$ is downwards convex with the proof of absolute minimum $\varphi(0) = 0$. Moreover, for any λ_0 and λ_1 with $\lambda_- < \lambda_0 \leqslant \lambda_1 < \infty$ it is obvious that

$$0 < C_1 \leqslant \varphi''(\lambda) \leqslant C_2 \quad \text{for all } \lambda \in [\lambda_0, \lambda_1]. \tag{27}$$

2) *Properties of* $f(t)$. By virtue of our assumptions the derivative $f'(t)$ is nonnegative and decresing in case 1°), and nonpositive and increasing in case 2°). It is clear that without loss of generality we can assume in what follows that

$$f(t) = f(t_0) \quad \text{for } 0 \leqslant t \leqslant t_0, \tag{28}$$

because in case 1°) the value of $\mathbf{P}\{\tau > T\}$ can only increase in passing from the boundary $f(t)$ to the boundary $\bar{f}(t) = \max(f(t_0), f(t))$, while in case 2°) it can only decrease in passing to the boundary $\underline{f}(t) = \min(f(t_0), f(t))$.

In the first case we choose t_0 large enough that in every case

$$f'(t_0) < \varphi'(+\infty). \tag{29}$$

This can always be done because $f'(t)$ is decreasing. (Other restrictions will be imposed on t_0 in the process of the proof.) In case 2°) the conditions on t_0 are stipulated in the statement of the theorem.

3) *Definition and properties of* $g(t)$. An essential role in what follows is played by the function $g(t)$, which is a solution of the equation

$$\varphi'(g(t)) = f'(t), \qquad t \geqslant 0. \tag{30}$$

By the assumed condition (28), $g(t) = 0$ for $0 \leqslant t < t_0$, and by the fact that the functions $f'(t)$ and $\varphi'(\lambda)$ are monotone, condition (29) and the condition $f'(t_0) > \varphi'(\lambda_-)$ ensure that (30) is solvable for all $t > t_0$. Clearly, $g(t)$ is nonnegative and decreasing (for $t > t_0$) in case 1°) and, respectively, nonpositive and increasing in case 2°), moreover, since $\varphi''(\lambda)$ is bounded below (see (27)), the estimate

$$|g(t)| \leqslant C|f'(t)|, \qquad t \geqslant 0, \tag{31}$$

also holds. For definiteness the functions $f'(t)$ and $g(t)$ are assumed to be right-continuous in what follows.

4) *Change of the measure*. Let us consider the process

$$Z_t = \exp\left\{\int_0^t g(u)\, dS_u - \int_0^t \varphi(g(u))\, du\right\},$$

where $g(u)$ is a solution of (30). From the theory of stochastic integrals with respect to martingales (see, for example, Jacod's monograph [26]) we know that if $g(t)$ has bounded variation (which holds in the case under consideration), then the usual formula for integration by parts holds:

$$\int_0^t g(u)\,dS_u = g(t)S_t - \int_0^t S_u\,dg(u). \tag{32}$$

(In our case this formula could be taken as the definition of the stochastic integral with respect to the martingale S_t.)

The possibility of introducing a new probability measure with the help of the process Z_t is based on the following lemma.

LEMMA 2. *Suppose that the conditions of Theorem 5 hold and that the function $g(t)$ is a solution of* (30). *Then the process Z_t is a uniformly integrable martingale with respect to the measure* **P**.

PROOF. The martingale property of the process Z_t can be proved in at least two ways. The first is to verify directly that Z_t has the property for step functions $g(t)$, and then to justify passing to the limit. The other way is to represent the process S_t in terms of Poisson point measures and then to use the Itô formula [26].

The uniform integrability of Z_t follows from the obvious estimate

$$\mathbf{E}Z_t^p \leqslant \exp\left(\int_0^t \varphi(pg(u))\,du\right), \qquad p > 1.$$

Since $g(t)$ is bounded, p can be chosen close enough to 1 that $pg(t) \in [\lambda_0, \lambda_1]$ for all $t \geqslant 0$, where $\lambda_- < \lambda_0 \leqslant \lambda_1 < \infty$. If we then use (27) and (31), we get

$$\int_0^\infty \varphi(pg(u))\,du \leqslant C\int_0^\infty g^2(u)\,du \leqslant C\int_0^\infty (f'(t))^2\,dt < \infty.$$

Thus, $\sup_{t\geqslant 0}\mathbf{E}Z_t^p < \infty$ for some $p > 1$ and, hence, Z_t is a uniformly integrable martingale. Lemma 2 is proved.

Since Z_t is a uniformly integrable martingale, $\lim_{t\to\infty} Z_t = Z_\infty$ exists (**P**-a.s.), and

$$\mathbf{E}Z_\infty = 1, \tag{33}$$

$$\mathbf{E}(Z_\infty \mid \mathfrak{F}_t) = Z_t, \qquad t \geqslant 0\ (\mathbf{P}\text{-a.s.}). \tag{34}$$

We now introduce a new measure on the σ-algebra $\mathfrak{F}$ by setting

$$\tilde{\mathbf{P}}\{A\} = EI\{A\}Z_\infty, \qquad A \in \mathfrak{F}.$$

By (33), this equality really does specify a probability measure, and, by (34), the restriction of this measure to the σ-algebra $\mathfrak{F}_t = \sigma\{S_u, u \leqslant t\}$ is determined by the formula

$$\tilde{\mathbf{P}}\{A\} = \mathbf{E}I\{A\}Z_t, \qquad A \in \mathfrak{F}_t. \tag{35}$$

Integration with respect to the new probability measure will be denoted by the letter **M**.

Using the independence of the increments of the process S_t (with respect to **P**) and the formula

$$\mathbf{M}(\eta_T \mid \mathfrak{F}_t) = Z_t^{-1}\mathbf{E}(\eta_T Z_T \mid \mathfrak{F}_t), \qquad t \leqslant T\ (\tilde{\mathbf{P}}\text{-a.s.}),$$

where η_T is any $\mathfrak{F}_T$-measurable random variable such that $\mathbf{M}|\eta_T|$ exists (see, for example, Liptser and Shiryaev [27], Lemma 6.6), we get that ($\tilde{\mathbf{P}}$-a.s.)

$$\mathbf{M}\{\exp(\lambda(S_T - S_t)) | \mathfrak{F}_t\} = \exp\left\{\int_t^T [\varphi(g(u) + \lambda) - \varphi(g(u))]\,du\right\} \tag{36}$$

for all λ such that $\inf_{u\geqslant 0} g(u) + \lambda > \lambda_-$ (for example, for $\lambda \geqslant 0$). This implies that S_t also has independent increments with respect to the measure $\tilde{\mathbf{P}}$, and

$$\mathbf{M}S_t = \int_0^t \varphi'(g(u))\,du. \tag{37}$$

The last integral is always equal to $f(t) - f(0)$ by the choice of the function $g(t)$.

Let

$$\tilde{S}_t = S_t - \int_0^t \varphi'(g(u))\,du.$$

Because the increments of S_t are independent and because of (37), the process $\tilde{S}_t$ is a martingale with respect to the measure $\tilde{\mathbf{P}}$. In terms of $\tilde{S}_t$ the stopping time τ can be written as follows:

$$\tau = \inf\{t \geqslant 0 \colon \tilde{S}_t > f(0)\}.$$

Since Z_T^{-1} is finite and positive ($\mathbf{P}$-a.s. and $\tilde{\mathbf{P}}$-a.s.), the relation (35) with $A = \{\tau > T\}$ implies that

$$\begin{aligned}\mathbf{P}\{\tau > T\} &= \mathbf{M} I\{\tau > T\} \\ &\quad \times \exp\left\{-\int_0^T g(u)\,d\tilde{S}_u - \int_0^T [g(u)\varphi'(g(u)) - \varphi(g(u))]\right\} du.\end{aligned} \tag{38}$$

This relation will give us the desired inequalities for $\mathbf{P}\{\tau > T\}$. But first we must get estimates for $\tilde{\mathbf{P}}\{\tau > T\}$.

5) *Lower and upper estimates for* $\tilde{\mathbf{P}}\{\tau > T\}$.

LEMMA 3. *Suppose that the conditions of Theorem* 5 *hold and that* $g(t)$ *is a solution of* (30). *Then*

$$T^{-1/2} \ll \tilde{\mathbf{P}}\{\tau > T\} \ll T^{-1/2}.$$

PROOF. We first establish the upper estimate. With this in mind consider the process

$$Z_t(\lambda) = \exp\{\lambda \tilde{S}_t - Q(\lambda, t)\} \tag{39}$$

with

$$Q(\lambda, t) = \int_0^t [(\varphi(g(u)) + \lambda) - \varphi(g(u)) - \lambda\varphi'(g(u))]\,du$$

and λ such that $\inf_{u\geqslant 0} g(u) + \lambda > \lambda_-$.

Since $g(t)$ is a bounded function and (27) holds for $\varphi''(\lambda)$, we have

$$C\lambda^2 \geqslant \varphi(g(u) + \lambda) - \varphi(g(u)) - \lambda\varphi'(g(u)) = \varphi''(g(u) + \lambda\theta)\lambda^2/2 \geqslant C\lambda^2$$

for the indicated values of λ and for all u, where $0 \leqslant \theta \leqslant 1$, and, hence,

$$C_1\lambda^2 t \leqslant Q(\lambda, t) \leqslant C_2\lambda^2 t, \qquad t \geqslant 0. \tag{40}$$

It is not hard to see that, by (36) and the independence of the increments $\tilde{S}_t$, the process $Z_t(\lambda)$ is a martingale with respect to $\tilde{\mathbf{P}}$. But since $\tilde{S}_{t\wedge\tau} \leqslant f(0) + L$ for all t, the process $Z_{t\wedge\tau}(\lambda)$ is a bounded martingale, and, hence, $\mathbf{M}Z_\tau(\lambda) = 1$, i.e.,

$$\mathbf{M}\exp\{\lambda\tilde{S}_\tau - Q(\lambda,\tau)\} = 1. \tag{41}$$

Using the Tchebycheff inequality, (41), and the estimate $\tilde{S}_\tau \leqslant f(0) + L$, we get

$$\tilde{\mathbf{P}}\{\tau > T\} = \tilde{\mathbf{P}}\{1 - e^{-Q(\lambda,\tau)} \geqslant 1 - e^{-Q(\lambda,T)}\} \leqslant (1 - e^{-\lambda(f(0)+L)})(1 - e^{-Q(\lambda,T)})^{-1}.$$

Here we take $\lambda = T^{-1/2}$. Then the inequality $1 - e^{-x} \leqslant x$ and (40) give us the desired upper estimate:

$$\tilde{\mathbf{P}}\{\tau > T\} \leqslant C(f(0) + L)T^{-1/2}. \tag{42}$$

We note for subsequent purposes that the same device can be used with conditional expectations to get the following estimate:

$$\tilde{\mathbf{P}}\{\tau > T \mid \mathfrak{F}_t\} \leqslant C(f(0) + L - \tilde{S}_t)(T - t)^{-1/2}, \qquad 0 \leqslant t < T, \tag{43}$$

on $\{\tau > t\}$ ($\tilde{\mathbf{P}}$-a.e.).

It is rather more complicated to get the lower estimate for $\tilde{\mathbf{P}}\{\tau > T\}$. The identity (41) implies that for $\lambda \geqslant 0$

$$\begin{aligned}1 - \mathbf{M}\exp&\left\{-\frac{\lambda^2}{2}\int_0^\tau \varphi''(g(u))\,du\right\} \\ &= \lambda\mathbf{M}\tilde{S}_\tau\exp\{-Q(\lambda,\tau)\} + \mathbf{M}\left(e^{\lambda\tilde{S}_\tau} - \lambda\tilde{S}_\tau - 1\right)\exp\{-Q(\lambda,\tau)\} \\ &\quad + \mathbf{M}\left[\exp\{-Q(\lambda,\tau)\} - \exp\left\{-\frac{\lambda^2}{2}\int_0^\tau \varphi''(g(u))\,du\right\}\right].\end{aligned} \tag{44}$$

Since $f(0) \leqslant \tilde{S}_\tau \leqslant f(0) + L$, the first term on the right-hand side obviously has order $\lambda\mathbf{M}\tilde{S}_\tau$, while the second has order $O(\lambda^2)$ as $\lambda \downarrow 0$. Before estimating the third term we observe that the third derivative of $\varphi(\lambda)$ satisfies the estimate $|\varphi'''(\lambda)| \leqslant C$, $\lambda \in [\lambda_0, \lambda_1]$, $\lambda_0 > \lambda_-$. Therefore, for all $t \geqslant 0$ and for sufficiently small λ

$$\left|Q(\lambda,t) - \frac{\lambda^2}{2}\int_0^t \varphi''(g(u))\,du\right| = \frac{\lambda^3}{6}\left|\int_0^t \varphi'''(g(u) + \lambda\theta)\,du\right| \leqslant C\lambda^3 t.$$

Applying the inequalities $|e^x - 1| \leqslant e^{|x|} - 1$ and (40), and the estimate just established to the third term in (44), we get

$$\begin{aligned}&\left|\mathbf{M}\exp\{-Q(\lambda,\tau)\} - \exp\left\{-\frac{\lambda^2}{2}\int_0^\tau \varphi''(g(u)\,du)\right\}\right| \\ &\qquad \leqslant \mathbf{M}\exp\{-C_1\lambda^2\tau\}\left(\exp\{C_2\lambda^3\tau\} - 1\right) \leqslant C\lambda^3\mathbf{M}\tau\exp\{-C_1\lambda^2\tau + C_2\lambda^3\tau\}.\end{aligned}$$

The last expression is certainly finite for sufficiently small λ. Using integration by parts and (42), we can estimate it above by the following expression (for sufficiently small $\lambda \geqslant 0$):

$$C\lambda^3\int_0^\infty \tilde{\mathbf{P}}\{\tau > t\}\exp\{-C_1\lambda^2 t + C_2\lambda^3 t\}\,dt \leqslant C\lambda^2\int_0^\infty u^{-1/2}\exp\{-cu\}\,du.$$

This implies that the third term in (44) also has order $O(\lambda^2)$ as $\lambda \downarrow 0$. Therefore,

$$1 - \mathbf{M}\exp\left\{-\frac{\lambda^2}{2}\int_0^\tau \varphi''(g(u))\,du\right\} \sim \lambda\mathbf{M}\tilde{S}_\tau, \qquad \lambda \downarrow 0.$$

Applying here a Tauberian theorem (in the same way as in §3), we get

$$\tilde{\mathbf{P}}\left\{\int_0^T \varphi''(g(u))\,du > T\right\} \sim CT^{-1/2}, \qquad T \to \infty.$$

By the boundedness of $g(u)$ and (27), $\varphi''(g(u)) \leqslant C$. Therefore, this implies that $\tilde{\mathbf{P}}\{\tau > T\} \gg T^{-1/2}$. Lemma 3 is proved.

6) *Proof of* (25). Since $g\varphi'(g) \geqslant \varphi(g)$, if we discard the last integral in (38) and use the formula (32) for integration by parts, then we get

$$\mathbf{P}\{\tau > T\} \leqslant \mathbf{M}I\{\tau > T\}\exp\left\{-g(T)\tilde{S}_\tau + \int_0^T \tilde{S}_u\,dg(u)\right\}. \tag{45}$$

We introduce a measure $\mu(t)$ on $[0, T]$ by setting

$$\mu(dt) = g(T)\sqrt{T}\,\delta(t-T)\,dt - \sqrt{t}\,dg(t),$$

where $\delta(t)$ is the Dirac function. In our case $g(t) \geqslant 0$ and $g(t)$ does not increase for $t > t_0$; therefore, $\mu(t)$ is indeed a measure. With the measure $\mu(t)$ the expression after exp in (45) can be written as follows:

$$-g(T)\tilde{S}_T + \int_0^T \tilde{S}_u\,dg(u) = -\int_{t_0}^T \tilde{S}_u u^{-1/2}\mu(du) + g(t_0)\tilde{S}_{t_0}.$$

On the set $\{\tau > T\}$ we have $\tilde{S}_{t_0} \leqslant f(0)$ for $T > t_0$, so $g(t_0)\tilde{S}_{t_0}$ does not exceed a constant. If we now apply Jensen's inequality in (45), we have for $T > t_0$ that

$$\mathbf{P}\{\tau > T\} \geqslant C\int_{t_0}^T \mathbf{M}I\{\tau > T\}\exp\{-\tilde{S}_u u^{-1/2}\mu(T)\}\mu(du). \tag{46}$$

Note that, by (31),

$$\mu(T) = g(T)\sqrt{T} - \int_0^T u^{1/2}\,dg(u) = g(t_0)\sqrt{t_0} + \int_{t_0}^T g(u)u^{-1/2}\,du$$
$$\leqslant C_1 + C_2\int_1^T f'(t)t^{-1/2}\,dt.$$

Application of the Schwarz inequality here gives $\mu(T) \ll (\log T)^{1/2}$. Moreover, using integration by parts and the fact that under the conditions of the theorem $f'(T)T^{1/2} \to 0$ and, hence, $f(T)T^{-1/2} \to 0$ as $T \to \infty$, it is easy to establish that

$$\mu(T) \ll \int_1^T |f(t)|\,t^{-3/2}\,dt.$$

For the rest of the proof we need the following result.

LEMMA 4. *Suppose that the conditions of Lemma* 2 *hold. Then for* $0 \leqslant \gamma < -\lambda_-$ *and* $t > t_0$

$$\mathbf{M}I\{\tau > T\}\exp\{-\gamma\tilde{S}_t\} \leqslant \left(C_1\gamma + C_2\gamma^3 t + C_3 t^{-1/2}\right)e^{C\gamma^2 t}.$$

PROOF. Using the notation in (39), we have

$$\mathbf{M}I\{\tau > T\}\exp\{-\gamma\tilde{S}_t\} = \exp\{Q(-\gamma, t)\}\mathbf{M}I\{\tau > t\}Z_t(-\gamma).$$

The estimate (40) gives us that $Q(-\gamma, t) \leqslant C\gamma^2 t$. Since the process $Z_t(-\gamma)$ is a martingale for $0 \leqslant \gamma < -\lambda_-$, and $\mathbf{M}Z_t(-\gamma) = 1$, it follows that

$$\mathbf{M}I\{\tau > t\}Z_t(-\gamma) = 1 - \mathbf{M}I\{\tau > t\}Z_\tau(-\gamma). \tag{47}$$

Note that

$$Z_\tau(-\gamma) = \exp\{-2\gamma\tilde{S}_\tau + Q(\gamma,\tau) - Q(-\gamma,\tau)\}Z_\tau(\gamma), \tag{48}$$

where, as established earlier, $\mathbf{M}Z_\tau(\gamma) = 1$ for $\gamma \geqslant 0$ (see (41)).

It follows from the definition (39) that

$$Q(\gamma,t) - Q(-\gamma,t) = \int_0^t [\varphi(g(u)+\gamma) - \varphi(g(u)-\gamma) - 2\gamma\varphi'(g(u))]\,du.$$

By Taylor's formula,

$$\varphi(g(u)+\gamma) - \varphi(g(u)-\gamma) - 2\gamma\varphi'(g(u)) = [\varphi'''(g(u)+\theta\gamma) + \varphi'''(g(u)-\theta\gamma)]\gamma^3/6 \qquad (0 \leqslant \theta \leqslant 1).$$

Since $\varphi'''(\lambda)$ is a bounded function of $\lambda \in [\lambda_0, \lambda_1]$ ($\lambda_0 > \lambda_-$) this implies that on the set $\{\tau \leqslant t\}$

$$Q(\gamma,\tau) - Q(-\gamma,\tau) \geqslant -C\gamma^3 t.$$

Therefore, (47) and (48) give us the following estimate:

$$\mathbf{M}I\{\tau > t\}Z_t(-\gamma) \leqslant 1 - \exp\{-2\gamma(f(0)+L) - C\gamma^3 t\}\mathbf{M}I\{\tau > t\}Z_\tau(\gamma).$$

But $\mathbf{M}I\{\tau \leqslant t\}Z_\tau(\gamma) = 1 - \mathbf{M}I\{\tau > t\}Z_t(\gamma)$. Applying the inequality $1 - e^{-x} \leqslant x$ ($x \geqslant 0$), we now have

$$\mathbf{M}I\{\tau > t\}Z_t(-\gamma) \leqslant C_1\gamma + C_2\gamma^3 t + \mathbf{M}I\{\tau > t\}Z_t(\gamma).$$

Since $\tilde{S}_t \leqslant f(0)$ on the set $\{\tau > t\}$, (42) implies that

$$\mathbf{M}I\{\tau > t\}Z_t(\gamma) \leqslant C\tilde{\mathbf{P}}\{\tau > t\} \leqslant C_3 t^{-1/2}.$$

Collecting all these estimates, we get the desired inequality of Lemma 4.

We pass to the concluding stage of the proof. Let us break up the integral in (46) into two parts:

$$\int_{t_0}^{T}(\cdot)dt = \int_{t_0}^{T/2}(\cdot)dt + \int_{T/2}^{T}(\cdot)dt, \qquad T > 2t_0. \tag{49}$$

In the second integral we use Lemma 4 with $\gamma = t^{-1/2}\mu(T)$. (Since $t \geqslant T/2$ here while $\mu(T) \ll (\log T)^{1/2}$, γ is sufficiently small for sufficiently large T, and Lemma 4 is indeed applicable.) Then

$$\begin{aligned}\int_{T/2}^{T}(\cdot)dt &\leqslant C\int_{T/2}^{T}\mathbf{M}I\{\tau > t\}\exp\{-\tilde{S}_t t^{-1/2}\mu(T)\}\mu(dt)\\ &\leqslant (C_1\mu(T) + C_2\mu^3(T) + C_3)\exp\{C\mu^2(T)\}T^{-1/2}\int_{T/2}^{T}\mu(dt)\\ &\leqslant C\exp\left\{C\left(\int_1^T f(t)t^{-3/2}\,dt\right)^2\right\}T^{-1/2}.\end{aligned} \tag{50}$$

The first integral in (49) is a little more complicated to estimate. We first note that

$$I\{\tau > T\} = I\{\tau > t\}I\{\tilde{S}_u - \tilde{S}_t \leqslant f(0) - \tilde{S}_t, t \leqslant u \leqslant T\}. \tag{51}$$

The following estimate holds for the second factor on the set $\{\tau > t\}$ (see (43)):

$$\tilde{\mathbf{P}}\{\tilde{S}_u - \tilde{S}_t \leqslant f(0) - \tilde{S}_t, t \leqslant u \leqslant T \mid \mathfrak{F}_t\} \leqslant C(f(0) + L - \tilde{S}_t)(T-t)^{-1/2}. \tag{52}$$

Therefore, the first integral in (49) can be estimated as follows:

$$\int_{t_0}^{T/2}(\cdot)\,dt \leqslant CT^{-1/2}\int_{t_0}^{T/2}\mathbf{M}I\{\tau>t\}\exp\{-\tilde{S}_t t^{-1/2}\mu(t)\}(f(0)+L-\tilde{S}_t)\mu(dt)$$
$$\leqslant CT^{-1/2}\int_{t_0}^{T}\mathbf{M}I\{\tau>t\}\exp\{-\tilde{S}_t t^{-1/2}(\mu(T)+1)\}t^{1/2}\mu(dt).$$

Let us now apply Lemma 4 with $\gamma=(\mu(T)+1)t^{-1/2}$. (By choosing the number $t_0=t_0(T)$ sufficiently large we can make γ sufficiently small, and Lemma 4 really is applicable.) In the same way as in (50) it then follows that the first integral in (49) does not exceed

$$C\exp\left\{C\left(\int_1^T f(t)t^{-3/2}\,dt\right)^2\right\}T^{-1/2}.$$

Combining this with (50), we arrive at the desired inequality.

7) *Proof of* (26). Since

$$\lambda\varphi'(\lambda)-\varphi(\lambda)=\int_0^\lambda u\varphi''(u)\,du\leqslant C\lambda^2$$

for $\lambda\geqslant\lambda_0>\lambda_-$ (see (28)), the estimate (31) gives us that

$$\int_0^T[g(u)\varphi'(g(u))-\varphi(g(u))]\,du\leqslant C\int_0^T g^2(u)\,du\leqslant C\int_0^T(f'(t))^2\,dt\leqslant C.$$

Therefore, replacing the last integral in (38) by constant and applying Jensen's inequality, we get

$$\mathbf{P}\{\tau>T\}\geqslant C\tilde{\mathbf{P}}\{\tau>T\}\exp\left\{-\tilde{\mathbf{P}}^{-1}\{\tau>T\}\mathbf{M}I\{\tau>T\}\int_0^T g(u)\,d\tilde{S}_u\right\}.$$

By Lemma 3, to obtain the desired estimate it suffices to show that

$$\mathbf{M}I\{\tau>T\}\int_0^T g(u)\,d\tilde{S}_u\leqslant T^{-1/2}\left(C_1+C_2\int_1^T|f(t)|t^{-3/2}\,dt\right). \tag{53}$$

Integration by parts yields

$$\begin{aligned}\mathbf{M}I\{\tau>T\}\int_0^T g(u)\,d\tilde{S}_u=\mathbf{M}I\{\tau>T\}g(t_0)(f(0)-\tilde{S}_{t_0})&\\ -g(T)f(0)\tilde{\mathbf{P}}\{\tau>T\}+g(T)\mathbf{M}I\{\tau>T\}\tilde{S}_T&\\ +\int_{t_0}^T\mathbf{M}I\{\tau>T\}(f(0)-\tilde{S}_t)\,dg(t).&\end{aligned} \tag{54}$$

Since $g(t)\leqslant 0$ in our case, the first term in (54) is nonpositive. The second term has order $o(T^{-1/2})$, $T\to\infty$, by virtue of (31) and (42). To estimate the third term we use the identity $\mathbf{M}\tilde{S}_{t\wedge\tau}=0$, which implies that

$$|\mathbf{M}I\{\tau>T\}\tilde{S}_T|=|\mathbf{M}I\{\tau\leqslant T\}\tilde{S}_\tau|\leqslant C. \tag{55}$$

Therefore, by (31) and the estimate $f'(T)=o(T^{-1/2})$, $T\to\infty$, the third term in (54) has the required order.

To estimate the last term in (54) we break up the integral into two parts (as in (49)). In the first integral we again use (51) and (52). Then

$$\int_{t_0}^{T/2}(\cdot)\,dt = \int_{t_0}^{T/2}\mathbf{M}I\{\tau > T\}\big(f(0) - \tilde{S}_t\big)dg(t)$$
$$\leqslant CT^{-1/2}\int_{t_0}^{T/2}\mathbf{M}I\{\tau > t\}\big(f(0) + L - \tilde{S}_t\big)^2\,dg(t).$$

Here the inequality $(a+b)^2 \leqslant 2(a^2+b^2)$ gives us that

$$\mathbf{M}I\{\tau > t\}\big(f(0) + L - \tilde{S}_t\big)^2 \leqslant C^2 + 2\mathbf{M}\tilde{S}^2_{t\wedge\tau}.$$

Since $\tilde{S}_t$ is a square-integrable martingale and (see (36))

$$\mathbf{M}\tilde{S}_t^2 = \int_0^t \varphi''(g(u))\,du,$$

the familiar martingale identity implies that

$$\mathbf{M}\tilde{S}^2_{t\wedge\tau} = \mathbf{M}\int_0^{t\wedge\tau}\varphi''(g(u))\,du$$

and, by (27) and (42),

$$\mathbf{M}\int_0^{t\wedge\tau}\varphi''(g(u))\,du \leqslant C\mathbf{M}t\wedge\tau \leqslant C\int_0^t\tilde{\mathbf{P}}\{\tau > u\}\,du \leqslant Ct^{1/2}.$$

Combining our estimates, we have

$$\int_{t_0}^{T/2}(\cdot)\,dt \leqslant CT^{-1/2}\int_{t_0}^{T/2}t^{1/2}\,dg(t) \leqslant T^{-1/2}\Big(C_1 + C_2\int_1^T|f(t)|t^{-3/2}\,dt\Big).$$

The second integral can be estimated as follows by virtue of (55):

$$\int_{T/2}^{T}(\cdot)\,dt \leqslant \int_{T/2}^{T}\mathbf{M}I\{\tau > t\}\big(f(0) - \tilde{S}_t\big)\,dg(t) \leqslant C\int_{T/2}^{T}dg(t)$$
$$\leqslant T^{-1/2}\Big(C_1 + C_2\int_1^T|f(t)|t^{-3/2}\,dt\Big).$$

As a result we get (53) and, hence, also the desired estimate (26).

4. Proof of Theorem 3. Suppose that condition 1°) holds. Then obviously $\mathbf{E}S_\tau > 0$. Since $f(t)$ is concave and bounded (for finite t), it is absolutely continuous, and, moreover, the derivative $f'(t)$ is a nonnegative function that decreases to zero, and there is a number t_0 such that $f'(t_0) < \varphi'(+\infty)$. Instead of the boundary $f(t)$ we now consider the boundary $\bar{f}(t) = \max(f(t_0), f(t))$ and let $\bar{\tau}$ be the time at which this boundary is reached. Obviously, $\mathbf{P}\{\tau > T\} \leqslant \mathbf{P}\{\bar{\tau} > T\}$. On the other hand,

$$\int_1^\infty \frac{|f(t)|}{t^{3/2}}dt < \infty \Rightarrow \int_0^\infty |f'(t)|^2\,dt < \infty, \tag{56}$$

and, hence, according to estimate (25) in Theorem 5, $\mathbf{P}\{\bar{\tau} > T\} \ll T^{-1/2}$. From this and Theorem 2 it follows that $\mathbf{E}S_\tau < \infty$. The converse, namely

$$\mathbf{E}S_\tau < \infty \Rightarrow \int_1^\infty \frac{f(t)}{t^{3/2}}dt < \infty,$$

was established under more general conditions in Remark 2 of §3.

Suppose now that condition 2°) holds. The implication (56) again is valid, and, therefore, the estimate (26) in Theorem 5 gives us that

$$\mathbf{P}\{\tau > T\} \ll T^{-1/2}.$$

From this and Theorem 2 it now follows that $\mathbf{E}S_\tau > 0$. The converse, namely

$$\mathbf{E}S_\tau > 0 \Rightarrow \int_1^\infty \frac{|f(t)|}{t^{3/2}}\,dt < \infty,$$

was established earlier under more general conditions in Remark 2 of §3.

5. Remark. For the case of a Wiener process Uchiyama [15] obtained lower and upper estimates for $\mathbf{P}\{\tau > T\}$ that have the same order of decrease also in the case where the integral $\int_1^\infty |f(t)|t^{-3/2}\,dt$ diverges. Unfortunately, Uchiyama's method, which is based on comparison theorems for conditional Gaussian distributions has eluded our attempts to carry it over to the general case.

In conclusion the author expresses gratitude to A. N. Shiryaev for his attention to this work and for support, and to M. V. Burnashev, O. V. Viskov and V. I. Rotar′ for a useful discussion of the results.

Appendix

Here we derive some properties of the functions $H(\nu, \alpha, x)$ and $\nu_\alpha(x)$ needed for the proof of Theorem 1.

1. Analytic continuation of $H(\nu, \alpha, x)$. It follows from the integral representation (1) that $H(\nu, \alpha, x)$ is analytic in the domain $\operatorname{Re}\nu < 0$ ($1 < \alpha \leqslant 2$, $-\infty < x < \infty$). Integrating by parts in (1) and using the fact that $\Gamma(z+1) = z\Gamma(z)$, we get that

$$H(\nu, \alpha, x) = -\frac{1}{\Gamma(1-\alpha\nu)} \int_0^\infty y^{-\alpha\nu} \frac{d}{dy} \exp\left(xy - \frac{y^\alpha}{\alpha}\right) dy \qquad (\operatorname{Re}\nu < 0).$$

From this it follows, in particular, that $\lim_{\nu\uparrow 0} H(\nu, \alpha, x) = 1$. Moreover, if we differentiate on the right-hand side and again use the notation of (1), then we arrive at the following functional relation:

$$H(\nu, \alpha, x) = -xH\left(\nu - \frac{1}{\alpha}, \alpha, x\right) + \frac{\Gamma(\alpha - \alpha\nu)}{\Gamma(1-\alpha\nu)} H(\nu - 1, \alpha, x). \qquad \text{(A.1)}$$

This relation can be taken as a basis for the definition of $H(\nu, \alpha, x)$ when $\operatorname{Re}\nu \geqslant 0$. Since the right-hand side of (A.1) is analytic for $\operatorname{Re}\nu < 1$ (recall that $\Gamma^{-1}(z)$ is analytic for all z), $H(\nu, \alpha, x)$ is also analytic in this domain.

Next, it is clear that all the singularities of $H(\nu, \alpha, x)$ for $\operatorname{Re}\nu \geqslant 1$ are determined by the singularities of $\Gamma(\alpha - \alpha\nu)$. But since the singularities of this function are simple poles on the real axis, $H(\nu, \alpha, x)$ can have only simple poles on the real axis as singular points, and the first singularity is a pole of first order at $\nu = 1$ when $1 < \alpha < 2$ (for $\alpha = 2$ this is a removable singularity). We do not need the other poles here.

We require one more functional relation in what follows: for $\operatorname{Re}\nu < 1$ and $1 < \alpha < 2$ or $\alpha = 2$

$$\partial H(\nu, \alpha, x)/\partial x = -\alpha\nu H(\nu - 1/\alpha, \alpha, x). \qquad \text{(A.2)}$$

In the domain $\operatorname{Re}\nu < 0$ this relation follows immediately from (1), and it then carries over to the domain $\operatorname{Re}\nu < 1$ by the uniqueness of the analytic continuation.

2. The asymptotic behavior of $H(\nu, \alpha, x)$. For $x \to -\infty$ and $\operatorname{Re}\nu < 1$

$$H(\nu, \alpha, x) \sim |x|^{\alpha\nu}. \tag{A.3}$$

This asymptotic behavior follows for $\operatorname{Re}\nu < 0$ from (1) if we make the change of variables $xy = -z$. In the case $0 \leqslant \operatorname{Re}\nu < 1$ it is necessary to use also (A.1).

For $x \to +\infty$ and $\operatorname{Re}\nu < 0$ application of the Laplace method [21] to the integral (1) gives the asymptotic behavior

$$H(\nu, \alpha, x) \sim \left(\frac{2\pi}{\alpha - 1}\right)^{1/2} \frac{1}{\Gamma(-\alpha\nu)} \exp\left\{\frac{\alpha - 1}{\alpha} x^{\alpha/(\alpha-1)}\right\} \left(x^{\alpha/(\alpha-1)}\right)^{-\nu-1/2}. \tag{A.4}$$

3. The real zeros of $H(\nu, \alpha, x)$. By analyticity,

$$H(0, \alpha, x) = \lim_{\nu\uparrow 0} H(\nu, \alpha, x) = 1$$

(the last equality was established above). If we now set $\nu = 1/\alpha$ in (A.1), we get that $H(1/\alpha, \alpha, x) = -x$. Therefore, for $x > 0$ the function $H(\nu, \alpha, x)$ has different signs at the endpoints of the interval $[0, 1/\alpha]$, and, hence, $0 < \nu_\alpha(x) < 1/\alpha$.

For $x = 0$ the function under consideration reduces to the gamma function

$$H(\nu, \alpha, 0) = \alpha^{-\nu-1} \frac{\Gamma(-\nu)}{\Gamma(-\alpha\nu)}.$$

This follows from (1) for $\operatorname{Re}\nu < 0$, and for $\operatorname{Re}\nu \geqslant 0$ we can take it as a definition. It follows from the properties of $\Gamma(z)$ that $\nu_\alpha(0) = 1/\alpha$ is the only zero of $H(\nu, \alpha, 0)$ in the domain $\operatorname{Re}\nu < 1/\alpha$.

From (A.1) it is not hard to derive that $H(\nu, \alpha, x)$ takes positive values for $\nu \in [0, 1/\alpha]$ when $x < 0$. But since $\lim_{\nu\uparrow 1} H(\nu, \alpha, x) = \infty$ when $1 < \alpha < 2$ and $x < 0$, it certainly follows that $1/\alpha < \nu_\alpha(x) < 1$. Moreover, it is not hard to see that

$$\lim_{x\to-\infty} \nu_\alpha(x) = 1 \qquad (1 < \alpha < 2). \tag{A.5}$$

Indeed, (A.1) and (A.3) give us that for fixed ν and $x \to -\infty$

$$H(\nu, \alpha, x) \sim |x|^{\alpha\nu-1}\left(|x| - \frac{\Gamma(\alpha - \alpha\nu)}{\Gamma(1 - \alpha\nu)}\right), \qquad \frac{1}{\alpha} \leqslant \nu < 1.$$

From this it is clear that $H(\nu, \alpha, x)$ can vanish only for ν close to 1 when $-x$ is large, and the larger $-x$ is, the closer ν must be to 1.

The monotonicity of $\nu_\alpha(x)$ can be proved by the following arguments. From (A.2) it follows that for any $x \geqslant y$ and $\nu < 1$

$$H(\nu, \alpha, x) = H(\nu, \alpha, y) - \alpha\nu\int_y^x H\left(\nu - \frac{1}{\alpha}, \alpha, u\right) du. \tag{A.6}$$

We consider this relation for $0 \leqslant \nu < \nu_\alpha(x)$, and first let $x \geqslant y \geqslant 0$. Then $\nu_\alpha(x) \leqslant 1/\alpha$ and, certainly, $H(\nu - 1/\alpha, \alpha, u) > 0$ (see (1)). Therefore, $H(\nu, \alpha, y) \geqslant H(\nu, \alpha, x) > 0$ (the last inequality follows from the definition of $\nu_\alpha(x)$). This implies that $\nu_\alpha(y) \geqslant \nu_\alpha(x)$.

Suppose now that $0 \geqslant x \geqslant y$ and again consider (A.6) for $0 \leqslant \nu < \nu_\alpha(x)$. As we already know, $1/\alpha \leqslant \nu_\alpha(x) < 1$ for $1 < \alpha < 2$ (we consider only this case for the time being), and hence $H(\nu - 1/\alpha, \alpha, u) > 0$ (this was established above for $\nu \leqslant 2/\alpha$ and $u \leqslant 0$). Therefore, again $H(\nu, \alpha, y) \geqslant H(\nu, \alpha, x) > 0$ and, hence, $\nu_\alpha(y) \geqslant \nu_\alpha(x)$.

For $\alpha = 2$ the zeros of $H(\nu, \alpha, x)$ coincide with the zeros of the function $D_{-2\nu}(-x)$, which satisfies the familiar differential equation of Weber [19]. Therefore, the monotonicity of $\nu_2(x)$ can be established with the help of a comparison theorem of Sturm.

The continuity of $\nu_\alpha(x)$ can be reduced from the monotonicity of $\nu_\alpha(x)$ and the continuity of $H(\nu, \alpha, x)$ with the help of (A.6).

We note two asymptotic relations for $\nu_\alpha(a)$ that can be obtained by analyzing (A.1):

$$\nu_\alpha(x) = \frac{1}{\alpha} - \alpha^{(1-\alpha)/\alpha} \frac{\Gamma(\alpha - 1)}{\Gamma(1 - 1/\alpha)} x + o(x), \qquad x \to 0, \tag{A.7}$$

$$\nu_\alpha(x) \sim \left(\frac{\alpha - 1}{2\pi}\right)^{1/2} x^{\alpha/2(\alpha-1)} \exp\left\{-\frac{\alpha - 1}{\alpha} x^{\alpha/(\alpha-1)}\right\}, \qquad x \to +\infty.$$

The first of these can be obtained from (A.1) by setting $\nu = \nu_\alpha(x) + 1/\alpha$ and using the continuity of $\nu_\alpha(x)$. To derive the second one it is necessary to set $\nu = \nu_\alpha(x) + 1/\alpha$ in (A.1) and to use (A.4). Both asymptotic relations were obtained in [15] for $\alpha = 2$.

BIBLIOGRAPHY

1. A. A. Borovkov, *Stochastic processes in queueing theory*, "Nauka", Moscow, 1972; English transl., Springer-Verlag, 1976.

2. V. S. Korolyuk, *Boundary problems for compound Poisson processes*, "Naukova Dumka", Kiev, 1975. (Russian)

3. A. V. Skorokhod, *Random processes with independent increments*, "Nauka", Moscow, 1964; English transls., Clearing House Federal Sci. Tech. Information, Springfield, Va., 1966; and *Theory of random processes*, Nat. Lending Library Sci. Technol., Boston Spa, Yorkshire, England, 1971.

4. Lajos Takàcs, *Combinatorial methods in the theory of stochastic processes*, Wiley, 1967.

5. A. A. Borovkov, *Boundary problems for random walks and large deviations in function spaces*, Teor. Veroyatnost. i Primenen. **12** (1967), 635–654; English transl. in Theor. Probab. Appl. **12** (1967).

6. V. I. Rotar', *Moments of the value and time of the first passage over a curvilinear boundary*, Teor. Veroyatnost. i Primemen. **12** (1967), 755–762; English transl. in Theor. Probab. Appl. **12** (1967).

7. Stephen Portnoy, *Probability bounds for first exits through moving boundaries*, Ann. Probab. **6** (1978), 106–117.

8. T. L. Lai and R.A. Wijsman, *First exit time of a random walk from the bounds $f(n) \pm cg(n)$, with applications*, Ann. Probab. **7** (1979), 672–692.

9. A. A. Borovkov, *Probability theory*, 2nd. rev. ed., "Nauka", Moscow, 1976; German transl., Birkhäuser, 1976.

10. A. A. Mogul'skiĭ and E. A. Pecherskiĭ, *On the first exit time out of a semigroup in R^m for a random walk*, Teor. Veroyatnost. i Primenen. **22** (1977), 837–844; English transl. in Theor. Probab. Appl. **22** (1977).

11. ______, *On the time of the first hit in a domain with a curved boundary*, Sibirsk. Mat. Zh. **19** (1978), 824–841; English transl. in Siberian Math. J. **19** (1978).

12. L. A. Shepp, *A first passage problem for the Wiener process*, Ann. Math. Statist. **38** (1967), 1912–1914.

13. Herbert Robbins and David Siegmund, *Boundary crossing probabilities for the Wiener process and sample sums*, Ann. Math. Statist. **41** (1970), 1410–1429.

14. A. A. Novikov, *The stopping times of a Wiener process*, Teor. Veroyatnost. i Primenen. **16** (1971), 458–465; English transl. in Theor. Probab. Appl. **16** (1971).

15. Kōhei Uchiyama, *Brownian first exit from and sojourn over one-sided moving boundary and application*, Z. Wahrsch. Verw. Gebiete **54** (1980), 75–116.

16. Jürgen Gärtner, *Upper and lower bounds for brownian first exit densities and propagation of wave fronts*, Preprint, Zentralinst. Math. Mech. Akad. Wiss. DDR, Berlin, 1980. Zbl. **443** #60079.

17. A. A. Novikov, *On estimates and the asymptotic behavior of the probability of nonintersection of moving boundaries by sums of independent random variables*, Izv. Akad. Nauk SSSR Ser. Mat. **44** (1980), 868–885; English transl. in Math. USSR Izv. **17** (1981).

18. I. S. Gradshteĭn and I. M. Ryzhik, *Tables of integrals, series and products*, 5th ed., "Nauka", Moscow, 1971; English transl. of 4th ed., Academic Press, 1965.

19. Milton Abramowitz and Irene A. Stegun (editors), *Handbook of mathematical functions, with formulas, graphs, and mathematical tables*, U. S. Govt. Printing Office, Washington, D. C., 1964; reprint, Dover, New York, 1966.

20. A. Erdélyi et al., *Higher transcendental functions*, Vols. 1–3, McGraw-Hill, 1953, 1955.

21. F. W. J. Olver, *Introduction to asymptotics and special functions*, Academic Press, 1974.*

22. A. A. Novikov, *On estimates and the asymptotic behavior of nonexit probabilities of a Wiener process to a moving boundary*, Mat. Sb. **110 (152)** (1979), 539–550; English transl. in Math. USSR Sb. **38** (1981).

23. Shunsuke Sato, *Evaluation of the first-passage time probability to a square root boundary for the Wiener process*, J. Appl. Probab. **14** (1977), 850–856.

24. V. Matskyavichus [Mackevičius], *On some problems of the optimal stopping of stable stochastic processes*, Litovsk. Mat. Sb. **12** (1972), no. 1, 173–180; English transl. in Selected Transl. Math. Statist. and Probab., Vol. 14, Amer. Math. Soc., Providence, R. I., 1977.

25. William Feller, *An introduction to probability theory and its applications*, Vol. II, Wiley, 1966.

26. Jean Jacod, *Calcul stochastique et problèmes de martingales*, Lecture Notes in Math., vol. 714, Springer-Verlag, 1979.

27. R. Sh. Liptser and A. N. Shiryaev, *Statistics of random processes*, "Nauka", Moscow, 1974; English transl., Vols. I, II, Springer-Verlag, 1977.

28. A. A. Novikov, *A martingale approach to first passage problems and a new condition for Wald's identity*, Stochastic Differential Systems (Proc. Third IFIP WG7/1 Working Conf., Visegrád, Hungary, 1980), Lecture Notes in Control and Information Sci., Vol. 36, Springer-Verlag, 1981, pp. 146–156.

Translated by H. H. McFADEN

**Editor's note*. This book consists of the first seven chapters of his book *Asymptotics and special functions*, Academic Press, 1974.

Труды Матем. Инст. Стеклов.
158(1981)

Proceedings of the Steklov
Institute of Mathematics
1983, Issue 4

ON NEW METHODS OF STUDYING THE ASYMPTOTIC PROPERTIES OF PARABOLIC EQUATIONS

UDC 517.9

A. A. SAMARSKIĬ

ABSTRACT. This is a survey of papers by the author and his coworkers concerned with the study of unbounded solutions of linear and quasilinear parabolic equations arising in connection with the problem of localization of diffusion processes. Its main attention is directed toward the problem of the asymptotic properties of parabolic equations. It is shown, in particular, in specific cases, that the solutions of parabolic equations may be described approximately by self-similar solutions of certain first order nonlinear equations.

Bibliography: 32 titles.

§1. Introduction

1. In recent years there has been considerable interest in the study of unbounded (not existing "in the large") solutions of parabolic equations. In papers by the author and his coworkers (see [1]-[15], [18], [21], [23] and [29]) several unusual properties were observed (properties, for example, such as the localization (inertia) of heat) relating to the diffusion of heat in a continuous medium in connection with the appearance in this medium of regimes with peaking. By a *regime with peaking* we usually mean a process some characteristic parameters of which increase without bound over a finite time (peaking time). Regimes with peaking arise in various, essentially nonstationary, physical processes (see [1], [2], [10] and [23]).

The effect of heat localization was observed in studying the quasilinear heat equation

$$\frac{\partial u}{\partial t}=\frac{\partial}{\partial x}\left(k(u)\frac{\partial u}{\partial x}\right),\quad k(u)=u^\sigma,\sigma>0,u>0;\quad u(0,x)=0,\quad x\in R^1_+, \quad (1)$$

with the boundary regime with peaking

$$u(t,0)=(T-t)^n,\qquad 0<t<T, n=\text{const}<0. \qquad (1')$$

Heat localization means that the temperature as $t\to T^-$ increases without bound in a finite domain (heat localization domain) and in the remaining portion of space it is bounded uniformly with respect to t.

1980 *Mathematics Subject Classification*. Primary 35K05, 35K55, 35K60, 35B40, 35-02; Secondary 80A20.

In [7]-[12] a study was made of the effect of heat localization for the Cauchy problem

$$\frac{\partial u}{\partial t} = \frac{\partial}{\partial x}\left(k(u)\frac{\partial u}{\partial x}\right) + Q(u), \qquad t > 0, x \in R^1,$$

$$k(u) = u^\sigma, \quad Q(u) = u^\beta, \quad \sigma > 0, \qquad \beta > 1; \qquad u(0, x) = u_0(x), \quad x \in R^1. \quad (2)$$

The effect of heat and combustion localization ("inertia") makes it possible, in particular, to concentrate an arbitrary amount of energy in a bounded volume and to retain it over a finite time with practically no propagation from the localization zone; this situation can have a number of important physical applications [10], [23]. Localization is also an internal source of the complex interaction in a nonlinear medium of various kinds of nonstationary dissipative structures (see [7]–[9] and [12]).

2. In studying heat localization processes for the equations (1) and (2) with the power-law dependencies $k(u) = u^\sigma$ and $Q(u) = u^\beta$, self-similar solutions were employed (see [1], [4]–[6], [7]–[9] and [13]). With the aid of these self-similar solutions lower and upper estimates were obtained for the non-self-similar solutions of (1) and (2). Asymptotic properties of these solutions were also studied.

Self-similar solutions, as we know [16], [17], exist only for $k(u) = u^\sigma$ and $k(u) = \exp u$. In [18]–[20] a mathematical tool of comparison theorems was developed, making it possible to carry out a study of regimes with peaking for the equations (1) and (2) with $k(u)$ and $Q(u)$ not of power-law form. Traditional comparison theorems for parabolic equations establish a continuous dependence of the solution of an equation on its boundary conditions. A method was developed in [18] allowing the comparison of solutions of problems corresponding not only to different boundary conditions but also to different parabolic equations (in the case of (1) and (2), to different $k(u)$ and $Q(u)$). With the aid of generalized comparision theorems, for example, heat localization was established for equation (1) with an arbitrary function $k(u)$, and corresponding estimates were given in [21].

3. In this paper, which contains the results from a number of papers of the author and his coworkers, we consider two problems. The first of these is connected with the localization of the heat diffusion process described by the linear heat equation

$$\partial u/\partial t = \partial^2 u/\partial x^2, \qquad 0 < t < T, x \in R^1_+ = (0, +\infty), \tag{3}$$

under the assumption that the boundary regime with peaking is given at $x = 0$. The second problem corresponds to a combustion process under the action of a volumetric heat liberation nonlinearly dependent on the temperature:

$$\partial u/\partial t = \partial^2 u/\partial x^2 + Q(u), \qquad t > 0, x \in R^1 = (-\infty, +\infty) \tag{4}$$

(semilinear parabolic equation).

We consider here a source term $Q(u)$ having the special form

$$Q(u) = (1 + u)\ln^\beta(1 + u), \qquad \beta > 1. \tag{5}$$

The equations (1) and (2) have been studied by many authors, usually from different points of view. We remark here that the semilinear equation (4) with the source term $Q(u) \geqslant 0$ bounded from above and satisfying the conditions $Q(0) = Q(1) = 0$ and $Q(u) \equiv 0$ for $u > 1$ was investigated in [25] and also in a large number of subsequent papers devoted to the study of travelling waves in biological, chemical and other media. In contrast to this group of papers there is, in our papers [7]–[12], [14], [15] and [17], as well as in the present paper, no restriction on the growth of $Q(u)$; this leads, in a number of cases, to the appearance of unbounded solutions (regimes with peaking).

Conditions for heat localization have been found for both of these problems. However, our main interest here is in the methods for studying the asymptotic properties of the solutions of parabolic equations in the presence of regime with peaking.

4. In this paper, using these two problems by way of example, we present a new method for studying the asymptotic properties (including the localization effect) of solutions of parabolic equations. We show that if an equation has no suitable self-similar solutions, the solution $u(t, x)$ of the problem in question may converge asymptotically to self-similar solutions $v_a(t, x)$ of altogether different equations. We speak of such solutions $v_a(t, x)$ as *approximate self-similar solutions* (a.s.-s.s.).

In §2 we establish the fact that a solution of the boundary value problem (3) with a boundary regime with peaking of exponential type, possessing the localization property, converges in a special norm to an a.s.-s.s. satisfying a first order equation of Hamilton-Jacobi type:

$$\frac{\partial v}{\partial t} = \frac{1}{1+v}\left(\frac{\partial v}{\partial x}\right)^2, \qquad 0 < t < T, x \in R^1_+ .$$

We obtain an estimate of the rate of convergence in $C(R^1_+)$.

In §3 we study the semilinear equation (4) with the source term (5). We show that the space-time structure of unbounded (existing over a finite time) solutions of the Cauchy problem for the equation (4)–(5) is determined by the properties of a.s.-s.s. satisfying the first order nonlinear equation

$$\frac{\partial v}{\partial t} = \frac{1}{1+v}\left(\frac{\partial v}{\partial x}\right)^2 + (1+v)\ln^{\beta}(1+v), \qquad t > 0, x \in R^1.$$

The a.s.-s.s. of these equations constructed in §§2 and 3 furnish additional information concerning the asymptotic properties of the processes studied.

§2. Localization and asymptotic properties of boundary regimes with peaking for the linear heat equation

1. In this section we examine the problem concerning the effect of a boundary regime with peaking on the propagation of heat in a medium with a constant coefficient of thermal conductivity:

$$\partial u/\partial t = \partial^2 u/\partial x^2, \qquad (t, x) \in S_T = (0, T) \times R^1_+ , \tag{1}$$

$$u(t, 0) = \varphi(t) \xrightarrow[t \to T^-]{} +\infty, \quad 0 < t < T, \qquad u(0, x) = 0, \quad x \in R^1_+ . \tag{2}$$

We say that in the problem (1), (2) there exists an (effective) *localization of heat* if the temperature $u(t, x)$ increases without bound as $t \to T^-$ on a set of finite measure, i.e.,

$$\operatorname{meas}\{x\colon u(T^-, x) = +\infty\} = l < +\infty. \tag{3}$$

We call the set $0 \leqslant x \leqslant l$ on which $u(t, x) \to +\infty$ as $t \to T^-$ (with the exception, possibly, of the point $x = l$) the *localization domain* and the quantity l the *localization depth*.

But if $\operatorname{meas}\{x: u(T^-, x) = +\infty\} = +\infty$, i.e., if the temperature becomes infinite for all $x \in R^1_+$ as $t \to T^-$, then heat localization is not present.

2. The solution of problem (1), (2) is given by the following formula (heat potential of a double layer) [24], [28]:

$$u(t, x) = \frac{x}{2\sqrt{\pi}} \int_0^t \exp\left[-\frac{x^2}{4(t-\tau)}\right](t-\tau)^{3/2}\varphi(\tau)\, d\tau, \qquad (t, x) \in S_T. \quad (4)$$

Letting $t \to T^-$ in (4), we obtain

$$u(T^-, x) = \frac{x}{2\sqrt{\pi}} \int_0^T \exp\left[-\frac{x^2}{4(T-\tau)}\right](T-\tau)^{-3/2}\varphi(\tau)\, d\tau, \qquad x \in R^1_+. \quad (5)$$

It follows directly from this that the regimes with localization are to be sought in the one-parameter family of exponential boundary functions

$$\varphi(t) = \exp((T-t)^n), \qquad 0 < t < T, n = \text{const} < 0, \quad (6)$$

depending on the parameter n.

Substituting (6) into (5), we obtain

$$u(T^-, x) = \frac{x}{2\sqrt{\pi}} \int_0^T \exp\left\{(T-\tau)^{-1}\left[(T-\tau)^{n+1} - \frac{x^2}{4}\right]\right\} \frac{d\tau}{(T-\tau)^{3/2}}. \quad (7)$$

From this formula we obtain the following classification of exponential regimes.[1]

a) If $n = -1$, we have heat localization in an S-regime with depth $l = 2$. Then (7) takes the form

$$u(T^-, x) = \frac{x}{2\sqrt{\pi}} \int_0^T \exp\left\{\frac{4-x^2}{4(T-\tau)}\right\} \frac{d\tau}{(T-\tau)}^{3/2}, \qquad x \in R^1_+.$$

In the localization domain $0 \leqslant x \leqslant l = 2$ the temperature $u(t, x)$ increases without bound as $t \to T^-$, while outside this domain it is bounded from above uniformly in t by the limiting curve

$$u(T^-, x) = \frac{1}{\sqrt{\pi}}\left[1 - \left(\frac{2}{x}\right)^2\right]^{-1/2} \int_s^\infty \exp(-\eta)\eta^{-1/2}\, d\eta, \quad x > 2, s = \frac{x^2-4}{4T}. \quad (8)$$

b) If $-1 < n < 0$, the condition (3) is satisfied for $l = 0$ (LS-regime); the temperature becomes infinite only at the point $x = 0$ while for $x > 0$ is it everywhere bounded from above, uniformly in t:

$$u(T^-, x) = \frac{1}{\sqrt{\pi}} \int_s^\infty \exp\left(-\eta + \frac{x^{2n}}{4^n}\eta^{-n}\right)\eta^{-1/2}\, d\eta, \qquad s = \frac{x^2}{4T}. \quad (9)$$

c) If $n < -1$, we see from (7) that $u(T^-, x) = +\infty$ for any $x > 0$ (HS-regime), i.e., heat localization is not present.

---

[1]A more general two-parameter family of exponential regimes with peaking was considered in [13] and [29].

Using a multidimensional analog of the potential (4), we can make a study of a multidimensional heat problem and clarify the possible forms of localization domains. In particular, as the boundary of a localization domain we can have an arbitrary second order surface. The principle of superposition makes it possible to combine localization domains corresponding to different boundary regimes and to obtain, for example, domains with nonsmooth boundaries.

3. The formula (7) gives us only limit ($t = T^-$) values of the temperature. Of interest here are the asymptotic (as $t \to T^-$) distributions of the temperature and, in particular, the time variation of the half-width $x_{\mathrm{eff}}(t)$ of the thermal wave penetration; the quantity $x_{\mathrm{eff}}(t)$ is determined by the condition

$$u(t, x_{\mathrm{eff}}(t)) = \tfrac{1}{2}u(t,0). \tag{10}$$

Equation (1) with the condition (6) has no self-similar solutions (it admits self-similar solutions only in the case of power-law regimes with peaking ($u(t,0) = (T-t)^n$, $n = \mathrm{const} < 0$); see [13]).

However, for our problem with an exponential regime, we are able to construct an a.s.-s.s. of (1). It is convenient to replace (6) by the condition

$$u(t,0) = \exp\left[(T-t)^n\right] - 1, \qquad 0 < t < T. \tag{11}$$

To (1) we make correspond a first order nonlinear equation of Hamilton-Jacobi type,

$$\partial v/\partial t = 1/(1+v)(\partial v/\partial x)^2, \qquad (t,x) \in S_T. \tag{12}$$

It has the self-similar solution

$$v_a(t,x) = \exp\left[(T-t)^n \theta_a(\xi)\right] - 1, \qquad \xi = x(T-t)^{-(1+n)/2}, \tag{13}$$

where $\theta_a(\xi) \geqslant 0$ is a solution of the boundary value problem

$$(\theta_a')^2 - (1+n)\xi\theta_a'/2 + n\theta_a = 0, \qquad \xi \in R_+^1,\ \theta_a(0) = 1,\ \theta_a(+\infty) = 0. \tag{14}$$

The boundary condition (11) for $v_a(t,x)$ is satisfied by virtue of the condition $\theta_a(0) = 1$.

The equation (14) is integrable in quadratures and the function $\theta_a(\xi)$, at those points where it is positive, is determined from the relations (at the remaining points $\theta_a = 0$)

$$\begin{gathered}[X - (1+n)/4]^{(1+n)/2}[X + (1-n)/4]^{(1-n)/2} = (-n)^{1/2}/\xi, \\ X = \left[((1+n)/4)^2 - n\theta_a(\xi)/\xi^2\right]^{1/2}.\end{gathered} \tag{15}$$

In the S-regime case, i.e., when $n = -1$, problem (14) has the simple solution

$$\theta_a(\xi) = (1-\xi/2)^2 \quad \text{for } 0 \leqslant \xi \leqslant 2, \qquad \theta_a(\xi) = 0 \quad \text{for } \xi > 2, \tag{16}$$

where $\xi \equiv x$.

A detailed study of the properties of the solution of (14) is given in [13]. Below we require only the estimate

$$\sup_{\xi \in R_+^1} |\theta_a''(\xi)| < +\infty. \tag{17}$$

The function (13) is an a.s.-s.s. for problem (1), (11).

THEOREM 1. *The function*

$$\theta(t, x) = (T - t)^{-n} \ln(1 + u(t, x)), \qquad (t, x) \in S_T,$$

where $u(t, x)$ *is the solution of the boundary value problem* (1), (11), *converges as* $t \to T^-$ *to the function* $\theta_a(\xi)$, *i.e., to the solution of* (14), *where, moreover,*

$$\sup_{x \in R^1_+} |\theta(t, x) - \theta_a(\xi)| = O\left[(T - t)^{-n} |\ln(T - t)|\right] \xrightarrow[t \to T^-]{} 0. \tag{18}$$

For the proof we introduce the functions $U = \ln(1 + u)$, $v = \ln(1 + v_a)$ and $z = U - V$. For z we obtain in S_T the equation

$$\frac{\partial z}{\partial t} = \frac{\partial^2 z}{\partial x^2} + \frac{\partial^2 V}{\partial x^2} + \frac{\partial z}{\partial x}\frac{\partial}{\partial x}(U + V) \tag{19}$$

with the condition $z(t, 0) = 0$, $0 < t < T$, where

$$\sup_x |z(0, x)| = \sup_x |\ln(1 + u_0(x)) - \theta_a(x)| < +\infty.$$

By virtue of [6] and [13] we can formally assume that $z \in C^{1,2}_{t,x}(S_T)$. Letting $Y(t) = \sup_x |z(t, x)|$ and using the maximum principle for (19), we find that as $t \to T^-$

$$\frac{dY(t)}{dt} \leqslant \sup_x \frac{\partial^2 V(t, x)}{\partial x^2}. \tag{20}$$

Since $V = \ln(1 + v_a)$ and v_a is given by (13), we have $\partial^2 V/\partial x^2 = (T - t)^{-1}\theta_a''(\xi)$. Hence it follows from (17) and (20) that

$$Y(t) = O\left[|\ln(T - t)|\right], \qquad t \to T^-. \tag{21}$$

Then, taking account of the fact that

$$\sup_{x \in R^1_+} |\theta(t, x) - \theta_a(\xi)| \leqslant (T - t)^{-n} Y(t),$$

we obtain the desired estimate (18), whence we also have the convergence $\theta \to \theta_a$ as $t \to T^-$.

4. By virtue of Theorem 1 the asymptotic behavior of the solution of the problem (1), (11) may be described by the a.s.-s.s. (13):

a) If $n = -1$ (S-regime), then, according to (16),

$$u(t, x) \sim v_a(t, x) = \exp\left[(T - t)^{-1}(1 - x/2)^2\right] - 1, \qquad 0 < x < 2; \tag{22}$$

the localization depth $l = 2$ coincides with the calculated from (7).

b) For $-1 < n < 0$ (LS-regime) the temperature is everywhere bounded in R^1_+ by a limit distribution which coincides with (9) as $x \to 0$:

$$u(t, x) \to \exp\left[C(n) x^{2n/(n+1)}\right] - 1, \qquad t \to T^- \quad \text{for } x \to 0,$$

where

$$C(n) = -[(n + 1)/2n] 2^{-2n/(n+1)} (-n)^{1/(n+1)}.$$

c) When $n < -1$ (HS-regime) the temperature in the domain

$$0 \leqslant x < (T - t)^{(n+1)/2} 2(-n)^{n/2}(-n - 1)^{-(n+1)/2}$$

changes (increases without bound) in accordance with the law

$$u(t, x) \sim \exp\{(T-t)^n - x(-n)^{1/2}(T-t)^{(n-1)/2}\}, \qquad t \to T^-.$$

The effective thermal wave depth $x_{\text{eff}}(t)$, i.e., the solution of (10), decreases in all three cases in accordance with the law

$$x_{\text{eff}}(t) = (T-t)^{(1-n)/2} \ln 2(-n)^{1/2} + o\left((T-t)^{(1-n)/2}\right), \qquad t \to T^-. \tag{23}$$

We remark that the estimate usually given for the thermal wave penetration depth, namely,

$$x_{\text{eff}}(t) \sim (T-t)^{1/2}, \tag{24}$$

which obtains from dimensional considerations, describes the asymptotic stage of the process incorrectly in our case. This is evident from a comparision of (24) with (23). The value of $x_{\text{eff}}(t)$, according to (23), is determined by the structure of the corresponding a.s.-s.s. (13).

§3. On conditions for unboundedness and localization of the solutions of the Cauchy problem for a semilinear equation

In this section we investigate the properties of the solutions of the Cauchy problem ($\beta = \text{const} > 1$)

$$\partial u/\partial t = \partial^2 u/\partial x^2 + (1+u)\ln^\beta(1+u), \qquad t > 0, x \in R^1; \tag{1}$$

$$u(0, x) = u_0(x) \geqslant 0, \qquad x \in R^1;\ u_0 \in C(R^1),\ \sup u_0 < \infty, \tag{2}$$

and, in particular, we construct the a.s.-s.s. of (1) to which the solutions of the problem considered converge asymptotically (we note that (1) has no suitable self-similar solutions [17]). We assume that either the initial function $u_0(x)$ has compact support ($\operatorname{meas}\operatorname{supp} u_0(x) < \infty$), or $u_0(x) \to 0$ as $|x| \to \infty$.

Following [7]–[15], we say that a solution of problem (1), (2) is *unbounded* (or is regime with peaking) if for some $T_0 < +\infty$ we have

$$\max_{x \in R^1} u(t, x) \to +\infty, \qquad t \to T_0^-. \tag{3}$$

We call an unbounded solution *localized* if $u(t, x) \to +\infty$ as $t \to T_0^-$ on a set of finite measure, i.e.,

$$\operatorname{meas}\{x: u(T_0^-, x) = +\infty\} = L < +\infty. \tag{4}$$

But if $u(T_0^-, x) = +\infty$ everywhere in R^1, we consider the solution to be *nonlocalized*. Problem (1), (2) is equivalent to the Cauchy problem

$$Au = \frac{\partial U}{\partial t} - \frac{\partial^2 U}{\partial x^2} - \left(\frac{\partial U}{\partial x}\right)^2 - U^\beta = 0, \qquad t > 0, x \in R^1, \tag{5}$$

$$U(0, x) = U_0(x) = \ln[1 + u_0(x)] \geqslant 0, \qquad x \in R^1, \tag{6}$$

a fact which can be checked by putting

$$U(t, x) = \ln(1 + u(t, x)).$$

1. We first determine the conditions for unboundedness of a solution of problem (1), (2). Let D_T denote the domain $(0, T) \times \{x: |x| < \zeta(t)\}$, where $\zeta \in C((0, T))$ is a nonnegative function, and let $P_T = (0, T) \times R^1$.

LEMMA. *Let $U(t,x)$ be a solution of problem* (5), (6), *and let the function* z: $D_T \to R^1_+$, $z \in C^{1,2}_{t,x}(D_T) \cap C(\bar{D}_T)$, $z \equiv 0$ *everywhere in* $P_T \setminus D_T$, *satisfy in* D_T *the inequality* $\mathbf{A} z \leqslant 0$. *Then if* $z(0,x) \leqslant U_0(x)$ *for all* $x \in R^1$, *it follows that* $U \geqslant z$ *everywhere in* P_T.

This lemma is proved on the basis of the maximum principle for parabolic equations (see, for example, [30]). Following [31], we call $z(t,x)$ a *lower solution* of the problem (5), (6).

We now consider in the domain

$$D_T = (0,T) \times \{x : |x| < a \times (T-t)^{(\beta-2)/2(\beta-1)}\},$$

where $T > 0$ and $a > 0$ are fixed constants (to be determined), the function

$$z(t,x) = (T-t)^{-1/(\beta-1)}\theta(\xi), \qquad \xi = x(T-t)^{-(\beta-2)/2(\beta-1)}, \tag{7}$$

where $\theta \in C^2(-a,a)$ is an even function. For (7) to be a lower solution of the problem it is sufficient that

$$(T-t)^{+1/(\beta-1)}\theta'' + (\theta')^2 - \frac{\beta-2}{2(\beta-1)}\theta'\xi - \frac{1}{\beta-1}\theta + \theta^\beta \geqslant 0,$$

$$\xi \in (0,a) \left(\theta' = \frac{d\theta}{d\xi}\right), \tag{8}$$

and that $z(0,x) \leqslant U_0(x)$ in R^1. We seek the function $\theta(\xi)$ in the form $\theta(\xi) = A(a^2 - \xi^2)^+$, where $A = \text{const} > 0$. Then (8) will certainly be satisfied if the constants T, a and A are such that

$$-[2T^{1/(\beta-1)} + a^2/(\beta-1)] + \xi^2(4A+1) + A^{\beta-1}(a^2-\xi^2)^\beta \geqslant 0$$

for any $\xi \in (0,a)$. Upon analyzing this inequality, we affirm the validity of the following assertion.

THEOREM 2. *Let the initial function* u_0 *be such that*

$$u_0(x) \geqslant \exp\{T^{-1/(\beta-1)}A[a^2 - |x|^2 T^{-(\beta-2)/(\beta-1)}]\} - 1, \qquad |x| < aT^{(\beta-2)/2(\beta-1)},$$

where the constants T, a and A satisfy the inequalities

$$A > (2-\beta)/4(\beta-1),$$

$$A^{\beta-1}a^{2\beta}\left[1 - \frac{1}{(1+4A)(\beta-1)}\right]^\beta \geqslant 2T^{1/(\beta-1)} + \frac{a^2}{\beta-1}.$$

Then the solution of problem (1), (2) *is unbounded, and condition* (3) *is satisfied for some* $T_0 < T$.

REMARK. From the estimates of Theorem 2 it is not difficult to derive the known result [15], [32] concerning the unboundedness of any solution of problem (1), (2) for $1 < \beta \leqslant 3$ and $u_0(x) \equiv 0$.

2. Let $U(t,x)$ be a solution of problem (5), (6) under the conditions of Theorem 2. Then as $T \to 0$ we can neglect the term $\partial^2 U/\partial x^2$ in (5), since the product

$$(T-t)^{1/(\beta-1)}\theta''(\xi) \quad \text{in (8)},$$

with $(T-t)^{1/(\beta-1)} = T^{1/(\beta-1)} \to 0$ as $T \to 0$, corresponds to $\partial^2 U/\partial x^2$. Thus (by analogy with §2) in the asymptotic stage ($t \to T_0^-$) of the process the second derivative $\partial^2 U/\partial x^2$ becomes negligibly small in comparison with other terms of (5), and as $t \to T_0^-$ the solution of the problem converges to the self-similar solution

$$V(t, x) = (T_0 - t)^{-1/(\beta-1)} \theta_a(\xi), \qquad \xi = x(T_0 - t)^{-(\beta-2)/2(\beta-1)} \tag{9}$$

of the first order equation

$$\partial V/\partial t = (\partial V/\partial x)^2 + V^\beta, \qquad t > 0, x \in R^1. \tag{10}$$

In (9) the function $\theta_a \geqslant 0$ is a solution of the boundary value problem

$$\begin{gathered}(\theta_a)^2 - \frac{\beta-2}{2(\beta-1)}\theta_a \xi - \frac{1}{\beta-1}\theta_a + \theta_a = 0, \qquad \xi \in R^1_+, \\ \theta_a(0) = 0, \qquad \theta_a(+\infty) = 0.\end{gathered} \tag{11}$$

The existence of a solution of this problem for any $\beta > 1$ was established in [15], where it was shown that $\theta_a(0) = (\beta-1)^{1/(\beta-1)}$, $\beta > 1$.

When $\beta = 2$, we have $\xi \equiv x$, and the solution of (11) is obtained in the explicit form

$$\theta_a(\xi) = \cos^2(\xi/2), \quad |\xi| \leqslant \pi; \qquad \theta_a(\xi) \equiv 0, \quad |\xi| > \pi. \tag{12}$$

The convergence of the unbounded solution of problem (5), (6) to the a.s.-s.s. (9) is understood, by analogy with §2 (see also [5], [6], [8] and [15]), as the proximity of their self-similar representations:

$$\lim_{t \to T_0^-} \sup_{\xi \in R^1} |\theta(t, \xi) - \theta_a(\xi)| = 0, \tag{13}$$

where $\theta(t, \xi)$ is the self-similar representation of the solution of problem (5), (6), calculated from the formula (see [8] and [15])

$$\theta(t, \xi) = A\Big[\max_x U(t, x)\Big]^{-1} U\Big(t, A^{\beta/2-1}\Big[\max_x U(t, x)\Big]^{1-\beta/2}\xi\Big),$$

where $A = \theta_a(0)$.

The validity of the limit equation (13) has been confirmed in a series of direct calculations (see [15]).

Going now, by way of the reverse substitutions $U = \ln(1+u)$ and $v = \ln(1+v)$, to the variables u and v, we see that an a.s.-s.s. of the initial equation (1) satisfies the first order Hamilton-Jacobi type equation

$$\frac{\partial v}{\partial t} = \frac{1}{(1+v)}\left(\frac{\partial v}{\partial x}\right)^2 + (1+v)\ln^\beta(1+v), \qquad t > 0, x \in R^2. \tag{14}$$

3. The asymptotic convergence (13) of the a.s.-s.s. (9) to the solution of problem (5), (6) guarantees in the limit the coincidence of their properties, so that as $t \to T_0^-$ their space-time distributions coincide.

Therefore, analyzing (9), we obtain the following classification of unbounded solutions of problem (1), (2), depending on the values of the parameter β:

a) If $\beta = 2$, we have an S-regime localization in which the solution increases without bound on a set with measure $L = 2\pi$ (see (12)).

b) If $\beta > 2$ (LS-regime), the temperature becomes infinite on a set of measure zero ($L = 0$).

c) If $1 < \beta < 2$ (HS-regime), then $u(t, x) \to +\infty$ as $t \to T_0^-$ everywhere in R^1, and localization is not present.

REMARK. In [15] it was shown that for $\beta > 3$ and sufficiently small initial functions $u_0(x)$ the problem (1), (2) has a solution, defined for arbitrary $t > 0$, with $\max_x u(t, x) \to 0$ as $t \to +\infty$. These global solutions of the problem converge asymptotically ($t \to +\infty$) to the a.s.-s.s.

$$v_a(t, x) = \exp\left[(T + t)^{-1/(\beta-1)} f_a(\eta)\right] - 1, \qquad \eta = x(T + t)^{-1/2}, \tag{15}$$

satisfying the parabolic equation

$$\frac{\partial v}{\partial t} = \frac{\partial^2 v}{\partial x^2} - \frac{1}{(1 + v)}\left(\frac{\partial v}{\partial x}\right)^2 + (1 + v)\ln^\beta(1 + v), \qquad t > 0, x \in R^1. \tag{16}$$

Substituting (15) into (16), we find that the positive function $f_a(\eta)$ is a solution of the problem

$$f_a'' + \frac{1}{2} f_a' \eta + \frac{1}{\beta - 1} f_a + f_a^\beta = 0, \quad \eta \in R_+^1, \qquad f_a'(0) = 0, \quad f_a(+\infty) = 0.$$

Thus, depending on which solutions, unbounded or global, are being investigated, equation (1) has, as its approximate self-similar solutions, functions satisfying equations of differing orders, namely, equations (14) or (16), respectively.

We note that unbounded solutions of the equation

$$\frac{\partial u}{\partial t} = \frac{\partial^2 u}{\partial x^2} + u^\beta, \qquad t > 0, x \in R^1; \beta > 1,$$

are always localized and converge asymptotically to its exact self-similar solutions [15].

BIBLIOGRAPHY

1. A. A. Samarskiĭ et al., *Effect of metastable localization of heat in a medium with nonlinear thermal conductivity*; Dokl. Akad. Nauk SSSR **223** (1975), 1344–1347; English transl. in Soviet Phys. Dokl. **20** (1975).

2. A. A. Samarskiĭ and I. M. Sobol′, *Example of numerical calculation of temperature waves*, Zh. Vychisl. Mat. i Mat. Fiz. **3** (1963), 702–719; English transl. in USSR Comput. Math. and Math. Phys. **3** (1963).

3. A. P. Mikhaĭlov, *Metastable localization of thermal perturbations in a medium with nonlinear heat conductivity*, Preprint No. 64, Keldysh Inst. Appl. Math. Acad. Sci. USSR, Moscow, 1977. (Russian) MR **58** #14417.

4. V. A. Galaktionov and A. P. Mikhaĭlov, *A self-similar problem for the equation on nonlinear heat conduction*, Preprint No. 53, Keldysh Inst. Appl. Math. Acad. Sci. USSR, Moscow, 1977. (Russian) MR **57** #13192

5. V. A. Galaktionov et al., *Asymptotic phase of modes with peaking and effective localization of heat in problems of nonlinear heat conduction*, Differentsial′nye Uravneniya **16** (1980), 1196–1204; English transl. in Differential Equations **16** (1980).

6. V. A. Galaktionov, *Approximate self-similar solutions of equations of heat conduction type*, Differentsial′nye Uravneniya **16** (1980), 1660–1676; English transl. in Differential Equations **16** (1980).

7. A. A. Samarskiĭ et al., *Thermal structures and fundamental length in a medium with nonlinear heat conduction and volumetric heat sources*, Dokl. Akad. Nauk SSSR **227** (1976), 321–324; English transl. in Soviet Phys. Dokl. **21** (1976).

8. G. G. Elenin and S. P. Kurdyumov, *Conditions that complicate the organization of a nonlinear dissipative medium*, Preprint No. 106, Keldyh Inst. Appl. Math. Acad. Sci. USSR, Moscow, 1977. (Russian) RZh Mekh. **1978** #8Б634.

9. A. A. Samarskiĭ et al, *The burning of a nonlinear medium in the form of complex structures*, Dokl. Akad. Nauk SSSR **237** (1977), 1330–1333; English transl. in Soviet Phys. Dokl. **22** (1977).

10. N. V. Zmitrenko et al., *Localization of thermonuclear combustion in a plasma with electronic thermal conductivity*, Pis'ma v Zh. Èksper. Teoret. Fiz. **26** (1977), 620–624; English transl. in JETP Letters **26** (1977).

11. S. P. Kurdyumov et al., *Interaction of dissipative thermal structures in nonlinear media*, Dokl. Akad. Nauk SSSR **251** (1980), 836–839; English transl. in Soviet Phys. Dokl. **25** (1980).

12. S. P. Kurdyumov, *Eigenfunctions for combustion of a nonlinear medium, and constructive laws for determining the organization of the medium*, Preprint No. 29, Keldysh Inst. Appl. Math. Acad. Sci. USSR, Moscow, 1979. (Russian) RZh Mekh. **1979** #8Б725.

13. V. A. Galaktionov et al., *The effect of boundary modes with aggravation on a medium with constant thermal conductivity*, Preprint No. 28, Keldysh Inst. Appl. Math. Acad. Sci. USSR, Moscow, 1979. (Russian) MR **80i**:80004

14. ______, *On unbounded solutions of the Cauchy problem for the parabolic equation* $u_t = \nabla(u^\sigma \nabla u) + u^\beta$, Dokl. Akad. Nauk SSSR **252** (1980), 1362–1364; English transl. in Soviet Phys. Dokl. **25** (1980).

15. ______, *Unbounded solutions of semilinear parabolic equations*, Preprint No. 161, Keldysh Inst. Appl. Math. Akad. Sci. USSR, Moscow, 1979. (Russian) MR **81c**:35060

16. L. V. Ovsyannikov, *Group relations of the equation of nonlinear heat conductivity*, Dokl. Akad. Nauk SSSR **125** (1959), 492–495. (Russian)

17. V. A. Dorodnitsyn, *Group properties and invariant solutions of an equation of nonlinear heat transport with a source or a sink*, Preprint No. 57, Keldysh Inst. Appl. Math. Acad. Sci. USSR, Moscow, 1979. (Russian) MR **80e**:80004.

18. V. A. Galaktionov et al., *An approach to the comparison of solutions of parabolic equations*, Zh. Vychisl. Mat. i Mat. Fiz **19** (1979), 1451–1461; English transl. in USSR Comput. Math. and Math. Phys. **19** (1979).

19. ______, *The comparison of solutions to parabolic equations*, Dokl. Akad. Nauk SSSR **248** (1979), 586–589; English transl. in Soviet Phys. Dokl. **24** (1979).

20. V. A. Galaktionov, *Two methods for comparing solutions of parabolic equations*, Dokl. Akad. Nauk SSSR **251** (1980), 832–835; English transl. in Soviet Phys. Dokl. **25** (1980).

21. V. A. Galaktionov, S. P. Kurdyumov and A. P. Mikhaĭlov, *Metastable localization of disturbances in problems for equations of nonlinear heat conduction type*, Preprint No. 181, Keldysh Inst. Appl. Math. Acad. Sci. USSR, Moscow, 1979. (Russian) MR **81g**:35050.

22. L. I. Sedov, *Similarity and dimensional methods in mechanics*, 6th ed., "Nauka", Moscow, 1967; English transl. of 4th ed., Academic Press, 1959.

23. N. V. Zmitrenko and S. P. Kurdyumov, *N- and S-modes of self-similar compression of a finite mass of plasma and properties of the modes with peaking*, Zh. Prikl. Mekh. i Tekhn. Fiz. **1977**, no. 1(107), 3–23; English transl. in J. Appl. Mech. Tech. Phys. **18** (1977).

24. O. A. Ladyzhenskaya, V. A. Solonnikov and N. N. Ural'tseva, *Linear and quasilinear equations of parabolic type*, "Nauka", Moscow, 1967; English transl., Amer. Math. Soc., Providence, R.I., 1968.

25. A. N. Kolmogorov, I. G. Petrovskiĭ and N. S. Piskunov, *Etude de l'équation de la diffusion avec croissance de la quantité et son application à un problème biologique*, Bull. Univ. Moscou Sér. Internat. Sect. A Math. Méch. **1** (1937/38), no. 6.

26. G. I. Barenblatt, *On some unsteady motions of a liquid and gas in a porous medium*, Prikl. Mat. Mekh. **16** (1952), 67–78. (Russian)

27. Ya. B. Zel'dovich and A. S. Kompaneets, *On the theory of propagation of heat with the heat conductivity depending upon the temperature*, Collection in Honor of the Seventieth Birthday of Academician A. F. Ioffe, Izdat. Akad. Nauk SSSR, Moscow, 1950, pp. 61–71. (Russian)

28. A. N. Tikhonov and A. A. Samarskiĭ, *The equations of mathematical physics*, 4th ed., "Nauka", Moscow, 1972; English transl. of 2nd ed., Pergamon Press, Oxford, and Macmillan, New York, 1963; also Vols. 1, 2, Holden-Day, San Francisco, Calif., 1964, 1967.

29. A. A. Samarskiĭ et al., *Localization of diffusion processes in media with constant properties*, Dokl. Akad. Nauk SSSR **247** (1979), 349–353; English transl. in Soviet Phys. Dokl. **24** (1979).

30. Avner Friedman, *Partial differential equations of parabolic type*, Prentice-Hall, Englewood Cliffs, N. J., 1964.

31. D. H. Sattinger, *Monotone methods in nonlinear elliptic and parabolic boundary value problems*, Indiana Univ. Math. J. **21** (1971/72), 979–1000.

32. Hiroshi Fujita, *On some nonexistence and nonuniqueness theorems for nonlinear parabolic equations*, Nonlinear Functional Analysis, Proc. Sympos. Pure Math., Vol. XVIII, Part 1, Amer. Math. Soc., Providence, R.I., 1970, pp. 105–113.

Translated by J. F. HEYDA

Труды Матем. Инст. Стеклов.
158(1981)

Proceedings of the Steklov
Institute of Mathematics
1983, Issue 4

THE STRUCTURE OF THE SPECTRUM OF THE TWO-DIMENSIONAL SCHRÖDINGER OPERATOR WITH PERIODIC POTENTIAL, AND SOME ARITHMETIC PROPERTIES OF TWO-DIMENSIONAL LATTICES

UDC 517

M. M. SKRIGANOV

ABSTRACT. In this paper the connection between the structure of the spectrum of the two-dimensional Schrödinger operator with periodic potential and certain purely arithmetic properties of period lattices is studied. These properties concern the approximation of the set of lengths of vectors of a plane lattice by rational numbers.

For a large class of period lattices of the potential it is proved that the number of connected components of the spectrum of the two-dimensional Schrödinger operator is finite.

Bibliography: 16 titles.

In this paper we prove a theorem relating the structure of the spectrum of the two-dimensional Schrödinger operator with periodic potential to purely arithmetic properties of the period lattice. As a corollary of this result, for a large class of periodic lattices it is established that the number of lacunae in the spectrum of such operators is finite. The suggestion that such an assertion is valid for the spectrum of the multidimensional Schrödinger operator with periodic potential forms the content of the Bethe-Sommerfeld conjecture advanced at the beginning of the thirties in solid-state quantum theory (see [8] and [9]).

The well-known estimate of the number of lattice points in two-dimensional domains given by I. M. Vinogradov in his first paper [1] plays an essential role in the proof of our theorem. Quite generally, insofar as can be seen, arithmetic considerations are basic for the investigation of the spectra of multidimensional operators with periodic coefficients (see [16]).

The paper consists of four sections. In the first section the main results are formulated and discussed. The required facts concerning the spectrum are presented in §2. The remaining sections are devoted to the proof of the main theorem: §3 contains the spectral part of the proof, and §4 contains the arithmetic part.

The approach to the investigation of the spectrum of the Schrödinger operator with periodic potential used in this paper was outlined in the author's note [13]. Other methods can be found in [15] and [16].

§1. The main definitions and results

We shall study the two-dimensional Schrödinger operator

$$H = -\Delta + v(x), \qquad x \in \mathbf{R}^2, \tag{1.1}$$

1980 *Mathematics Subject Classification.* Primary 35J10, 35P99; Secondary 10E05.

with a periodic potential $v(x)$. Here Δ is the two-dimensional Laplace operator defined with respect to the standard inner product $\langle\cdot,\cdot\rangle$ in $\mathbf{R}^2$, and the potential $v(x)$ is assumed to be a real, bounded, periodic function. Periodicity means that for some plane lattice Γ, called the period lattice of the potential, the following relations are satisfied:

$$v(x+\gamma)=v(x), \qquad \gamma\in\Gamma.$$

We recall that a lattice in $\mathbf{R}^2$ is a set of vectors of the form

$$\gamma=n_1\gamma_1+n_2\gamma_2, \tag{1.2}$$

where γ_1, γ_2 is some basis in $\mathbf{R}^2$, called the basis of the lattice Γ, and n_1 and n_2 are arbitrary integers.

In the Hilbert spaces $L_2(\mathbf{R}^2)$ the operator H is selfadjoint and lower semibounded on its natural domain, the Sobolev space $W_2^2(\mathbf{R}^2)$. Its spectrum is purely continuous and consists of a number of segments of the real axis separated by lacunae (see [10] and [11]).

We shall be interested in the question of the number of lacunae in the spectrum of H or, equivalently, the number of connected components of the spectrum of H. Before formulating our main result regarding the spectrum of H, we give several definitions concerning the properties of period lattices.

The set of real numbers of the form $\lambda=|\gamma|$, $\gamma\in\Gamma$, we call the spectrum of the lattice Γ, i.e., the spectrum of Γ is the set of lengths of vectors of Γ.

From (1.2) we immediately find that the spectrum of Γ consists of numbers

$$\lambda=\lambda(n_1,n_2)=\left[a_1n_1^2+2bn_1n_2+a_2n_2^2\right]^{1/2}, \tag{1.3}$$

where

$$a_1=|\gamma_1|^2, \quad b=\langle\gamma_1,\gamma_2\rangle, \quad a_2=|\gamma_2|^2.$$

The spectrum of a lattice is an important characteristic of it. A number of interesting problems are connected with the spectrum, for example, the problem of recovering the lattice from its spectrum, first considered by Delone [2].

For our purposes the question of simultaneous approximation of points of the spectrum by rational numbers is of interest. We recall the definition (see [5]).

A point $A=(\alpha_1,\dots,\alpha_m)$ of an m-dimensional real space is called an *extremal point* (also called *poorly approximable*) if for any $\varepsilon>0$ the system of inequalities

$$|n_i-\alpha_i q|<q^{-1/m-\varepsilon}, \qquad i=1,\dots,m, \tag{1.4}$$

has no more that a finite number of solutions of integers $n_1,\dots,n_m$, q with $q>0$.

We introduce the following classes of lattices.

The class $\mathcal{A}_m$, where m is a natural number, consists of lattices with spectra containing $m+1$ numbers $\lambda_0,\lambda_1,\dots,\lambda_m$ such that the point

$$A=(\lambda_1/\lambda_0,\dots,\lambda_m/\lambda_0) \tag{1.5}$$

is extremal. The class $\mathcal{A}_\infty$ is the intersection of all classes A_m.

We point out that the classes A_m and A_∞ are invariant under dilations and rotations of the lattices.

We can now formulate our main result.

THEOREM 1.1. *If the period lattice of the potential belongs to the class A_m for some $m \geqslant 7$, then the spectrum of the corresponding Schrödinger operator H consists of a finite number of connected components.*

It is important to emphasize that this assertion admits a direct generalization to the case of nonlocal periodic perturbations. We shall discuss this in somewhat more detail at the end of §2. We remark that the number 7 in the formulation of Theorem 1.1 is related to the order of the remainder in the estimate of I. M. Vinogradov.

ADDED IN TRANSLATION. Finiteness of the number of lacunae in the spectrum can be proved for two-dimensional operators of general type, in particular for the two-dimensional Schrödinger operator with any period lattice of the potential. Similar results can also be obtained for the periodic Schrödinger operator in arbitrary dimension (see M. M. Shriganov, *The structure of the spectrum of the multidimensional Schrödinger operator with periodic potential*, Dokl. Akad. Nauk SSSR **262** (1982), 847–850; English transl. in Soviet Phys. Dokl. **27** (1982), no. 2).

It is not hard to construct specific examples of lattices belonging to the classes A_m. For this purpose we use the theorem of Schmidt on the simultaneous approximation of algebraic numbers by rational numbers (see [3]).

THEOREM 1.2. *Let $\alpha_1,\dots,\alpha_m$ be real algebraic numbers and suppose that $1, \alpha_1,\dots,\alpha_m$ are linearly independent over the field of rational numbers. Then the point $A = (\alpha_1,\dots,\alpha_m)$ is extremal.*

Suppose, for example, that Γ is a rational lattice, i.e., $\langle \gamma_1, \gamma_2 \rangle = \beta r_{12}$, for any $\gamma_1, \gamma_2 \in \Gamma$, where β is a constant and r_{12} are rational numbers. In this case the spectrum of Γ consists of the numbers (see (1.3))

$$\lambda = \beta_1 [f(n_1, n_2)]^{1/2},$$

where β_1 is a positive constant, and $f(n_1, n_2)$ is a positive definite quadratic form which may be assumed integral and primitive (purely fundamental). It is known that such a form represents infinitely many distinct prime numbers (see [4]). We denote these primes by $p_0, p_1, \dots$. Taking into account the invariance of the classes A_m noted above and also the invariance of the definition of a rational lattice with respect to lattice dilations, we assume that $\beta_1 = 1$. Thus, for any m the spectrum of Γ contains the numbers $\sqrt{p_0}, \sqrt{p_1}, \dots, \sqrt{p_m}$. These numbers are linearly independent over the field of rational numbers (a special case of the theorem of Besicovitch; see [6]). Hence, the collection of algebraic numbers

$$\sqrt{p_1/p_0}, \dots, \sqrt{p_m/p_0}$$

satisfies the conditions of Schmidt's theorem. We have proved the following result.

LEMMA 1.1. *The class A_∞ contains the rational lattices.*

Using Schmidt's theorem in a similar way, we can easily increase the number of such examples.

There is a reason to suppose that the following conjecture is true: for any m almost all lattices belong to the class A_m.

We shall discuss this assertion in more detail. Let $\mathfrak{M}$ be the quotient space of the set of all two-dimensional lattices modulo the equivalence relations generated by lattice dilations and rotations. We introduce the following parametrization on $\mathfrak{M}$. For each lattice Γ we choose a reduced basis (see [2]). It may be assumed that this basis has the form

$$\gamma_1 = (1, 0), \qquad \gamma_2 = (t_2, t_2), \tag{1.6}$$

where t_1 and t_2 satisfy the reduction conditions

$$\tfrac{1}{2} \geqslant t_1 \geqslant 0, \qquad t_1^2 + t_2^2 \geqslant 1.$$

Thus, $\mathfrak{M}$ is parametrized by the vector γ_2 belonging to the right half of the modular figure. As concerns the measure on $\mathfrak{M}$, its specific choice is clearly of no consequence for our purposes; only the class of measures which are absolutely continuous with respect to one another is of importance. The measure $d\tau = dt_1\, dt_2/t_2^2$, which is naturally connected with the modular figure, can be taken as a representative of this class.

For any m we choose $m+1$ pairs of relative by prime integers $(n_{10}, n_{20}) = (1, 0)$, (n_{1i}, n_{2i}), $i = 1, \ldots, m$. The spectrum of the lattice $\Gamma = \Gamma(t_1, t_2)$ with basis (6.16) contains the $m+1$ numbers $\lambda_0 = 1$, $\lambda_i = \lambda(n_{1i}, n_{2i})$, $i = 1, \ldots, m$. Hence, the coordinates of the corresponding point $A(t_1, A_2)$ (see (1.5)) have the form

$$\alpha_i = \left[n_{1i}^2 + 2t_1 n_{1i} n_{2i} + \left(t_1^2 + t_2^2\right) n_{2i}^2\right]^{1/2}. \tag{1.7}$$

The above conjecture asserts that for $d\tau$-almost all $\gamma_2 = (t_1, t_2)$ the point $A(t_1, t_2)$ with coordinates (1.7) is extremal.

For $m = 1$ and 2 this assertion is actually true, since in this case the variables α_i are independent. For $m \geqslant 3$ the problem of approximating dependent quantities arises. In m-dimensional Euclidean space the relations (1.7) determine a two-dimensional algebra variety $\mathcal{M}$. Our conjecture is equivalent to the conjecture that almost every point of $\mathcal{M}$ (in the sense of measure on $\mathcal{M}$) is extremal, i.e., to the conjecture that the variety $\mathcal{M}$ is extremal. Apparently, results now known in the theory of Diophantine approximations suffice only to prove this conjecture for $m = 3$. More details concerning this circle of problems can be found in Sprindzhuk's book [5] where, in particular, very general conjectures regarding the extremal property of smooth varieties are formulated.

With this we complete the discussion of questions to which Theorem 1.1 leads. We now turn to its proof.

§2. General properties of the spectrum of the operator H

The operator H commutes with the commutative family of unitary operators generated in $L_2(\mathbf{R}^2)$ by translations by vectors of the period lattice of the potential. This circumstance, which follows directly from the periodicity of the potential, makes it possible to partially diagonalize H and reduce its study to the study of a certain family of operators $H(k)$ with purely discrete spectrum.

Let $L_2(\mathcal{F}_\Gamma)$ be the space of periodic functions with period lattice Γ which are square-integrable on the fundamental domain $\mathcal{F}_\Gamma = \mathbf{R}^2/\Gamma$ of the lattice Γ. The Sobolev space $W_2^2(\mathcal{F}_\Gamma)$ is defined similarly. A concrete choice of $\mathcal{F}_\Gamma$ is of no consequence for our purposes. It may be assumed that $\mathfrak{F}_\Gamma$ is the parallelogram spanned by the basis vectors of Γ. We remark that with the identification of opposite sides $\mathfrak{F}_\Gamma$ may be considered a two-dimensional torus.

Let Γ' be the lattice inverse to Γ. We recall that Γ' is defined as the lattice of all vectors γ' such that, for any $\gamma \in \Gamma$, $\langle \gamma, \gamma' \rangle$ is an integer. A basis of Γ' is the basis γ'_1, γ'_2 biorthogonal to the basis γ_1, γ_2 of the lattice Γ: $\langle \gamma_i, \gamma'_j \rangle = \delta_{ij}$. We set $\Omega = 2\pi\Gamma'$, i.e., the lattice Ω is obtained by a 2π-dilation of the lattice Γ'.

In $L_2(\mathfrak{F}_\Gamma)$, we consider on $W_2^2(\mathfrak{F}_\Gamma)$ the following family of operators depending on a parameter $k \in \mathbf{R}^2$:

$$\begin{aligned} H(k) &= H^0(k) + v(x), \\ H^0(k) &= \langle i\nabla + k, i\nabla + k \rangle = -\Delta + 2i\langle k, \nabla \rangle + |k|^2. \end{aligned} \tag{2.1}$$

Here ∇ is the gradient operator in $\mathbf{R}^2$, and $i = \sqrt{-1}$.

We cite the main properties of the family $H(k)$.

LEMMA 2.1. 1. *For each k the operator $H(k)$ is selfadjoint and lower semibounded.*

2. *For each k the resolvent of $H(k)$ is compact; the spectrum of $H(k)$ consists of isolated eigenvalues of finite multiplicity with the sole accumulation point at infinity.*

3. *On its domain the family of operators $H(k)$ depends on the parameter $k \in \mathbf{R}^2$ in a strongly continuous manner.*

4. *The operators $H(k_1)$ and $H(k_2)$ are unitarily equivalent if $k_1 - k_2 \in \Omega$; the corresponding unitary operator consists in multiplication by the periodic function $\exp(i\langle k_1 - k_2, x \rangle)$.*

Assertions 1, 3 and 4 of Lemma 2.1 follow from the explicit form (2.1) of the operators $H(k)$. Assertion 2 follows from the fact that the $H(k)$ can be considered elliptic operators on a compact manifold—the torus $\mathfrak{F}_\Gamma$ (the Sobolev space $W_2^2(\mathfrak{F}_\Gamma)$ is compactly imbedded in $L_2(\mathfrak{F}_\Gamma)$).

For a natural number n we denote by $E_n(k)$ the eigenvalues of $H(k)$ enumerated, counting multiplicity, in nondecreasing order: $E_n(k) \leq E_{n+1}(k)$. We denote by $E_n^0(k)$ the eigenvalues of $H^0(k)$, also enumerated in nondecreasing order. In general, characteristics pertaining to the operator $H^0(k)$ we henceforth denote by a superscript 0.

LEMMA 2.2. 1. *The functions $E_n(k)$ are continuous in k.*

2. *The functions $E_n(k)$ are periodic in k with period lattice Ω:*

$$E_n(k + \omega) = E_n(k), \qquad \omega \in \Omega.$$

3. *The estimates*

$$|E_n(k) - E_n^0(k)| \leq \|v\| \tag{2.2}$$

hold, where $\|v\| = \sup_x |v(x)|$ is the norm of the operator of multiplication by $v(x)$ in the spaces $L_2(\mathfrak{F}_\Gamma)$ and $L_2(\mathbf{R}^2)$.

Assertion 1 of Lemma 2.2 follows from assertions 2 and 3 of Lemma 2.1; see Kato's book [12], where the role of the ordering of the eigenvalues for their continuous dependence on parameters is also discussed. Assertion 2 follows immediately from assertion 4 of Lemma 2.1. The estimate (2.2) expresses the well-known fact that under perturbation of an operator with discrete spectrum its eigenvalues are shifted by no more than the norm of the perturbation (see [12]).

The connection between the spectrum of H and the spectra of the operators $H(k)$ can be formulated as follows.

THEOREM 2.1. *In order that a point λ belong to the spectrum of the operator H it is necessary and sufficient that λ belong to the spectrum of at least one of the operators $H(k)$; in other words, the spectrum of H coincides with the union of the ranges of all the functions $E_n(k)$, $k \in \mathfrak{F}_\Omega$, where $\mathfrak{F}_\Omega = \mathbf{R}^2/\Omega$ is the fundamental domain of the lattice Ω.*

This important theorem is well known (see, for example, [10] or [11]). We point out that in those books a somewhat different form of writing the operators $H(k)$ is used: the differential expression (2.1) defining the family $H(k)$ does not depend on the parameter k, but rather the space in which this family is considered depends on k. Our notation corresponds to the form adopted in the physics literature (see [8] and [9]). Of course, the two definitions differ only by a unitary transformation.

The part of Theorem 2.1 pertaining to sufficient conditions is the most essential for the sequel. It should be noted that this part of the theorem can be proved very simply. Indeed, if λ_0 is an eigenvalue of one of the operators $H(k)$ with $k = k_0$ and $\varphi(x) \in W_2^2(\mathfrak{F}_\Gamma)$ is the corresponding eigenfunctions, then the function

$$\psi(x) = \exp(-i\langle k_0, x\rangle)\varphi(x)$$

satisfies the Schrödinger equation $H\psi = \lambda_0\psi$. We introduce the sequence of "test" functions

$$\psi_n(x) = c_n^{-1}\chi_n(x)\psi(x), \qquad c_n = \|\chi_n\psi\|,$$

where $\chi_\eta(x)$ are smooth functions equal to 1 for $|x| \leqslant n$ and to 0 for $|x| \geqslant n+1$; $\|\cdot\|$ is the norm in $L_2(\mathbf{R}^2)$. The following assertions are easily verified.

1) *The sequence $\psi_n(x)$ is noncompact in $L_2(\mathbf{R}^2)$.*

2) $\psi_n(x) \in W_2^2(\mathbf{R}^2)$.

3) $\|\psi_n\| = 1$.

4) $\|H\psi_n - \lambda_0\psi_n\| \to 0$ *as* $n \to \infty$.

Now by definition this means that the point λ_0 is a point of the (continuous) spectrum of the operator H.

We shall not discuss Theorem 2.1 in more detail.

For what follows we shall need the following spectral characteristic:

$$N(\lambda, k) = \sum_{E_n(k) \leqslant \lambda} 1.$$

$N(\lambda, k)$ is equal to the number of eigenvalues of $H(k)$ on the semiaxis $(-\infty, \lambda]$. The analogous function for $H^0(k)$ we denote by $N^0(\lambda, k)$.

LEMMA 2.3. 1. *The function $N(\lambda, k)$ is periodic in k with period lattice Ω:*

$$N(\lambda, k+\omega) = N(\lambda, k), \qquad \omega \in \Omega.$$

2.

$$N^0(\lambda - \|v\|, k) \leqslant N(\lambda, k) \leqslant N^0(\lambda + \|v\|, k). \tag{2.3}$$

The assertions of Lemma 2.3 paraphrase assertions 2 and 3 of Lemma 2.2. The propositions presented above contain all the preliminary information we require concerning the spectrum of H.

To conclude this section we note the possibility of generalizing the results to the case of more general operators of the form $\mathcal{H} = -\Delta + V$, where V is a bounded, selfadjoint operator which commutes with translations by vectors of Γ. For such operators all the results of this section are valid with the exception of assertion 3 of Lemma 2.1 and assertion 1 of Lemma 2.2 concerning the properties of continuity in the parameter k. Nevertheless, if some "smoothness" of the operator V is required with respect to the operator $H^0 = -\Delta$, then these assertions also hold. For example, it can be assumed that V is a pseudodifferential operator with symbol $v(x, \xi)$ periodic in x which satisfies certain smoothness conditions in the variable ξ in addition to the continuous ensuring selfadjointness and boundedness. Our main assertion, Theorem 1.1, remains in force for operators $\mathcal{H}$ with such V.

§3. Reduction of the spectral problem to an arithmetic problem

The first step in reducing the original spectral problem to an arithmetic problem concerning period lattices is the following result.

LEMMA 3.1. *In order that a point λ_0 belong to the spectrum of the operator H it is sufficient that $N(\lambda_0, k)$ as a function of k not reduce identically to a constant.*

Indeed, if $N(\lambda_0, k_1) \neq N(\lambda_0, k_2)$, for some k_1 and k_2, then this implies that in the three-dimensional space of pairs (λ, k), $\lambda \in \mathbf{R}^1$, $k \in \mathbf{R}^2$, the plane $\lambda = \lambda_0$ intersects the graph of at least one of the functions $\lambda = E_n(k)$ (here we have used the continuity of the functions $E_n(k)$). Hence, there exist a point $k_0 \in \mathcal{F}_\Omega$ and an index n such that $\lambda_0 = E_n(k_0)$ (the fact that $k_0 \in \mathcal{F}_\Omega$ follows from the periodicity of the functions $E_n(k)$). By Theorem 2.1 the point λ_0 belongs to the spectrum of H.

We shall find the spectrum of $H^0(k)$. The complete system of orthonormal eigenfunctions of $H^0(k)$ does not depend on k and is formed by the functions

$$\varphi_\omega(x) = e^{i\langle \omega, x\rangle}/\sqrt{\mu_\Gamma}, \qquad \omega \in \Omega,$$

where μ_Γ is the area of the fundamental domain of Γ. Thus, the spectrum of $H^0(k)$ consists of all numbers of the form

$$\lambda_\omega(k) = |\omega - k|^2, \qquad \omega \in \Omega.$$

From what has been said we obtain the following result.

LEMMA 3.2. $N^0(\lambda, k) = 0$ *for* $\lambda < 0$, *while* $N^0(\lambda, k) = n(\sqrt{\lambda}, k)$ *for* $\lambda \geqslant 0$ *where*

$$n(R, k) = \sum_{|\omega - k|^2 \leqslant R^2} 1, \qquad \omega \in \Omega, \tag{3.1}$$

is the number of points of the lattice Ω in the closed disk of radius R with center at the point k. The function $n(R, k)$ is periodic in k with period lattice Ω:

$$n(R, k + \omega) = n(R, k), \qquad \omega \in \Omega.$$

We denote by $n_+(R)$ and $n_-(R)$ respectively the maximum and minimum number of points of the lattice Ω that can be contained in the closed disk of radius R under all possible translations of it:

$$n_+(R) = \max_{k\in\mathfrak{F}_\Omega} n(R,k), \qquad n_-(R) = \min_{k\in\mathfrak{F}_\Omega} n(R,k). \tag{3.2}$$

We shall formulate conditions that a point λ belong to the spectrum of H in terms of the functions $n_+(R)$ and $n_-(R)$.

LEMMA 3.3. *In order that a point $\lambda_0 \geqslant \|v\|$ belong to the spectrum of H it suffices that*

$$n_-\left(\sqrt{\lambda_0 + \|v\|}\right) < n_+\left(\sqrt{\lambda_0 - \|v\|}\right). \tag{3.3}$$

Indeed, from Lemma 3.2 and (2.3) it follows that in $\mathfrak{F}_\Omega$ there are two points k_1 and k_2 such that

$$N(\lambda_0, k_1) \geqslant n_+\left(\sqrt{\lambda_0 - \|v\|}\right), \qquad N(\lambda_0, k_2) \leqslant n_-\left(\sqrt{\lambda_0 + \|\lambda\|}\right).$$

From (3.3) we obtain

$$N(\lambda_0, k) > N(\lambda_0, k_2),$$

i.e., the function $N(\lambda_0, k)$ does not reduce identically to a constant. By Lemma 3.1 the point λ_0 belongs to the spectrum of H.

As $R \to \infty$, there are the following asymptotic formulas for the functions $n_\pm(R)$:

$$n_\pm(R) = \pi R^2/\mu_\Omega + r_\pm(R), \tag{3.4}$$

where μ_Ω is the area of the fundamental domain $\mathfrak{F}_\Omega$ and $r_\pm(R)$ are remainders of order at most $O(R^{2/3}\log^{2/3} R)$ (see §4, Lemma 4.5).

THEOREM 3.1. *Suppose that the remainders $r_+(R)$ and $r_-(R)$ as $R \to \infty$ admit the estimate*

$$r_+(R) \geqslant cR^\alpha, \qquad r_-(R) \leqslant -cR^\alpha \tag{3.5}$$

with $c > 0$ and $\alpha > 0$. Then the spectrum of H consists of no more than a finite number of connected components.

By Theorem 2.1 any semiaxis $(-\infty, \lambda]$ contains no more than a finite number of connected components of the spectrum of H. It therefore suffices to prove that in a neighborhood of infinity the spectrum of H contains only one connected component. From (3.4) and the hypothesis of Theorem 3.1 we find that as $\lambda \to \infty$

$$n_+\left(\sqrt{\lambda - \|v\|}\right) - n_-\left(\sqrt{\lambda + \|v\|}\right) \geqslant -2\pi\|v\|/\mu_\Omega + 2c\lambda^{\alpha/2} + O(\lambda^{\alpha/2-1}) \to \infty.$$

By Lemma 3.3 all sufficiently large λ belong to the spectrum of H. The proof of Theorem 3.1 is complete.

As a result, the question of the number of connected components of the spectrum of the Schrödinger operator has been reduced to the question of estimating the oscillation of the number of lattice points in a disk under translation of its center. This question is studied in the next section.

§4. Oscillation of the number of lattice points in a disk under translation of the center of the disk

This section is devoted to the proof of the following assertion.

THEOREM 4.1. *Suppose that the lattice Ω' inverse to Ω belongs to the class $\mathcal{C}_m$ with $m \geqslant 3$. Then the remainders $r_+(R)$ and $r_-(R)$ in the asymptotic formulas (3.4) as $R \to \infty$ admit the estimates*

$$r_+(R) \geqslant c_\delta R^{1/3-2/m-\delta}, \qquad r_-(R) \leqslant -c_\delta R^{1/3-2/m-\delta} \tag{4.1}$$

with some $c_\delta > 0$ for any $\delta > 0$.

REMARK. Our main assertion, Theorem 1.1, follows from Theorem 3.1 and 4.1. Indeed, the period lattice of the potential Γ and the lattice Ω involved in the formulation of Theorem 3.1 are connected by the relation $\Omega = 2\pi\Gamma'$ (see §2). Therefore, $\Omega' = (2\pi)^{-1}\Gamma$, i.e., the lattice Ω' belongs to the class $\mathcal{C}_m$ if the period lattice Γ belongs to this class. Finally, for $m \geqslant 7$ the exponent in (4.1) is positive.

We shall prove Theorem 4.1 following the method expounded by A. I. Vinogradov and the author in [14].

As noted in Lemma 3.2 $n(R, k)$ is a periodic function of k with period lattice Ω. We expand it in a Fourier series.

LEMMA 4.1. *The function $n(R, k)$ can be expanded in the Fourier series*

$$n(R, k) = \frac{\pi R^2}{\mu_\Omega} + \frac{R}{\mu_\Omega} \sum_{\omega \in \Omega' \setminus \{0\}} |\omega|^{-1} J_1(2\pi|\omega|R) e^{2\pi i \langle \omega, k \rangle}, \tag{4.2}$$

where $J_1(\cdot)$ is the Bessel function. The series (4.2) converges in mean square on the fundamental domain $\mathcal{F}_\Omega$.

Formulas of the type (4.2) expressing the number of lattice points in a region in terms of a series with cylinder functions were first obtained by Voronoĭ and were investigated in detail by Hardy (see, for example, [6] and [7]).

The expansion (4.2) is most simply obtained from the Poisson summation formula

$$\mu_\Omega \sum_{\omega \in \Omega} \varphi(\omega) = \sum_{\omega \in \Omega'} \tilde{\varphi}(\omega). \tag{4.3}$$

Here $\varphi(\xi)$ is any smooth, rapidly decreasing function, and $\tilde{\varphi}(\xi)$ is its Fourier transform.

If we formally substitute into (4.3) the characteristic function of a disk of radius R with center at the point k, we immediately arrive at (4.2). To rigorously justify this procedure it suffices to choose a sequence of smooth, compactly supported functions converging to the indicated characteristic function, to write the Poisson summation formula for each of these functions, and to verify that the Fourier coefficients of the corresponding expansions converge to the Fourier coefficients of (4.2). We take the liberty of omitting the standard details.

We set

$$n(R, k) = \pi R^2/\mu_\Omega + r(R, k). \tag{4.4}$$

The remainders $r_\pm(R)$ in (3.4) can be expressed in terms of $r(R, k)$ as follows:

$$r_+(R) = \max_{k \in \mathcal{F}_\Omega} r(R, k), \qquad r_-(R) = \min_{k \in \mathcal{F}_\Omega} r(R, k). \tag{4.5}$$

We further set

$$M(R)=\max_{k\in\mathfrak{F}_\Omega}|r(R,k)|. \tag{4.6}$$

Comparing (4.4) with (4.2), we obtain the Fourier series for $r(R,k)$:

$$r(R,k)=\frac{R}{\mu_\Omega}\sum_{\omega\in\Omega'\setminus\{0\}}|\omega|^{-1}J_1(2\pi|\omega|R)e^{2\pi i\langle\omega,k\rangle}.$$

Taking into account the Parseval equality, we arrive at the following assertion.

LEMMA 4.2. *The function $r(R,k)$ satisfies the relations*

$$\int_{\mathfrak{F}_\Omega}r(R,k)\,dk=0,\qquad \int_{\mathfrak{F}_\Omega}r^2(R,k)\,dk=\mu_\Omega\theta(R), \tag{4.7}$$

where

$$\theta(R)=\frac{R^2}{\mu_\Omega^2}\sum_{\omega\in\Omega'\setminus\{0\}}|\omega|^{-2}J_1^2(2\pi|\omega|R). \tag{4.8}$$

The relations (4.7) make it possible to estimate the remainders $r_\pm(R)$. We shall need a general proposition from function theory.

Suppose that $\mathbf{K}$ is a space with measure dk, and the total measure of the space is $\mu<\infty$. On $\mathbf{K}$ there is given a real-valued function $f(k)$, $k\in\mathbf{K}$. Suppose that

$$\begin{gathered}M=\max|f(k)|<\infty,\\ \int f(k)\,dk=0,\qquad \int f^2(k)\,dk=\mu\theta.\end{gathered} \tag{4.9}$$

The integrals here extend over all of $\mathbf{K}$. We set

$$f_+=\max f(k),\qquad f_-=\min f(k).$$

LEMMA 4.3. *Under the above assumptions,*

$$f_+\geqslant\theta/2M,\qquad f_-\leqslant-\theta/2M. \tag{4.10}$$

We denote by $f_+(k)$ the positive part of the function $f(k)$ and by $f_-(k)$ the negative part, so that $f(k)=f_+(k)-f_-(k)$. From the first relation (4.9) we obtain

$$\int f_+(k)\,dk=\int f_-(k)\,dk. \tag{4.11}$$

The second relation of (4.9) leads to the estimate

$$M\left(\int f_+(k)\,dk+\int f_-(k)\,dk\right)\geqslant\int f_+^2(k)\,dk+\int f_-^2(k)\,dk=\mu\theta.$$

Using (4.11), we arrive at (4.10). The proof of Lemma 4.3 is completed.

If we set $\mathbf{K}=\mathfrak{F}_\Omega$ and $f(k)=r(R,k)$, and use Lemmas 4.3 and 4.2 we obtain the following assertion (here we correct an error in the formulation of an analogous assertion in [14]).

LEMMA 4.4. *The remainders $r_+(R)$ and $r_-(R)$ admit the estimates*

$$r_+(R)\geqslant\frac{\theta(R)}{2M(R)},\qquad r_-(R)\leqslant-\frac{\theta(R)}{2M(R)}. \tag{4.12}$$

Theorem 4.1 follows immediately from Lemma 4.4, and Lemmas 4.5 and 4.6 presented below.

LEMMA 4.5. *As* $R \to \infty$

$$M(R) = O(R^{2/3} \log^{2/3} R). \tag{4.13}$$

LEMMA 4.6. *Suppose that the lattice* Ω' *belongs to the class* $\mathcal{C}_m$, $m \geqslant 3$; *then as* $R \to \infty$

$$\theta(R) \geqslant c_\delta R^{1-2/m-\delta} \tag{4.14}$$

with some $c_\delta > 0$ *for any* $\delta > 0$.

Lemma 4.5 is contained as a special case in the following theorem of I. M. Vinogradov [1].

THEOREM 4.2. *Suppose that a domain* Σ *of area* S *is bounded by a closed contour on which the ratio of the largest radius of curvature* R *to the smallest does not exceed a given positive number. If* n *is the number of integral points (i.e., points with integral coordinates) belonging to* Σ, *then*

$$n = S + O(R^{2/3} \log^{2/3} R).$$

To prove Lemma 4.5 it suffices to observe that each lattice Ω is obtained from the lattice of integral points by some linear transformation; in particular, the measure μ_Ω in the asymptotic formula (3.4) arises as the Jacobian of this transformation.

We proceed to the proof of Lemma 4.6. By the hypothesis of Lemma 4.6 the spectrum of the lattice Ω' contains $m + 1$ numbers $\lambda_0, \ldots, \lambda_m$ such that the point

$$A = (\lambda_1/\lambda_0, \ldots, \lambda_m/\lambda_0) \tag{4.15}$$

is extremal (see §1).

We estimate the function $\theta(R)$ as follows:

$$\theta(R) \geqslant \frac{R^2}{\mu_\Omega^2} \sum_{i=0}^{m} \lambda_i^{-2} J_1^2(2\pi\lambda_i R).$$

Substituting here the asymptotics of the Bessel function

$$J_1(z) = \sqrt{\frac{2}{\pi z}} \cos\left(z - \frac{3\pi}{4}\right) + O(z^{-3/8}), \qquad z \to \infty,$$

we obtain

$$\theta(R) \geqslant cR\xi(R) + O(1), \qquad R \to \infty, \tag{4.16}$$

where c is a positive constant, and

$$\xi(R) = \sum_{i=0}^{m} \cos^2 2\pi\left(\lambda_i R - \frac{3}{8}\right). \tag{4.17}$$

To prove Lemma 4.6 it now suffices to show that under our assumptions

$$\xi(R) \geqslant\geqslant c_\delta R^{-2/m-\delta} \tag{4.18}$$

with some $c_\delta > 0$ for any $\delta > 0$. We recall that by the hypothesis of the lemma $m \geqslant 3$, and hence the first term in (4.16) is the leading term.

The estimates (4.18), in turn, will have been proved as soon as we show that for any sufficiently large R in the sum (4.17) there is at least one term having all zeros at distance at least $c_\varepsilon R^{-1/m-\varepsilon}$ from the point R for some $c_\varepsilon > 0$ and any $\varepsilon > 0$. We shall prove this assertion.

The zeros of the term with index i have the form

$$R_i(p) = (5+4p)/8\lambda_i;$$

here p is an integer (the index of the zero).

Suppose $\varepsilon > 0$ is given. We consider the sequence of intervals

$$T_p = \left(R_0(p) - c_\varepsilon[R_0(p)]^{-1/m-\varepsilon}, R_0(p) + c_\varepsilon[R_0(p)]^{-1/m-\varepsilon}\right),$$

where c_ε is so far an arbitrary positive constant.

If R does not belong to any of the intervals T_p, then all zeros of the term with index $i = 0$ are at distance at least $cR^{-1/m-\varepsilon}$ from R. We must thus prove that for any sufficiently large p there is a term with index i, $i = 1,\dots,m$, with zeros not belonging to the interval T_p. We suppose otherwise that for some infinite sequence of integers p the intervals T_p contain the zeros of all terms with indices $i = 1,\dots,m$. In this case the system of inequalities

$$|R_i(d_i) - R_0(p)| < c_\varepsilon[R_0(p)]^{-1/m-\varepsilon}, \qquad i = 1,\dots,m, \tag{4.19}$$

has an infinite number of solutions in integers d_i and p. Setting

$$n_i = 5 + 4d_i, \qquad q = 5+4p, \qquad \alpha_i = \lambda_i/\lambda_0$$

and suitably choosing the constant c_ε in the definition of the intervals T_p, we can rewrite (4.19) in the form

$$|n_i - \alpha_i q| < q^{-1/m-\varepsilon}, \qquad i = 1,\dots,m.$$

The assumption of an infinite number of solutions in integers n_i and q of this system contradicts the extremal property of the point (4.15) (see §1). Our assertion regarding the zeros has been proved. With this the proof of Lemma 4.6 is completed.

The proof of Theorem 4.1 has now been completed.

In conclusion we call attention to a circumstance connected with the questions we have considered.

We call the quantity

$$\nu(R) = n_+(R) - n_-(R) = r_+(R) - r_-(R)$$

the oscillation of the number of points of the lattice Ω in a disk a radius R.

It is of interest to obtain a lower bound for the oscillation $\nu(R)$. Suppose for simplicity that Ω (and hence also Ω') is a rational lattice. Then by Lemma 1.1 it is possible to set $m = \infty$ in (4.1), and as $R \to \infty$ we obtain

$$\nu(R) \geqslant c_\delta R^{1/3-\delta}$$

with some $c_\delta > 0$ for any $\delta > 0$. It turns out, however, that a minor modification of Lemma 4.3 makes it possible to obtain a considerable improvement of this estimate, namely,

$$\nu(R) \geqslant c_\delta R^{1/2-\delta}$$

with some $c_\delta > 0$ for any $\delta > 0$. For the proof it suffices to note that

$$\nu(R) \geqslant \sqrt{\theta(R)}\,.$$

BIBLIOGRAPHY

1. I. M. Vinogradov, *A new method of obtaining the asymptotic expression of arithmetic functions*, Izv. Rossiĭsk. Akad. Nauk (6) **1917**, 1347–1378; reprinted in his *Selected works*, Izdat. Akad. Nauk SSSR, Moscow, 1952. (Russian)

2. B. N. Delone, *Geometry of positive quadratic forms*. II, Uspekhi Mat. Nauk **4** (1938), 102–164. (Russian)

3. Wolfgang M. Schmidt, *Simultaneous approximation to algebraic numbers by rationals*, Acta Math. **125** (1970), 189–201.

4. H. Weber, *Bewers des Satzes, dass jede eigentlich primitive quadratische Form unendlich viele Primzahlen darzustellen fähig ist*, Math. Ann. **20** (1882), 301–329.

5. V. G. Sprindzhuk, *The metric theory of diophantine approximations*, "Nauka", Moscow, 1977; English transl., Wiley, 1979.

6. K. Chandrasekharan, *Arithmetical functions*, Springer-Verlag, 1970.

7. Hua Loo-keng, *Die Abschätzung von Exponentialsummen und ihre Anwendung in der Zahlen theorie*, Enzyklopädie der Math. Wissenschaften, Band I, 2, Heft 13, Teil 1, Teubner, Leipzig, 1959.

8. A. Sommerfeld and H. Bethe, *Elektronentheorie der Metalle*, Handbuch der Physik, 2nd ed., Band XXIV/2, Springer-Verlag, 1933, Chap. 3, pp. 333–622; reprint (as a separate book), Springer-Verlag, 1967.

9. A. H. Wilson, *The theory of metals*, Cambridge Univ. Press, 1936.

10. E. C. Titchmarsh. *Eigenfunction expansions associated with second-order differential equations*, Part 2, Clarendon Press, Oxford, 1958.

11. Michael C. Reed and Barry Simon, *Methods of modern mathematical physics*. Vol IV, Academic Press, 1978.

12. Tosio Kato, *Perturbation theory for linear operators*, Springer-Verlag, 1966.

13. M. M. Skriganov, *Proof of the Bethe-Sommerfeld conjecture in dimension two*, Dokl. Akad. Nauk SSSR **248** (1979), 39–42; English transl. in Soviet Math. Dokl. **20** (1979).

14. A. I. Vinogradov and M. M. Skriganov, *The number of lattice points inside the sphere with variable center*, Zap. Nauehn. Serv. Leningrad. Otdel. Mat. Inst. Steklov. (LOMI) **91** (1979), 25–30; English transl. in J. Soviet Math. **17** (1981), no. 5.

15. M. M. Skriganov, *Finiteness of the number of lacunae in the spectrum of the multidimensional polyharmonic operator with a periodic potential*, Mat. Sb. **113 (155)** (1980), 133–145; English transl. in Math. USSR Sb. **41** (1982).

16. ______, *General properties of the spectrum of differential and pseudodifferential operators with periodic coefficients and some problems of the geometry of numbers*, Dokl. Akad. Nauk SSSR **256** (1981), 47–51; English transl. in Soviet Math. Dokl. **23** (1981).

Translated by J. R. SCHULENBERGER

Труды Матем. Инст. Стеклов.
158(1981)

Proceedings of the Steklov
Institute of Mathematics
1983, Issue 4

ON THE PRODUCT OF n LINEAR FORMS IN n VARIABLES

UDC 511.9

B. F. SKUBENKO

ABSTRACT. In this paper an isolation theorem for decomposable forms is proved. It is shown that if $F(x)$ is a decomposable form in n variables ($n > 2$) corresponding to a full module of a purely real algebraic field of degree n, then any form in its ε-neighborhood integrally represents an arbitrarily small number ε_1 (ε_1 depends on ε).

Bibliography: 3 titles.

Notation and definitions

Elements of R^n will be called points or vectors. A point (vector) will be written as a row: $X = (x_1,\ldots,x_n)$.

Suppose $A = (a_{ij})$ is a matrix. The symbol $\|A\|$ will denote the quantity

$$\|A\| = \max_{1 \leqslant i,j \leqslant n} |a_{ij}|.$$

The symbol $N(X)$ will stand for $N(X) = x_1 \cdots x_n$.

Suppose Λ is a lattice in R^n. The symbol $N(\Lambda)$ will denote the quantity

$$N(\Lambda) = \inf |N(X)|, \qquad X \in \Lambda, X \neq 0.$$

We will call two lattices Λ and Λ' *similar*, and write $\Lambda \sim \Lambda'$ (or $\Lambda \nsim \Lambda'$ otherwise), if $\Lambda D = \Lambda'$, where D is a diagonal matrix with $\det D \neq 0$. The symbol $\Lambda(\varepsilon)$ will denote the ε-neighborhood of Λ:

$$\Lambda(\varepsilon) = \{\Lambda_i \mid \Lambda_i \nsim \Lambda;\ \Lambda A_i = \Lambda_i, \|A_i - I\| \leqslant \varepsilon\},$$

where I is the identity matrix.

A lattice Λ will be called *algebraic* if it corresponds to a full module M of a purely real algebraic field K. (To each $\alpha \in K$ there corresponds a vector Z of R^n in accordance with the rule $\alpha \leftrightarrow Z = (\alpha^{(1)},\ldots,\alpha^{(n)})$, where $\alpha^{(1)},\ldots,\alpha^{(n)}$ are the conjugates of α.) For brevity, such a lattice will be denoted by Λ_0.

The rest of the notation is fairly standard or will be introduced as needed.

Introduction

We will restate in the language of lattices some isolation theorems that appear in the literature in the language of forms (for a certain class of forms). Rogers (see [1]) proved

1980 *Mathematics Subject Classification*. Primary 10E15.

THEOREM A. *If a two-dimensional lattice* Λ_0 *with* $N(\Lambda_0) = \lambda$ *contains in each orthant (quadrant) a point* X *with* $|N(X)| = \lambda$, *then for each* $\Lambda_i \in \Lambda_0(\varepsilon)$, *where* $\varepsilon > 0$ *is sufficiently small,*

$$N(\Lambda_i) < \lambda' < \lambda.$$

This theorem was generalized by Barnes [2].

THEOREM B. *If a lattice* Λ_0 *of arbitrary dimension with* $N(\Lambda_0) = \lambda$ *contains in each orthant a point* X *with* $|N(X)| = \lambda$, *then for each* $\Lambda_i \in \Lambda_0(\varepsilon)$, *where* $\varepsilon > 0$ *is sufficiently small,*

$$N(\Lambda_i) < \lambda' < \lambda.$$

The conditions imposed on the lattice are "severe", but such lattices exist; for example, lattices corresponding to ideals of the fields $K(2\cos(2\pi/m))$, $m \neq 12$. Somewhat unusual is the "exceptional" isolation property for three-dimensional algebraic lattices, which was discovered by Cassels and Swinnerton-Dyer [1].

THEOREM C. *If* Λ_0 *is a three-dimensional lattice, then for a given* $\varepsilon_1 > 0$ *there exists* $\varepsilon > 0$ *such that for each* $\Lambda_i \in \Lambda_0(\varepsilon)$

$$N(\Lambda_i) < \varepsilon_1.$$

Theorem C can be found in [3], Chapter X, Theorem IX.

In this note we prove the following assertion.

THEOREM. *If* Λ_0 *is a lattice of dimension greater than* 2, *then for a given* $\varepsilon_1 > 0$ *there exists* $\varepsilon > 0$ *such that for each* $\Lambda_i \in \Lambda_0(\varepsilon)$

$$N(\Lambda_i) < \varepsilon_1.$$

The proof will be preceded by several remarks and lemmas.

We begin with six remarks in the form we need. Let $\Phi(\Lambda)$ denote a complete set of representatives of the cosets in R^n/Λ (exactly one representative from each coset).

REMARK 1. For the set $\Phi(\Lambda)$ to be contained in $S \subset R^n$ it is necessary and sufficient that for any $Y \in R^n$ the set $\Lambda \cap (Y + S)$ be nonempty.

This implies

REMARK 2. If $\Lambda' \subset \Lambda$ and $\Phi(\Lambda') \subset S$, then $\Phi(\Lambda) \subset S$.

The symbol R_m will denote a linear subspace of R^n of dimension m that contains the origin, Λ_m will denote a lattice in R_m, and Λ_m^* ($\Lambda_m^* \subset R_m$) the reciprocal (dual) of Λ_m.

REMARK 3. Suppose $Z \in R_m$. In order that $(Z, X^*) = 0$ for some $X^* \in \Lambda_m^*$ it is necessary and sufficient that there exist linearly independent vectors $X_1, \dots, X_{m-1} \in \Lambda_m$ such that $Z = \lambda_1 X_1 + \cdots + \lambda_{m-1} X_{m-1}$.

The symbol $S(L, r, h)$ will denote the cylinder

$$S(L, r, h) = \{Z \mid Z \in R^n, |L|^2|Z|^2 - (L, Z)^2 \leqslant r^2|L|^2, (L, Z)^2 \leqslant h^2|L|^2\}. \quad (1)$$

Here L is a direction vector of the axis of the cylinder, r is the radius of the ball of the orthogonal section, and $2h$ is the height.

The symbol $S_m(L, r, h)$ will denote the following m-dimensional cylinder in R_m:

$$S_m(L, r, h) = R_m \cap S(L, r, h), \qquad L \in R_m. \quad (2)$$

REMARK 4. If $R_m \supset R_{m-1} \not\ni L$, then $R_{m-1} \cap S_m(L, r, \infty)$ is bounded.

REMARK 5. The number of points in $\Lambda_m \cap S_m(L, r, h)$ is at least

$$hr^{m-1}(2^m \det \Lambda_m \Gamma(m))^{-1} - 1.$$

Remarks 4 and 5 imply

REMARK 6. If all points of $\Lambda_m \cap S_m(L, r, h)$ lie in R_{m-1}, then L lies in R_{m-1}.

We will now prove three lemmas.

LEMMA 1. *Suppose there are given in R_m a countable set of vectors T: $\{X_i\}$ and a set V_{m-1} such that $V_{m-1} \subset R_{m-1} \subset R_m$ and V_{m-1} has nonzero $(m-1)$-dimensional volume. If T contains no vector orthogonal to R_{m-1}, then there exists a vector Z in V_{m-1} such that $(Z, X_i) \neq 0$ for each $X_i \in T$.*

PROOF. We prove the lemma by induction on m. For $m = 2$ the lemma is obvious. Assume the lemma is true for $m - 1 \geqslant 2$. To each X_i there corresponds the orthogonal subspace $R^{(i)}_{m-1} \subset R_m$. The set $\{R^{(i)}_{m-1}\}$ is countable, and no $R^{(i)}_{m-1}$ contains V_{m-1}. Consequently, there exists R'_{m-1} not contained in $\{R^{(i)}_{m-1}\}$ such that $R'_{m-1} \subset R_m$, $R'_{m-1} \neq R_{m-1}$, and the set $V_{m-2} = R'_{m-1} \cap V_{m-1}$ has nonzero $(m-2)$-dimensional volume. Each X_i can be written as $X_i = X'_i + \lambda_i X$, where $X \in R_m$, X is orthogonal to R'_{m-1}, and $X'_i \in R'_{m-1}$. Therefore, the countable set $\{X'_i\}$ in R'_{m-1} is such that no X'_i is orthogonal to V_{m-2}. By assumption, V_{m-2} contains a vector Z such that $(Z, X'_i) \neq 0$. Since $(Z, X'_i) = (Z, X_i)$, the lemma is proved.

LEMMA 2. *If $(L, X^*) \neq 0$ for each $X^* \in \Lambda^*_m$, then for any $r > 0$ there exists an h such that $S_m(L, r, h) \supset \Phi(\Lambda_m)$.*

PROOF. Consider $S' = (1/m)S_m(L, r, \infty)$. It follows from Remark 5 that S' contains infinitely many points of Λ_m. Since $(L, X^*) \neq 0$ for all X^* in Λ^*_m, it follows from Remarks 3 and 6 that the set $\Lambda_m \cap S'$ contains m linearly independent vectors. Consequently, there exists $h' = (1/m)h$ such that the cylinder $(1/m)S_m(L, r, h)$ contains m linearly independent vectors of Λ_m. The lattice Λ'_m spanned by these vectors is a sublattice of Λ_m. The parallelepiped spanned by these vectors contains $\Phi(\Lambda'_m)$ and is contained in $S_m(L, r, h)$. Therefore, $S_m(L, r, h) \supset e\Phi(\Lambda'_m)$. It follows from Remark 2 that $S_m(L, r, h) \supset \Phi(\Lambda_m)$. The lemma is proved.

COROLLARY. *If $S_m(L, r, h) \supset \Phi(\Lambda_m)$, then for any $Y = (y_1, \ldots, y_i, \ldots, y_j, \ldots, y_n)$ in R_m the lattice Λ_m contains a point $\bar{Z}(\bar{z}_1, \ldots, \bar{z}_i, \ldots, \bar{z}_j, \ldots, \bar{z}_n)$ whose coordinates satisfy the conditions*

I. $y_k + hl_k - r \leqslant \bar{z}_k \leqslant y_k + hl_k + r$ $(k = 1, \ldots, n)$;

II. $y_j - y_i + h(l_j - l_i) - 2r \leqslant \bar{z}_j - \bar{z}_i \leqslant y_j - y_i + h(l_j - l_i) + 2r$ $(i, j = 1, \ldots, n)$, *where* $(l_1, \ldots, l_n) = L$.

Note that I implies II.

Condition I follows from (1) and (2) with the aid of Remark 1, where $S = S_m(L, r, h)$ and $Y \in R_m$.

To prove Lemma 3 we introduce some additional notation. We denote by G a group of automorphisms of the module M corresponding to Λ_0 such that for each unit $\zeta \in G$ its conjugates $\zeta^{(1)}, \ldots, \zeta^{(n)}$ are positive. There exists such a group with $n - 1$ generators. The correspondence $\zeta \leftrightarrow Z = (\log \zeta^{(1)}, \ldots, \log \zeta^{(n)})$ allows us to represent G as a lattice Λ_{n-1} in the subspace R_{n-1} orthogonal to the bisector of the first orthant.

LEMMA 3. *For a given pair of indices i, j there exists in R_{n-1} a vector $L = (l_1, \ldots, l_i, \ldots, l_j, \ldots, l_n)$ with $l_i = l_j = 1$ such that $S_{n-1}(L, r, h)$ contains $\Phi(\Lambda_{n-1})$.*

PROOF. We may assume without loss of generality that $i = 1$ and $j = 2$. The full set of vectors $\{Y\}$ in R_{n-1} such that $y_1 = y_2$ is a subspace R_{n-2}. If no vector of the lattice Λ^*_{n-1} is orthogonal to R_{n-2}, then Lemma 1 guarantees the existence of such a vector L and the lemma is proved. We will show that Λ^*_{n-1} contains no vector orthogonal to R_{n-2}. Assume the contrary. Then Remark 3 implies the existence of linearly independent vectors $X_1, \ldots, X_{n-2} \in \Lambda_{n-1}$ such that any $Y \in R_{n-2}$ can be represented in the form $Y = \lambda_1 X_1 + \cdots + \lambda_{n-2} X_{n-2}$. It then follows immediately (since the first two coordinates of Y are equal) that $x_{11} = x_{21}$, $x_{12} = x_{22}, \ldots, x_{1n-2} = x_{2n-2}$. (Here $X_k = (x_{1k}, \ldots, x_{nk})$). These equalities are equivalent to $\zeta_1^{(1)} = \zeta_1^{(2)}$, $\zeta_2^{(1)} = \zeta_2^{(2)}, \ldots, \zeta_{n-2}^{(1)} = \zeta_{n-2}^{(2)}$, where $\zeta_k \leftrightarrow X_k = (x_{1k}, \ldots, x_{nk})$, $x_{ik} = \log \zeta_k^{(i)}$, and $\zeta_k^{(1)}, \ldots, \zeta_k^{(n)}$ are the conjugates of ζ_k. Therefore, $\zeta_1, \ldots, \zeta_{n-2}$ belong to some subfield K' of K, where the degree of K' is at most $n/2$, and thus are dependent in the sense of units. Therefore, $X_1, \ldots, X_{n-2}$ are linearly dependent, which leads to a contradiction.

COROLLARY. *For given indices i, j and given η, $\rho > 0$, the group G contains a unit ζ_0 whose conjugates satisfy*

$$1 - \eta \leqslant \left(\zeta_0^{(j)}/\zeta_0^{(i)}\right)/\rho \leqslant 1 + \eta, \quad \left(\sqrt{\rho}\, H\right)^{-1} \leqslant \zeta_0^{(i)}, \zeta_0^{(j)} \leqslant \sqrt{\rho}\, H, \tag{3}$$
$$H^{-1} \leqslant \zeta_0^{(k)} \leqslant H \quad \textit{for } k \neq i, j,$$

where H depends only on G and η.

This follows from Lemma 3 and the Corollary of Lemma 2 with $y_i = -y_j = \log \sqrt{\rho}$, $\log H = r^2 + \max(hl_i)^2$, $y_k = 0$ for $k \neq i, j$, and $\overline{Z} \leftrightarrow \zeta_0 \in G$.

We now turn to a proof of the theorem.

PROOF. We will prove the theorem by the method used by Cassels (see [3], Chapter X, proof of Theorem IX) to prove Theorem C.

Choose in each orthant exactly one point of Λ_0: $Z_1, \ldots, Z_{2^n}$. Suppose $\varepsilon_1 > 0$ is given. Consider any Λ_1 in the set $\Lambda_0(\varepsilon)$. We have $\Lambda_0 A_1 = \Lambda_1$, where $\|A_1 - I\| \leqslant \varepsilon$, $A_1 = (a_{rs})$. We assume without loss of generality that $a_{21} \geqslant \max|a_{rs}|$, $r \neq s$. From among $Z_1, \ldots, Z_{2^n}$ we choose a point $Z = (z_1, \ldots, z_n)$ for which $z_1 > 0$ and $z_2 < 0$. Consider the set Λ': $\{(z_1\zeta^{(1)}, \ldots, z_n\zeta^{(n)})\}$ of vectors in Λ_0, where ζ ranges over all units of G. Here $\zeta^{(1)}, \ldots, \zeta^{(n)}$ are the conjugates of ζ. We have $\Lambda_1 = \Lambda_0 A_1 \supset \Lambda' A_1$. Now use the Corollary of Lemma 3. Put $i = 1$, $j = 2$, and $\rho = -a_{21}z_2(a_{11}z_1)^{-1}$. We obtain

$$\Lambda_1 \ni \left(z_1\zeta_0^{(1)}, \ldots, z_n\zeta_0^{(n)}\right)A_1 = (\omega_1, \ldots, \omega_n),$$

where $\zeta_0^{(1)}$, $\zeta_0^{(n)}$ satisfy (3) with $i = 1$ and $j = 2$. We have

$$\omega_k = z_1\zeta_0^{(1)}a_{1k} + \cdots + z_k\zeta_0^{(k)}a_{kk} + \cdots + z_n\zeta_0^{(n)}a_{nk} \qquad (k = 1, \ldots, n).$$

It follows from (3) that

$$|\omega_1| \leq \zeta_0^{(1)} |(\eta z_1 a_{11} + c_1 H^2\sqrt{\varepsilon})|,$$

$$\vdots$$

$$|\omega_k| \leq \zeta_0^{(k)} |(z_k a_{kk} + c_k H^2\sqrt{\varepsilon})|,$$

$$\vdots$$

$$|\omega_k| \leq \zeta_0^{(n)} |(z_n a_{nn} + c_n H^2\sqrt{\varepsilon})|,$$

where $c_1, \ldots, c_n$ are constants depending only on n.

Since Norm $\zeta = 1$, for a suitable choice of η and ε we have $|\omega_1 \cdots \omega_n| < \varepsilon_1$. The theorem is proved.

We mention in conclusion that all of our theorems are true for lattices similar to algebraic ones.

BIBLIOGRAPHY

1. J. W. S. Cassels and H. P. F. Swinnerton-Dyer, *On the product of three homogeneous linear forms and indefinite ternary quadratic forms*, Philos. Trans. Roy. Soc. London Ser. A **248** (1955), 73–96.

2. E. S. Barnes, *Isolated minima of the product of n linear forms*, Proc. Cambridge Philos. Soc. **49** (1953), 59–62.

3. J. W. S. Cassels, *An introduction to the geometry of numbers*, Springer-Verlag, 1959.

Translated by G. A. KANDALL

Труды Матем. Инст. Стеклов.
158(1981)

Proceedings of the Steklov
Institute of Mathematics
1983, Issue 4

UDC 511

DIOPHANTINE EQUATIONS WITH UNKNOWN PRIME NUMBERS

V. G. SPRINDZHUK

ABSTRACT. Let $F(x, y)$ be an absolutely irreducible integral polynomial with $\deg_y F(x, y) = n \geqslant 2$. Suppose that $F(0, y)$ has a simple root of degree $k < n$, that a and b are integers, $(a, b) = 1$, and that for at least one prime p dividing a the p-component a_p is greater than $\max(|a|, |b|)^{1-1/nk+\delta}$, $0 < \delta < 1/nk$. Let $F(a/b, y) = F_1(y) \cdots F_r(y)$ be a decomposition into irreducible factors in $\mathbf{Q}[y]$ and $d_j = \deg F_j(y), j = 1, \ldots, r$. Then all the numbers $kd_j, j = 1, \ldots, r$, are divisible by n whenever

$$\max(|a|, |b|) \geqslant (H_F + 1)^{c\delta^{-2}},$$

where H_F is the height of $F(x, y)$ and the value of c is determined from the degree of $F(x, y)$.

Bibliography: 13 titles.

§1. Introduction

Let $F(x, y)$ be an irreducible integral polynomial. By a well-known theorem of Hilbert [1], there exist infinitely many rational numbers x_0 such that $F(x_0, y)$ is irreducible in $\mathbf{Q}[y]$. Hilbert, and later Emmy Noether [2], gave the first applications of this theorem to the solution of the inverse problem of Galois theory (in the case of special groups). This theorem has been generalized and sharpened repeatedly from various points of view, and several new proofs of it have been suggested (see [3], Chapter VIII). One such proof was suggested by Siegel [4], who based it on the theorem of the finiteness of the number of integral points on curves of genus greater than zero.

We denote by $\mathfrak{H}_F$ the set of all rational x_0 that satisfy the assertion of Hilbert's theorem (the "Hilbert" set of the polynomial $F(x, y)$). It is known that $\mathfrak{H}_F$ is "extensive" from various points of view (for example, $\mathfrak{H}_F$ contains "almost all" natural numbers, and its elements lie densely in $\mathbf{Q}$ and $\mathbf{Q}_p$ for any p). Schinzel [5] proved that $\mathfrak{H}_F$ contains an arithmetic progression (of course, the progression depends on the polynomial).

The proof of the irreducibility theorem given by Hilbert himself [1] is noneffective and does not make it directly possible to determine elements of the set $\mathfrak{H}_F$. Although Hilbert noticed that his reasoning can be supplemented by the necessary estimates in order to indicate an explicit bound for the first element of $\mathfrak{H}_F$ ([1], p. 129), nobody has yet done this, and his method of reasoning has almost been forgotten. Making

1980 *Mathematics Subject Classification*. Primary 10B15, 10B40, 12A20, 12E05; Secondary 12A55, 12B05, 12D05.

0081-5438/84 $1.00 + $.25 per page

Hilbert's theorem effective on the basis of Siegel's theorem is impossible, since the latter is noneffective. Just recently Fried [6] obtained a proof of Schinzel's [5] theorem that admits in principle the possibility of computing the first element and difference of the progression contained in $\mathfrak{H}_F$ from the coefficients of the polynomial $F(x, y)$, and Cohen([1]) has obtained estimates of these values (see also [7]).

In this article we give, in particular, a proof of the following assertion (see §6):

THEOREM 1. *Let $F(x, y)$ be an irreducible integral polynomial and*

$$x_m = \left[\exp\sqrt{\ln \ln m}\right] + m!2^{m^2} \qquad (m \geqslant e^e).$$

Then $F(x_m, y)$ is irreducible in $\mathbf{Q}[y]$ for all $m \geqslant m_0(F)$, where the value of $m_0(F)$ can be determined explicitly from the coefficients and degree of $F(x, y)$.

The interesting feature of Theorem 1 is not so much the specific form of the numbers x_m as its universal nature with respect to the set of polynomials $F(x, y) \in \mathbf{Z}[x, y]$ and the possibility of effectively determining $m_0(F)$ in all cases.

The fundamental basis of the proof of Theorem 1 is provided by the recently suggested method of investigating the multiplicative structure of the polynomials $F(x_0, y)$ in the ring $\mathbf{Q}[y]$ via the multiplicative structure of the numbers $x_0 \in \mathbf{Q}$ (see [8] and [9]). In this article we give a further development of this method, carrying it, in a certain sense, as far as it will go. On this path we obtain the following assertion:

THEOREM 2. *Let $F(x, y)$ be an absolutely irreducible integral polynomial with $\deg_y F(x, y) = n \geqslant 2$. Suppose that $F(0, y)$ has a simple root of degree $k < n$, that a and b are rational integers, $(a, b) = 1$, and that for at least one prime p dividing a the p-component a_p (the largest power of p in a) satisfies the inequality*

$$a_p > \left(\max(|a|, |b|)^{1-1/nk+\delta}\right), \qquad 0 < \delta < 1/nk. \tag{1.1}$$

Let

$$F(a/b, y) = F_1(y) \cdots F_r(y) \tag{1.2}$$

be a decomposition of the polynomial $F(a/b, y)$ into irreducible factors in $\mathbf{Q}[y]$ and $d_j = \deg F_j(y)$ $(1 \leqslant j \leqslant r)$. Then all the numbers kd_j $(j = 1,\ldots,r)$ are divisible by n whenever

$$\max(|a|, |b|) \geqslant (H_F + 1)^{c_1\delta^{-2}},$$

where H_F is the height of $F(x, y)$ and the value of c_1 can be determined explicitly from the degree of $F(x, y)$.

We see, in particular, that if $(k, n) = 1$, then $F(a/b, y)$ is irreducible in $\mathbf{Q}[y]$ (the case $k = 1$ corresponds to Theorem 1 in [8]). In any case $d_j \geqslant n/k$ $(j = 1,\ldots,r)$, and so $F(a/b, y)$ has no linear factors. This shows that if (x_1, y_1) is a rational point on the curve $F(x, y) = 0$, $x_1 = u/v$, $(u, v) = 1$ and $|u| \geqslant |v|$, then all the p-components u_p satisfy the inequalities $u_p \leqslant |u|^{1-1/nk+\delta}$ whenever

$$|u| \geqslant (H_F + 1)^{c_1\delta^{-2}},$$

([1])Personal communication.

where $\delta > 0$ is arbitrary. In particular, all the solutions of the diophantine equation

$$F(p^t, y) = 0 \tag{1.3}$$

is unknown primes p and integers t and y ($t > 0$) satisfy the inequality

$$\max(p^t, |y|) < (H_F + 1)^{c_2},$$

where c_2 can be determined explicitly from the degree of $F(x, y)$.

Under the conditions of Theorem 2, it is probably impossible to remove the assumption $k < n$ (i.e., to assume that $k = n$), since otherwise we would find that the number of solutions of (1.3) for $t = 1$ is finite, and it would follow that any nonlinear polynomial represents only a finite number of prime numbers (if the discussion involves the polynomial $p(y)$, then we take $F(x, y) = x - p(y)$). But the belief that there are infinitely many prime numbers of the form $y^2 + 1$ or $y^2 - 2$ is widespread. It can be expected that further work in the development and deepening of this method will make it possible to clarify this question.

§2. Preliminary facts and the main theorem

Henceforth we assume that the polynomial $F(x, y)$ has been reduced to the form

$$F(x, y) = y^n - a_1(x)y^{n-1} - \cdots - a_n(x), \qquad a_i(x) \in \mathbf{Z}[x].$$

This does not limit the generality, since we can easily achieve this by a linear (in y) substitution while preserving the assumptions on the irreducibility of the original and transformed polynomials (a consequence of Gauss' lemma).

Suppose that $F(x, y)$ satisfies the conditions of Theorem 2, that γ is a simple root of $F(0, y)$, and that $[\gamma : \mathbf{Q}] = k < n$ and $\mathbf{K} = \mathbf{Q}(\gamma)$. The analytic function $y = y(x)$ defined by $F(x, y) = 0$ and the initial conditions $x = 0$ and $y = \gamma$ expands in a neighborhood of $x = 0$ expands into a power series

$$f(x) = \sum_{\nu=0}^{\infty} f_\nu x^\nu, \qquad f_0 = \gamma, \tag{2.1}$$

which identically satisfies the equation

$$F(x, f(x)) = 0. \tag{2.2}$$

Obviously, $f'(x) \in \mathbf{Q}(x)[f(x)]$ and $\partial F(0, \gamma)/\partial y \neq 0$. Therefore, $f^{(\nu)}(x) \in \mathbf{Q}(x)[f(x)]$ and $f^{(\nu)}(0) \in \mathbf{K}$ for all ν, so that $f_\nu \in \mathbf{K}$ ($\nu = 0, 1, \ldots$).

By virtue of the well-known generalization of Eisenstein's theorem (see Lemma 3.2 below), there exists an effectively determinable natural number $E = E_f$ such that all numbers $E^\nu f_\nu$ are algebraic integers ($\nu = 0, 1, \ldots$).

Assuming that the rational integers a and b and the prime number p satisfy the conditions of Theorem 2, we denote by $|\cdots|_p$ a fixed extension of the p-adic metric in $\mathbf{Q}$ to the algebraic closure $\overline{\mathbf{Q}}$, $|p|_p = p^{-1}$. Letting σ denote isomorphisms of $\mathbf{K}$ over $\mathbf{Q}$, we consider, in conjunction with (2.1), the series

$$f^\sigma(x) = \sum_{\nu=0}^{\infty} f_\nu^\sigma x^\nu. \tag{2.4}$$

which are produced by the substitution

$$f_\nu \to f_\nu^\sigma \qquad (\nu = 0, 1, \ldots). \tag{2.3}$$

Regarding (2.2) as an infinite system of equations for the coefficients of (2.1) and applying the transformation (2.3) to this system of equations, we see that the series (2.4) satisfy the equations

$$F(x, f^{\sigma}(0)) = 0 \quad (\forall\sigma). \tag{2.5}$$

Setting $x_0 = a/b$, we denote by $\theta^{\sigma} = f^{\sigma}(x_0)$ the sum of (2.4) in the metric $|\cdots|_p$ (assuming that the series converges, i.e., let us say, $|x_0|_p < |E_f|_p$).

All the numbers θ^{σ} are different unless x_0 is a root of the equation

$$R(x) = R_y(F, \partial F/\partial y) = 0, \tag{2.6}$$

where $R(x)$ is the resultant of the polynomials F and $\partial F/\partial y$ with respect to y. This resultant is not identically zero, since $F(x, y)$ is irreducible and $\partial F(x, y)/\partial y$ has degree lower by one in y. The polynomial $R(x)$ has degree at most $(2n-1)n_1$, where $n_1 = \deg_x f(x, y)$, and height of order at most $H_F^{(2n-1)n_1}$, so that all rational roots of (2.6) have heights of no larger order. In subsequent discussions we exclude all the x_0 that satisfy (2.6).

By (2.5), the numbers θ^{σ} are roots of $F(x_0, y)$, and expansion (1.2) shows that they are distributed among the roots of the polynomials $F_j(y)$ $(j = 1,\ldots,r)$. The following sharpening of Theorem 2 is the central result of this article and indicates the direction of its proof.

THEOREM 3. *Suppose that the conditions of Theorem 2 hold. Let k_j denote the number of θ^{σ} that are roots of the polynomial $F_j(y)$. Then*

$$kd_j = nk_j \qquad (j = 1,\ldots,r).$$

§3. General lemmas concerning algebraic power series

Proceeding now with the proof of Theorem 3, we begin with a discussion of well-known facts, to which we give interpretations that are appropriate for our purposes.

LEMMA 3.1. *The coefficient of the series* (2.1) *satisfy the inequality*

$$|\overline{f_\nu}| \leqslant A \qquad (\nu = 0, 1, \ldots), \tag{3.1}$$

where $1 < A \leqslant c_3 H_F^{8(nn_1)^2}$, $n_1 = \deg_x F(x, y)$ and c_3 can be determined from n and n_1.

PROOF. Since

$$|\overline{f_\nu}| = \max_{(\sigma)} |f_\nu^{\sigma}| \qquad (\nu = 0, 1, \ldots),$$

it suffices to prove the inequality

$$|f_\nu^{\sigma}| \leqslant A \qquad (\nu = 0, 1, \ldots). \tag{3.2}$$

The power series (2.4) defines a regular function of the complex variable x in the disc $|x| < \rho_0$, where ρ_0 is the distance from the point $x = 0$ to the nearest (other than $x = 0$) singular point of the algebraic function $y(x)$ defined by the equation $F(x, y) = 0$ (see [10], §4.31). For $0 < \rho < \rho_0$ we find that

$$f_\nu^{\sigma} = \frac{1}{2\pi i} \int_{|z|=\rho} \frac{f^{\sigma}(x)}{z^{\nu+1}}\, dz,$$

and so

$$|f_\nu^\sigma| \leqslant \rho^{-\nu} \max_{|z|=\rho} |f^\sigma(z)| \qquad (\nu = 0, 1, \dots). \tag{3.3}$$

From (2.5), written in the form

$$(f^\sigma(z))^n = a_1(z)(f^\sigma(z))^{n-1} + \cdots + a_n(z),$$

for $|z| = \rho$ we obtain either $|f^\sigma(z)| \leqslant 1$ or

$$|f^\sigma(z)| \leqslant n \max\Big(\max_{|z|=\rho} |a_1(z)|, \dots, \max_{|z|=\rho} b\, |a_n(z)| \Big) \leqslant n \max(1, \rho)^{n_1}(n_1 + 1)H_F.$$

Therefore, from (3:3) we obtain (assuming $\rho \leqslant 1$)

$$|f_\nu^\sigma| \leqslant n(n_1 + 1)\rho^{-\nu} H_F \qquad (\nu = 0, 1, \dots). \tag{3.4}$$

Since the singular points of $y(x)$ are roots of the discriminant of $F(x, y)$ with respect to y, i.e., the roots of the integral polynomial

$$D(x) = D(1, -a_1(x), \dots, -a_n(x)),$$

to estimate the value of ρ_0 it suffices to estimate from below the absolute values of the complex roots $\kappa \neq 0$ of the polynomial $D(x)$. The height H of this polynomial does not exceed $c_4 H_F^{2(n-1)n_1}$, and the absolute values of its roots can be estimated from above by a similar quantity (c_4 depends only on n and n_1). If d is the smallest natural number such that $d\kappa$ is an algebraic integer, then $d \leqslant H$ and

$$1 \leqslant |\mathrm{Nm}(d\kappa)| \leqslant |\kappa| \big(c_4^2 H_F^{4(n-1)n_1} \big)^{\deg \kappa}.$$

Since the degree of $D(x)$ does not exceed $2(n - 1)n_1$, we obtain

$$|\kappa| > \big(c_4^2 H_F^{4(n-1)n_1} \big)^{-2(n-1)n_1},$$

which is also a lower bound for ρ_0. From (3.4) we now obtain

$$|f_\nu^\sigma| < n(n_1 + 1)\big(c_5 H_F^{8(n-1)2n_1^2} \big)^\nu H_F,$$

whence (3.1) follows.

LEMMA 3.2. *The number E_f defined for the series* (2.1) *divides* $\mathrm{Nm}(\partial F(0, \gamma)/\partial y)^2$.

PROOF. Making the substitution $x = \alpha^2 v$, $y = \alpha u + \gamma$ in the equation $F(x, y) = 0$, where

$$\alpha = \partial F(0, \gamma)/\partial y \neq 0,$$

we find that the power series

$$g(v) = \alpha^{-1}(f(\alpha^2 v) - \gamma)$$

satisfies the equation $G(v, g(v)) = 0$, in which the polynomial

$$G(v, u) = b_0(v)u^n + \cdots + b_{n-1}(v)u + b_n(v)$$

has coefficients lying in the ring $I_{\mathbf{K}}$ of all integers of the field $\mathbf{K}$; here $b_{n-1}(0) = 1$. Therefore, we have the formal expansion

$$\frac{1}{b_{n-1}(v)} = \sum_{\nu=0}^{\infty} d_\nu v^\nu, \qquad d_\nu \in I_{\mathbf{K}}\ (\nu = 0, 1, \dots). \tag{3.5}$$

Setting

$$g(v)=\sum_{\nu=1}^{\infty} g_\nu v^\nu \qquad (g_\nu = f_\nu \alpha^{2\nu-1}),$$

from the equation $G(v,n)=0$, written in the form

$$-u=\frac{b_n(v)}{b_{n-1}(v)}+\frac{b_{n-2}(v)}{b_{n-1}(v)}u^2+\cdots+\frac{b_0(v)}{b_{n-1}(v)}u^n,$$

by making the substitution $u=g(v)$, applying the expansion (3.5), and comparing the coefficients of identical powers of v on the left and right sides of the equality, we obtain the infinite system of equations ($m=1,2,\ldots$)

$$g_m=e_m+\sum_{\substack{\nu+\mu_1+\mu_2=m,\\ \mu_1\geq 1,\,\mu_2\geq 1}} e'_\nu g_{\mu_1}g_{\mu_2}+\cdots+\sum_{\substack{\nu+\mu_1+\cdots+\mu_n=m,\\ \mu_1\geq 1,\ldots,\mu_n\geq 1}} e_\nu^{(n-1)} g_{\mu_1}\cdots g_{\mu_{n-1}},$$

where all the numbers $e_m, e'_\nu,\ldots,e_\nu^{(n-1)}$ lie in $I_{\mathbf{K}}$ ($m=1,2,\ldots$ and $\nu=0,1,\ldots$). Thus, $g_1=e_1$, $g_2=e_2+e_0g_1^2,\ldots,g_m$ is a polynomial in $g_1,\ldots,g_{m-1}$ over $I_{\mathbf{K}}$, and we see that $g_m\in I_{\mathbf{K}}$ ($m=1,2,\ldots$). Since $g_m=\alpha^{2m-1}f_m$ and $\alpha\in I_{\mathbf{K}}$, we find that

$$\alpha^{2m}f_m\in I_{\mathbf{K}},\quad \left(\mathrm{Nm}^2_{\mathbf{K}/\mathbf{Q}}(\alpha)\right)^m f_m\in I_{\mathbf{K}}\qquad (m=1,2,\ldots),$$

which proves the lemma.

LEMMA 3.3. *If the polynomial $F(x,y)$ is absolutely irreducible, then for any prime number $q>c_6$ the reduced polynomial*

$$F(x,y)(\mathrm{mod}\,q)=\bar{F}(x,y)$$

is irreducible in $\mathbf{F}_{q^s}[x,y]$ for any s, where $\mathbf{F}_{q^s}$ is the finite field of q^s elements, and

$$c_6=(4H_F)^c,\quad c=k^{2^k},\quad k=(n+n_1-1)(n+n_1)/2,\quad n_1=\deg_x F(x,y).$$

The proof can be found in [11] (p. 193).

LEMMA 3.4. *Let $\mathfrak{q}$ be a prime ideal of the ring $L_{\mathbf{K}}$, $\mathbf{F}_{\mathfrak{q}}$ the residue class field of $I_{\mathbf{K}}$ (mod $\mathfrak{q}$), and q a prime number divisible by $\mathfrak{q}$, with $q>c_6$ and $q\nmid E_f$. Then the formal power series*

$$\bar{f}(x)=\sum_{\nu=0}^{\infty}\bar{f}_\nu x^\nu,$$

obtained from the series (2.1) *by the imbedding*

$$f_\nu\to\bar{f}_\nu \pmod{\mathfrak{q}}\qquad (\nu=0,1,\ldots), \tag{3.6}$$

is algebraic over $F_{\mathfrak{q}}[x]$, and

$$\left[\bar{f}(x):\mathbf{F}_{\mathfrak{q}}[x]\right]=n. \tag{3.7}$$

PROOF. Regarding (2.2) as an infinite system of equations obtained by the substitution of (2.1) in the left side of (2.2) and the vanishing of all the coefficients of the resulting power series, we perform the reduction (3.6) in these equations. We find that

$$\bar{F}(x,\bar{f}(x))=0,$$

and since the polynomial $\bar{F}(x,y)$ is irreducible in $\mathbf{F}_{\mathfrak{q}}[x,y]$ by Lemma 3.3, we obtain (3.7).

§4. An auxiliary power series

We now turn to the construction of an auxiliary power series that plays the main role in the proof of Theorem 3.

LEMMA 4.1. *Let* $\mathbf{K}$ *be an algebraic number field of degree* k *over* $\mathbf{Q}$, β_{ij} *integers in* $\mathbf{K}$ $(1 \leqslant i \leqslant N,\ 1 \leqslant j \leqslant M)$, $N > kM$, $B > 0$ *an integer, and*

$$\max_{(i,j)} \left|\overline{\beta_{ij}}\right| \leqslant B. \tag{4.1}$$

Then there exists rational integers $x_1, \ldots, x_N$ *with*

$$0 < \max_{(i)} |x_i| \leqslant (2^{1/k} NB)^{kM/(N-kM)},$$

satisfying the equations

$$\sum_{i=1}^{N} \beta_{ij} x_i = 0 \qquad (j = 1, \ldots, M).$$

The proof can be found in [12] (Lemma 1.2.2).

LEMMA 4.2. *For any numbers* h *and* ε, *where* $h \geqslant 4nk$ *is an integer and* $0 < \varepsilon < 1/n$, *there exist polynomials*

$$A_j(x) \in \mathbf{Z}[x] \qquad (j = 0, 1, \ldots, n-1)$$

of degrees at most h, *heights at most*

$$H_0 = (2E_j A)^{n(h+1)/\varepsilon}$$

and not all equal to zero, such that the power series

$$V(x) = A_0(x) + A_1(x) f(x) + \cdots + A_{n-1}(x) f^{n-1}(x) \tag{4.2}$$

has order

$$\operatorname{ord} V(x) \geqslant (1 - \varepsilon)(h + 1)n/k$$

in x (*the quantity* A *is defined in Lemma* 3.1).

PROOF. We set

$$A_j(x) = \sum_{i=0}^{k} a_{ij} x^i \qquad (0 \leqslant j \leqslant n-1), \tag{4.3}$$

where the a_{ij} are unknown rational integers. We find that

$$f^j(x) = \sum_{k=0}^{\infty} d_{kj} x^k,$$

$$d_{kj} = \sum_{\nu_1 + \cdots + \nu_j = k} f_{\nu_1} \cdots f_{\nu_j}, \qquad E_f^k d_{kj} \in I_{\mathbf{K}}.$$

By (3.1) we obtain

$$\left|\overline{d_{kj}}\right| < (k+1)^n A^k \qquad (1 \leqslant j \leqslant n-1;\ k = 0, 1, \ldots). \tag{4.4}$$

For the polynomials (4.3) the power series $V(x)$ assumes the form

$$V(x) = \sum_{l=0}^{\infty} b_l x^l, \qquad b_l = a'_{l0} + \sum_{j=1}^{n-1} \sum_{\substack{i+k=l,\\ 0 \leqslant i \leqslant h,\ k \geqslant 0}} a_{ij} d_{kj},$$

where $a'_{l0} = a_{l0}$ for $l \geqslant h+1$. We set $l_0 = [(1-\varepsilon)(h+1)n/k]$ and consider the system of equations $b_l = 0$ $(l = 0, 1, \ldots, l_0)$, linear in a_{ij}. The number of such equations equals l_0, and the number of unknowns is $(h+1)n > kl_0$. Multiplying each equation by $E_f^{l_0}$, we obtain equations with coefficients in $I_{\mathbf{K}}$, whose "size," by virtue of (4.4), does not exceed

$$B = l_0^n (E_f A)^{l_0}.$$

By Lemma 4.1, we find that this system of equations is satisfied by a collection of rational integers a_{ij}, not all zero, that in absolute value do not exceed

$$\left(2^{1/k}(h+1)nl_0^n(E_f A)^{l_0}\right)^{kl_0/((h+1)n-kl_0)} < (2E_f A)^{(k+1)n/\varepsilon}.$$

LEMMA 4.3. *Starting with the equality*

$$f^n(x) = a_1(x) f^{n-1}(x) + \cdots + a_n(x), \qquad a_i(x) \in \mathbf{Z}[x],$$

for $k = n, n+1$, *set*

$$f^k(x) = a_{1k}(x) f^{n-1}(x) + \cdots + a_{nk}(x), \qquad a_{ik}(x) \in \mathbf{Z}[x].$$

Then, with $|\overline{a(x)}|$ *denoting the height of the polynomial* $a(x)$,

$$\max_{(i)} \left|\overline{a_{ik(x)}}\right| < c_7^k H_F^{k-n+1},$$

$$\max_{(i)} \deg a_{ik}(x) \leqslant n_1(k-n+1) \qquad (k = n, n+1, \ldots),$$

where c_7 *can be determined from* $n_1 = \max \deg a_i(x)$ $(i = 1, \ldots, n)$.

The proof is by induction on k.

LEMMA 4.4. *Let*

$$S(x) = s_0(x) + s_1(x) f(x) + \cdots + s_{n-1}(x) f^{n-1}(x), \qquad s_i(x) \in \mathbf{Z}[x],$$

$$T(x) = t_0(x) + t_1(x) f(x) + \cdots + t_{n-1}(x) f^{n-1}(x), \qquad t_i(x) \in \mathbf{Z}[x],$$

where the heights and degrees of the polynomials $s_i(x)$ *do not exceed* S *and* s, *respectively, and the heights and degrees of the polynomials* $t_i(x)$ *do not exceed* T *and* t, *respectively. Then*

$$S(x)T(x) = u_0(x) + u_1(x) f(x) + \cdots + u_{n-1}(x) f^{n-1}(x), \quad u_i(x) \in \mathbf{Z}[x],$$

where the heights and degrees of the polynomials $u_i(x)$ *satisfy the inequalities*

$$\max_{(i)} \left|\overline{u_i(x)}\right| < c_8 \mu S T H_F^{n-1}, \qquad \max_{(i)} \deg u_i(x) \leqslant s + t + n_1(n-1),$$

where c_8 *can be determined from* n *and* n_1, *and* $\mu = \min(s, t)$.

The proof is by direct calculation.

LEMMA 4.5. *Suppose that a power series* B *is defined by the equality*

$$B = B(x, f(x)) = (\partial/\partial y) F(x, y)|_{y=f(x)},$$

that $D = d/dx$ *is the operator of differentiation with respect to* x, *and that* $V = V(x)$ *is defined by Lemma* 4.2. *Then for any integer* $s \geqslant 1$

$$D^s(B^{2s}V) = A_{0s}(x) + A_{1s}(x) f(x) + \cdots + A'_{n-1s}(x) f^{n-1}, \tag{4.5}$$

where the $A_{is}(x)$ *are integral polynomials of degrees at most* $h + 2nn_1 s$ *and heights at most*

$$|\overline{A_{is}(x)}| < c_9^s H_F^{2ns} H_0 \prod_{r=0}^{s-1} (s + h + 2nn_1 r),$$

where c_9 *can be expressed in terms of* n *and* n_1.

PROOF. Let

$$P(x, y) = A_0(x) + A_1(x)y + \cdots + A_{n-1}(x)y^{n-1},$$

where the polynomials $A_i(x)$ are defined in Lemma 4.2. Then for any l, $0 \leqslant l \leqslant s$,

$$D^l(B^{2s}P(x, f(x))) = B^{2s-2l}P_l(x, f(x)),$$

where

$$P_l(x, y) = A_{0l}(x) + A_{1l}(x)y + \cdots + A_{n-1l}(x)y^{n-1}, \qquad A_{il}(x) \in \mathbf{Z}[x].$$

This is proved by induction on l (see [7], Lemma 4). For $l = s$ we obtain (4.6). Application of Lemmas 4.3 and 4.4 yields the bounds for the degrees and heights of the polynomials $A_{is}(x)$ indicated in the lemma.

LEMMA 4.6. *If the original polynomial* $F(x, y)$ *is absolutely irreducible and* π_s *is the largest divisor of* $s!$ *made up of powers of prime numbers bigger than* c_6 *and not occurring in* E_f, *then all the coefficients of all the polynomials* $A_{is}(x)$ *defined by* (4.5) *are divisible by* π_s.

PROOF. All the coefficients of the formal power series $D^s(B^{2s}V)$ are divisible by π_s, since all the coefficients of the series $(B^{2s}V)(Ex)$ lie in $I_{\mathbf{K}}$ and the operator D^s provides each term of the series with a factor of the form $k(k-1)\cdots(k-s+1)$ divisible by $s!$. If q is a prime divisor of π_s and $\mathfrak{q}$ is a prime ideal of $I_{\mathbf{K}}$ occurring in q, then from (4.5) we obtain the following equality over $\mathbf{F}_{\mathfrak{q}}$:

$$\bar{A}_{0s}(x) + \bar{A}_{1s}(x)\bar{f}(x) + \cdots + \bar{A}_{n-1s}(x)\bar{f}^{n-1}(x) = 0,$$

where $\bar{A}_{is}(x)$ and $\bar{f}(x)$ are obtained from $A_{is}(x)$ and $f(x)$ by imbedding their coefficients in $\mathbf{F}_{\mathfrak{q}}$. By Lemma 3.4, the degree of $\bar{f}(x)$ over $\mathbf{F}_{\mathfrak{q}}[x]$ equals n, so that all the polynomials $\bar{A}_{is}(x)$ must be zero. This means that all the coefficients of all the $A_{is}(x)$ are divisible by $\mathfrak{q}$ and therefore by q as well. If q occurs in π_s to more than the first power, then, dividing all the coefficients of all the $A_{is}(x)$ by q, we continue the argument in a similar manner. In this way we find that all the coefficients of all the $A_{is}(x)$ are divisible by $q^{\mathrm{ord}_q \pi_s}$ and therefore by the product of all such factors, i.e., by π_s.

§5. Proof of Theorem 3

We denote by Ω_p the completion of the algebraic closure of the field of p-adic numbers $\mathbf{Q}_p$ in the metric $|\cdots|_p$. For $\omega \in \Omega_p$ such that $|\omega|_p < |E_f|_p$ the series (2.1) and (2.4) converge and define regular functions of ω. Using (4.2), we correspondingly define regular functions $V(\omega)$ and $V^\sigma(\omega)$ for the same ω.

LEMMA 5.1. *Each zero* ω_0 *of the function* $V(\omega)$ *such that* $0 < |\omega_0|_p < |E|_p$ *has multiplicity at most*

$$\varepsilon(h + 1)n + (n - 1)(n_1 - 1) - 1.$$

PROOF. Let $f_1(x),\dots,f_n(x)$ be all the power series expansions in a neighborhood of $x=0$ of the algebraic function $y=y(x)$ defined by

$$y^n = a_1(x)y^{n-1} + \cdots + a_n(x), \qquad a_i(x) \in \mathbf{Z}[x].$$

We can assume that $f_1(x),\dots,f_k(x)$ are the power series (2.4) corresponding to different σ; in general, $f_{k+1}(x),\dots,f_n(x)$ are Puiseux series.

We note that the elementary symmetric functions of $f_{k+1}(x),\dots,f_n(x)$ are integral polynomials in $x, f_1(x),\dots,f_k(x)$.

Indeed, in the field of formal power series we find that

$$F_1(x, y) = \frac{F(x, y)}{y - f_1(x)} = \frac{F(x, y) - F(x, f_1(x))}{y - f_1(x)}$$
$$= \sum_{r=1}^{n} a_r(x)\frac{y^r - f_1^r(x)}{y - f_1(x)} \in \mathbf{Z}[x, f_1(x); y].$$

Therefore, by repeated application of this we obtain

$$F_2(x, y) = \frac{F_1(x, y)}{y - f_2(x)} \in \mathbf{Z}[x, f_1(x), f_2(x); y],$$
$$\dots\dots\dots\dots\dots\dots\dots\dots\dots\dots$$
$$F_k(x, y) = \frac{F_{k-1}(x, y)}{y - f_k(x)} \in \mathbf{Z}[x, f_1(x),\dots,f_k(x); y].$$

Since

$$F_k(x, y) = \frac{F(x, y)}{(y - f_1(x)) \cdots (y - f_k(x))} = \sum_{r=0}^{n-k} (-1)^r s_r y^{n-k-r},$$

where the s_r are elementary symmetric functions of $f_{k+1}(x),\dots,f_n(x)$, we see that

$$s_r \in \mathbf{Z}[x, f_1(x),\dots,f_k(x)] \qquad (r = 1,\dots,n-k).$$

Now let $P(x, y)$ be as in the proof of Lemma 4.5. We represent the resultant $R_y(P, F)$ of the polynomials $P(x, y)$ and $F(x, y)$ with respect to y in the form

$$R_y(P, F) = \prod_{i=1}^{k} P(x, f_i) \prod_{i=k+1}^{n} P(x, f_i) = \prod_{\sigma} V^\sigma(x) \cdot M(x),$$

where $M(x)$, as a symmetric function of $f_{k+1}(x),\dots,f_n(x)$, is a polynomial in $x, f_1(x),\dots,f_k(x)$:

$$M(x) = M(x, f_1(x),\dots,f_k(r)). \tag{5.1}$$

By Lemma 4.2, the order with respect to x of each power series $V^\sigma(x)$ (obtained from (4.2) by the substitution $f(x) \to f^\sigma(x)$) is at least $(1-\varepsilon)(h+1)n/k$, whereas the order of $M(x)$ is at least zero. Therefore,

$$m = \operatorname{ord}_x R_y(P, F) \geqslant (1-\varepsilon)(h+1)n.$$

Since $R_y(P, F)$ is a polynomial in x of degree at most $nh + (n-1)n_1$, we have

$$R_y(P, F) = x^m R(x), \qquad R(x) \in \mathbf{Z}[x],$$
$$\deg R(x) \leqslant \varepsilon(h+1)n + (n-1)(n_1 - 1) - 1. \tag{5.2}$$

Passing now to p-adic functions, we note that

$$R_y(P, F) = \prod_{\sigma} V^{\sigma}(\omega) \cdot M(\omega).$$

where, by virtue of (5.1), $M(\omega)$ is regular in the disc $|\omega|_p < |E|_p$ since this is true of the functions $f_1(\omega), \dots, f_k(\omega)$ (which coincide with the functions $f^{\sigma}(\omega)$). Obviously, $V(\omega)$ is one of the functions $V^{\sigma}(\omega)$, and each zero $\omega_0 \neq 0$ of $V(\omega)$ is a root of the polynomial $R(\omega)$. Inequality (5.2) yields the lemma.

We now turn directly to the proof of Theorem 3.

Let a and b be integers. $(a, b) = 1$, p a prime number that divides a but not E_f, and $f(\omega)$ and $V(\omega)$ p-adic functions defined for $\omega \in \Omega_p$ with $|\omega|_p < 1$ by the power series (2.1) and equality (4.2), respectively. We set $x_0 = a/b$ and denote by $\theta = f(x_0)$ the sum of the series (2.1) in the metric $|\cdots|_p$.

Assume that

$$V^{(i)}(x_0) = 0 \quad (0 \leqslant i \leqslant s-1), \qquad V^{(s)}(x_0) \neq 0.$$

By Lemma 5.1, $s \leqslant \varepsilon(h+1)n + (n-1)(n_1-1) - 1 < 2\varepsilon(h+1)n$ if $h > n_1/\varepsilon$, which we assume together with the constraint $0 < \varepsilon < 1/3k$.

We set $V_s(x) = D^s(B^{2^s}V)$ in accordance with Lemma 4.5. Then

$$\operatorname{ord} V_s(x) \geqslant (1-\varepsilon)(h+1)n/k - s \geqslant (1-\varepsilon_1)(h+1)n/k, \quad \varepsilon_1 = (2k+1)\varepsilon.$$

We set $W(x) = (1/\pi_s)V_s(x)$, where the π_s are defined in Lemma 4.6. Then

$$W(x) = B_0(x) + B_1(x)f(x) + \cdots + B_{n-1}(x)f^{n-1}(x),$$
$$B_i(x) = 1/\pi_s A_{is}(x) \in \mathbf{Z}[x], \quad g = \max \deg B_i(x) \leqslant h + 2nn_1 s.$$

Let us not estimate π_s. It follows from Lemma 3.2 that

$$E_f \leqslant \left((n_1+1)H_F|\overline{\gamma}|^{n_1}\right)^{2k},$$

whereas $|\overline{\gamma}| \leqslant n_1 H_F$. Therefore,

$$E_f \leqslant \left((n_1+1)n_1^{n_1}H_F^{n_1+1}\right)^{2k}. \tag{5.3}$$

To estimate π_s it suffices to exclude from $s!$ the factor

$$\prod_{q \leqslant c_6} q^{[s/q]+[s/q^2]+\cdots} \leqslant \prod_{q \leqslant c_6} q^{s/(q-1)} < c_6^s e^{O(s)},$$

where the symbol O conceals an absolute constant, and we obtain

$$\pi_s \geqslant s!\,(c_{10}H_F^c)^{-s}, \qquad c_{10} = c_{10}(n, n_1).$$

We now find that

$$\max_{(i)} |\overline{B_i(x)}| < (c_9 H_F^{2n})^s H_0 \frac{(c_{10}H_F^c)^s}{s!} \prod_{r=0}^{s-1} (s + h + 2nn_1 r)$$
$$< (c_{11}H_F^{2n+c})^s H_0 \left(\frac{h}{s}\right)^s, \qquad c_{11} = c_{11}(n, n_1).$$

Bearing in mind the explicit form of H_0 presented in Lemma 4.2, Lemma 3.1 and (5.3), we obtain

$$H_0 < \left(c_{12}H_F^{9(nn_1)^2}\right)^{n(h+1)/\varepsilon}, \qquad c_{12} = c_{12}(n, n_1).$$

Therefore,

$$\max_{(i)} \left|\overline{B_i(x)}\right| < \left(c_{13} H_F^{9n3n_1^2+2n(2n+c)\varepsilon^2}\right)^{(h+1)/\varepsilon},$$

where $c_{13} = c_{13}(n, n_1)$. Setting $B_i = b^g B_i(x_0)$ $(i = 0, 1, \dots, n-1)$, we find that the B_i are rational integers, and that

$$\begin{aligned} B = \max_{(i)} |B_i| &< \left(c_{14} H_F^{9n3n_1^2+2n(2n+c)\varepsilon^2}\right)^{(h+1)/\varepsilon} \\ &\times \left(\max(|a|, |b|)\right)^{h(1+5n^2 n_1 \varepsilon)}, \qquad c_{14} = c_{14}(n, n_1). \end{aligned} \tag{5.4}$$

We further find that

$$V_s(x_0) = B^{2s}(x_0, f(x_0)) V^{(s)}(x_0) \neq 0,$$

and so $W(x_0) = (1/\pi_s) V_s(x_0) \neq 0$. Setting

$$L(y) = B_0 + B_1 y + \cdots + B_{n-1} y^{n-1},$$

we see that $L(\theta) \neq 0$. Since $f(Ex)$ is an integral power series, the coefficients of the series $W(x)$ are p-integral; and since

$$\operatorname{ord} W(x) \geqslant (1 - \varepsilon_1)(h+1)n/k, \tag{5.5}$$

we find that

$$0 \neq |L(\theta)|_p \leqslant p^{-\operatorname{ord}_p a \cdot \operatorname{ord} W(x)} \leqslant |x_0|_p^{(1-\varepsilon_1)(h+1)n/k}. \tag{5.6}$$

Obviously, (5.5) also holds for the power series $W^\sigma(x)$ obtained from $W(x)$ by the substitution $f(x) \to f^\sigma(x)$; and by analogy with the preceding, for all σ we obtain

$$|L(\theta^\sigma)|_p \leqslant |x_0|_p^{(1-\varepsilon_1)(h+1)n/k} \tag{5.7}$$

(it is not guaranteed that the left inequality in (5.6) holds).

We now turn to (1.2). We first note that in absolute value the complex roots ξ of the polynomial $F(x_0, y)$ do not exceed

$$n(n_1 + 1) H_F \max\left(1, |x_0|\right)^{n_1}. \tag{5.8}$$

By (2.2), θ is a root of $F(x_0, y)$ and therefore a root of one of the polynomials $F_j(y)$, say, $F_1(y)$. Since $L(\theta) \neq 0$ and $F_1(y)$ is irreducible, the resultant $R(L, F_1) \neq 0$. All the complex roots of $F_1(y)$, as roots of $F(x_0, y)$, do not exceed the quantity (5.8), and we obtain

$$|R(L, F_1)| = \prod_{\xi \in F_1} |L(\xi)| < c_{15}(n, n_1)\left(B H_F^{n-1} \max\left(1, |x_0|\right)^{n_1(n-1)}\right)^d. \tag{5.9}$$

Let $\theta_1, \dots, \theta_d$ be the roots of $F_1(y)$ in $\overline{\mathbf{Q}}_p$, where d is the degree of $F_1(y)$. Since the coefficients of the polynomial $F(x_0, y)$ are p-integral, so are the coefficients of $F_1(y)$; therefore, $|\theta_i|_p \leqslant$ $(i = 1, \dots, d)$. Suppose that k_1 roots θ^σ of $F(x_0, y)$ are contained among $\theta_1, \dots, \theta_d$. Then by (5.7) we obtain

$$|R(L, F_1)|_p = \prod_{i=1}^{d} |L(\theta_i)|_p \leqslant \prod_{\theta^\sigma \in F_1} |L(\theta^\sigma)|_p \leqslant |x_0|_p^{(1-\varepsilon)(h+1)nk_1/k}. \tag{5.10}$$

Regarding the resultant $R(L, F_1)$ as the determinant formed by the coefficients of the polynomials $L(y)$ and $F_1(y)$, we see that it is a rational number with a denominator that divides $b^{n_1(n-1)}$. Therefore, bearing (5.9) and (5.10) in mind, we obtain

$$1 \leqslant |b|^{n_1(n-1)} |R(L, F_1)| \, |R(L, F_1)|_p$$
$$< c_{15}(n, n_1) |b|^{n_1(n-1)} \left(H_F^{n-1} \max(1, |x_0|)^{n_1(n-1)} \right)^d B^d |x_0|_p^{(1-\varepsilon_1)(h+1)nk_1/k}.$$

Using (5.4) and the possibility of taking h arbitrarily large, we find that

$$1 \leqslant \left(c_{14} H_F^{9n3n_1^2 + 2n(2n+c)\varepsilon_2} \right)^{d/\varepsilon} \left(\max(|a|, |b|) \right)^{(1+5n^2 n_1 \varepsilon)d} |x_0|_p^{(1-\varepsilon_1)nk_1/k}.$$

Since $|x_0|_p = a_p^{-1}$, setting $h(x_0) = \max(|a|, |b|)$ and $\varepsilon_2 = 5n^2 n_1 \varepsilon$ we obtain

$$\frac{\ln a_p}{\ln h(x_0)} \leqslant \frac{dk}{nk_1} \frac{1+\varepsilon_2}{1-\varepsilon_1} + c_{16} \frac{\ln(H_F + 1)}{\varepsilon \ln h(x_0)}$$
$$\leqslant \frac{dk}{nk_1} + c_{17} \left(\varepsilon + \frac{\ln(H_F + 1)}{\varepsilon \ln h(x_0)} \right), \qquad c_{17} = c_{17}(n, n_1).$$

Assuming that (1.1) holds, we take $\varepsilon = \delta(16nc_{17})^{-1}$ and assume that

$$h(x_0) \geqslant (H_F + 1)^{16nc_{17}\delta^{-2}}.$$

We then have

$$\left(1 - \frac{1}{nk} + \delta \right) \frac{k_1}{k} \leqslant \frac{d}{n} + \frac{\delta}{2n}. \tag{5.11}$$

Since θ is one of the numbers θ^σ, for $k_1 \neq k$ we can find a θ^σ, not a root of $F_1(y)$, which can be taken as a new θ in the above arguments. This again leads to an inequality of type (5.11). Exhausting all the θ^σ in a similar manner, we obtain the inequalities

$$\left(1 - \frac{1}{nk} + \delta \right) \frac{k_j}{k} \leqslant \frac{d_j}{n} + \frac{\delta}{2n} \qquad (j = 1, \ldots, r'), \tag{5.12}$$

where $r' \leqslant r$ and we assume that the numbers θ^σ are roots only of the polynomials $F_j(y)$ for $j = 1, \ldots, r'$; $k_1 + \cdots + k_{r'} = k$. If $k = 1$, it follows from (5.11) that $d = n$, and if $k \geqslant 2$, then, summing (5.12) over all j, we find that $d_1 + \cdots + d_{r'} = n$, and so $r' = r$. Again summing (5.12) over all j except for the given value, we obtain

$$\left(1 - \frac{1}{nk} + \delta \right) \left(1 - \frac{k_j}{k} \right) \leqslant 1 - \frac{d_j}{n} + \frac{\delta}{2},$$

and, since $0 < \delta < 1/nk$,

$$\frac{d_j}{n} - \frac{k_j}{k} < \frac{1}{nk} - \frac{\delta}{2} \qquad (j = 1, \ldots, r).$$

Noting that (5.12) entails

$$\frac{d_j}{n} - \frac{k_j}{k} > -\frac{1}{nk} + \frac{\delta}{2} \qquad (j = 1, \ldots, r),$$

we find that

$$\left|\frac{d_j}{n}-\frac{k_j}{k}\right|<\frac{1}{nk}-\frac{\delta}{2}\qquad (j=1,\dots,r). \tag{5.13}$$

If we assume that $d_j k \neq k_j n$ for some j, we obtain a contradiction to (5.13). Therefore, Theorem 3 is proved in the case when p does not divide E_f.

In the case when p divides E_f, we consider the functions $f^\sigma(\omega)$ in the disc $|\omega|_p<|E_f|_p$ and require that x_0 belong to this disc. If $a_p>E_f$, then this requirement holds since in this case

$$|x_0|_p=1/a_p<1/E_j\leqslant|E_f|_p.$$

By virtue of (5.3), we can assume that $a_p>E_f$. The above arguments are now applicable almost unchanged.

§6. Universal subsets of Hilbert sets

We now consider some corollaries of Theorem 2 which bear on the solutions of diophantine equations and questions of the irreducibility of polynomials. We shall first verify the following assertion.

THEOREM 4. *Let $A=\{a_m\}_1^\infty$ be a sequence of natural numbers that satisfy the following two conditions*:

a) *For each m the number a_m has a prime divisor p such that the p-component $(a_m)_p$ of a_m satisfies the inequality*

$$(a_m)_p\geqslant a_m^{1-\chi(m)}, \tag{6.1}$$

where $\chi(m)>0$ and $\chi(m)\to 0\ (m\to\infty)$.

b) *For each prime number q all the numbers a_m such that $m\geqslant m_0(q)$ are divisible by q.*

Then for any integral polynomial $F(x,y)$, irreducible in $\mathbf{Q}[x,y]$, of degree at least 2 with respect to y and such that $F(0,y)$ has at least one simple root, the diophantine equation

$$F(x,y)=0,\qquad x\in A,\ y\in\mathbf{Z}, \tag{6.2}$$

has only a finite number of solutions. All the solutions can be determined effectively if the functions $\chi(m)$ and $m_0(q)$ are defined effectively.

LEMMA 6.1. *Let $F(x,y)$ be an integral primitive polynomial that is irreducible in $\mathbf{Q}[x,y]$ but reducible in $\overline{\mathbf{Q}}[x,y]$, where* $\deg F(x,y)=d\geqslant 2$. *Then all the solutions of* (6.2) *in integers x and y satisfy the inequality*

$$\max(|x|,|y|)\leqslant(H_F+1)^{c_{18}},$$

where c_{18} can be determined from d.

PROOF. Since $F(x,y)$ is reducible in $\overline{\mathbf{Q}}[x,y]$, we have the expansion

$$F(x,y)=F_1(x,y)F_2(x,y), \tag{6.3}$$

where $F_1(x,y)$ and $F_2(x,y)$ are nonconstant polynomials with coefficients lying in a finite extension of $\mathbf{Q}$. Using Kronecker's substitution $y\to x^d$, we note that the expansion of $F(x,y)$ in $\overline{\mathbf{Q}}[x,y]$ is parallel to the expansion of $F(x,x^d)$ in $\overline{\mathbf{Q}}[x]$ and conversely. Let F, F_1 and F_2 be the polynomials $F(x,x^d)$, $F_1(x,x^d)$ and $F_2(x,x^d)$, respectively, and let g, γ_1 and γ_2 be their leading coefficients. From (6.3) we note that

$$g^{-1}F = \gamma_1^{-1}F_1\gamma_2^{-1}F_2 \qquad (g = \gamma_1\gamma_2),$$

where the transformed polynomials have leading coefficients 1. The smallest field containing the coefficients of the polynomials $\gamma_1^{-1}F_1$ and $\gamma_2^{-1}F_2$ is contained in the splitting field of F and therefore has degree at most $(d^2)!$.

Since $F(x, y)$ is irreducible in $\mathbf{Q}[x, y]$, it follows that F is irreducible in $\mathbf{Q}[x]$, the heights of these polynomials coincide, and the roots of F have the same height. The coefficients of $\gamma_1^{-1}F_1$ and $\gamma_2^{-1}F_2$, as symmetric functions of the corresponding roots of F, and algebraic numbers of height at most $(H_F + 1)^{c_{19}}$, where c_{19} can be determined from d.

If x_0 and y_0 are integers that satisfy (6.2), then from (6.3) we obtain, say, $F_1(x_0, y_0) = 0$. We can assume that $F_1(x, y)$ depends on y (otherwise we immediately obtain a bound for x_0). Since $F(x, y)$ is irreducible, not all the coefficients of $F_1(x, y)$ are rational numbers and, replacing the coefficients of $F_1(x, y)$ by their conjugates, we obtain the new equality $\tilde{F}(x_0, y_0) = 0$. Regarding $F_1(x, y)$ and $\tilde{F}_1(x, y)$ as polynomials in y over $\mathbf{Q}(x)$, we note that they do not have a common root. Otherwise, $F(x, y) \in \mathbf{Q}(x)[y]$ would have a multiple root (both $F_1(x, y)$ and $\tilde{F}_1(x, y)$ divide $F(x, y)$). Therefore, the resultant $R_y[F_1(x, y), \tilde{F}_1(x, y)]$ is not identically zero, and x_0 is a root of the nontrivial equation

$$R_y\big[F_1(x_0, y), \tilde{F}_1(x_0, y)\big] = 0.$$

From this equation we obtain a bound for $|x_0|$, which yields the lemma.

LEMMA 6.2. *Let* $\mathbf{K}$ *be an algebraic number field of degree* $k \geqslant 2$ *and* $\pi_{\mathbf{K}}^{(0)}(x)$ *the number of prime number* $p \leqslant x$ *not divisible by prime ideals of the first degree of the ring* $I_{\mathbf{K}}$. *Then for* $x > \exp\{k^{-1}|D_{\mathbf{K}}|^{(k-1)!}\}$

$$\pi_{\mathbf{K}}^{(0)}(x) > \frac{k-1}{k!}\operatorname{li} x - c_{20}x\exp\left\{-c_{21}\sqrt{\frac{\ln x}{k!}}\right\}, \tag{6.4}$$

where c_{20} *and* c_{21} *are computable absolute constants.*

PROOF. We shall use well-known results on the distribution of prime ideals obtained on the basis of the theory of the Dedekind zeta function [13].

Let $P_{\mathbf{K}}(x)$ be the number of prime ideals of $I_{\mathbf{K}}$ whose norms do not exceed . Then

$$P_{\mathbf{K}}(x) < \operatorname{li} x + c_{22}xe^{-c_{23}\sqrt{k^{-1}\ln x}} \tag{6.5}$$

for $x > \exp\{k^{-1}(\ln|D_{\mathbf{K}}|)^2\}$, where c_{22} and c_{23} are computable absolute constants.

Let $\mathbf{G}$ be the smallest normal field over $\mathbf{K}$. Then

$$P_{\mathbf{G}}(x) > \operatorname{li} x - 2x^{1-c_{24}/\sqrt{|D_{\mathbf{G}}|}} - c_{22}xe^{-c_{23}\sqrt{g^{-1}\ln x}} \tag{6.6}$$

for $x > \exp\{g^{-1}(\ln|D_{\mathbf{G}}|)^2\}$, $g = \deg \mathbf{G}$, and c_{24} is a computable absolute constant.

We denote by $\pi_{\mathbf{K}}^{(r)}(x)$ the number of prime numbers $p \leqslant x$ divisible by exactly r distinct prime ideals $\mathfrak{p}$ of the first degree of the ring $I_{\mathbf{K}}$, $p \nmid D_{\mathbf{K}}$. Then for $x \geqslant |D_{\mathbf{K}}|$ we have

$$\pi(x) = \pi_{\mathbf{K}}^{(0)}(x) + \pi_{\mathbf{K}}^{(1)}(x) + \cdots + \pi_{\mathbf{K}}^{(k)}(x) + \nu, \tag{6.7}$$

where $\pi(x)$ is the number of all prime numbers $p \leqslant x$ and ν is the number of distinct prime divisors $D_{\mathbf{K}}$. Obviously,

$$P_{\mathbf{K}}(x) = \sum_{N(\mathfrak{p}) \leqslant x} 1 \geqslant \sum_{\substack{N(\mathfrak{p}) \leqslant x, \\ \deg \mathfrak{p} = 1}} - k^2 \pi(x^{1/2})$$

$$> k\pi_{\mathbf{K}}^{(k)}(x) + (k-1)\pi_{\mathbf{K}}^{(k-1)}(x) + \cdots + \pi_{\mathbf{K}}^{(1)}(x) - k^2\pi(x^{1/2}).$$

Using (6.7), we obtain

$$P_{\mathbf{K}}(x) > (k-1)\pi_{\mathbf{K}}^{(k)}(x) + \pi_{\mathbf{K}}^{(k)}(x) + \pi_{\mathbf{K}}^{(k-1)}(x) + \cdots + \pi_{\mathbf{K}}^{(1)}(x) - k^2 x^{1/2}$$
$$= (k-1)\pi_{\mathbf{K}}^{(k)}(x) + \pi(x) - \pi^{(0)}\mathbf{K}(x) - \nu - k^2 x^{1/2}.$$

Therefore, bearing in mind that $2^{\nu} \leqslant |D_{\mathbf{K}}|$, we find that

$$\pi_{\mathbf{K}}^{(0)}(x) > (k-1)\pi_{\mathbf{K}}^{(k)}(x) + \pi(x) - P_{\mathbf{K}}(x) - \frac{\ln|D_{\mathbf{K}}|}{\ln 2} - k^2 x^{1/2},$$

and (6.5) together with the known formula

$$\pi(x) = \operatorname{li} x + O(xe^{-c_{25}\sqrt{\ln x}})$$

leads to the estimate

$$\pi_{\mathbf{K}}^{(0)}(x) > (k-1)\pi_{\mathbf{K}}^{(k)}(x) - c_{26}xe^{-c_{23}\sqrt{(1/k)\ln x}}, \tag{6.8}$$

whenever $x > \exp\{k^{-1}(\ln|D_k|)^2\}$.

We now note that if a prime number p does not occur in $D_{\mathbf{G}}$ and is the norm of a prime ideal in $I_{\mathbf{G}}$, then p splits into exactly k distinct prime ideals of the first degree in $I_{\mathbf{K}}$. Therefore, $\pi_{\mathbf{K}}^{(k)}(x) \geqslant \pi_{\mathbf{G}}^{(g)}(x)$. Since

$$\pi_{\mathbf{G}}^{(g)}(x) = g^{-1}P_{\mathbf{G}}(x) + O(gx^{1/2}),$$

inequalities (6.6) and (6.8) lead to (6.4) if we bear in mind that $|D_{\mathbf{G}}| \leqslant |D_{\mathbf{K}}|^{(k-1)!}$ (which follows from the theory of differents).

We actually use much less than (6.4): to prove Theorem 1 it suffices to know that

$$\pi_{\mathbf{K}}^{(0)}(x) > 0 \quad \text{for } x > \max\left(c_{27}, \exp\left\{k^{-1}|D_{\mathbf{K}}|^{(k-1)!}\right\}\right), \tag{6.9}$$

where c_{27} is a computable absolute constant (to prove Theorem 4 it suffices that there exist an effectively determinable quantity x_1 such that $\pi_{\mathbf{K}}^{(0)}(x_1) > 0$).

PROOF OF THEOREM 4. We shall distinguish two cases, which correspond to conditions a) and b) imposed on the sequence A: the polynomial $F(0, y)$ is reducible or irreducible in $\mathbf{Q}[y]$. In the reducible case we would obtain the theorem as a corollary of Theorem 2 if $F(x, y)$ were absolutely irreducible. The possibility of such an assumption is based on Lemma 6.1: if (6.2) has a sufficiently large solution, then $F(x, y)$ is absolutely irreducible.

If $F(0, y)$ is an irreducible polynomial of degree at least 2, then by Lemma 6.2 there exists an effectively determinable prime number q such that $F(0, y) \pmod q$ has no linear factors. Then $F(a_m, y) \pmod q$ for $m \geqslant m_0(q)$ likewise has no linear factors, and the equality $F(a_m, y) = 0$ for integral y is impossible.

Finally, the case of a linear polynomial $F(0, y)$ is covered by Theorem 2.

THEOREM 5. *Let $F(x, y)$ be an integral polynomial that is irreducible in $\mathbf{Q}[x, y]$. Then for all integers a, except for a finite number of effectively determinable singular numbers a, and for all $m \geqslant m_0(a, F)$ the polynomials $F(a + a_m, y)$ are irreducible in $\mathbf{Q}[y]$, where the a_m are terms of the sequence A introduced in Theorem 4 and $m_0(a, F)$ is a quantity that is explicitly determinable from a and the parameters of F.*

We shall again assume that $F(x, y)$ has the form

$$F(x, y) = y^n + a_1(x)y^{n-1} + \cdots + a_n(x), \qquad a_i(x) \in \mathbf{Z}[x]. \tag{6.10}$$

LEMMA 6.3. *Let $y_1(a), \ldots, y_n(a)$ $(a \in \mathbf{Z})$ be the roots of the polynomial (6.10) for $x = a$, and suppose that for all $a \geqslant a_0 > 0$*

$$\max(|y_1(a)|, \ldots, |y_n(a)|) < A|a|^m. \tag{6.11}$$

Then

$$n_1 = \max_{(i)} \deg a_i(x) \leqslant mn, \qquad \max_{(i)} |\overline{a_i(x)}| < c_{28}A^n,$$

where c_{28} can be determined from n, n_1 and a_0.

PROOF. Since the $a_i(a)$ are symmetric functions of $y_1(a), \ldots, y_n(a)$, from condition (6.11) for $a \geqslant a_0$ we obtain

$$|a_i(a)| < c_{29}(A|a|^m)^i \qquad (i = 1, \ldots, n),\ c_{29} = c_{29}(n).$$

Since the $a_i(x)$ are polynomials, $\deg a_i(x) \leqslant mi$. Lagrange's interpolation formula now yields the estimate indicated in the lemma for the heights of the polynomials $a_i(x)$ $(i = 1, \ldots, n)$.

PROOF OF THEOREM 5. Let $a \in \mathbf{Z}$ and $F(a, y) = F_1(y)F_2(y)$ in $\mathbf{Z}[y]$, where we can assume that $F_1(y)$ and $F_2(y)$ have no common roots. Indeed, the resultant $R_y(F, \partial F/\partial y)$ is not identically zero, since F is irreducible and $\partial F/\partial y$ has degree one less than F with respect to y. If a is not a root of the resultant, then $F(a, y)$ has no multiple roots.

The roots $y_1, \ldots, y_n$ of the polynomial $F(x, y)$ over $\mathbf{Q}(x)$ can be derived into two groups, say, $y_1, \ldots, y_s$ and $y_{s+1}, \ldots, y_n$, such that the polynomials

$$F_1(x, y) = (y - y_1) \cdots (y - y_s), \qquad F_2(x, y) = (y - y_{s+1}) \cdots (y - y_n)$$

for $x = a$ coincide with $F_1(y)$ and $F_2(y)$, respectively. The coefficients $g_1, \ldots, g_s$ of $F_1(x, y)$ over $\mathbf{Q}(x)$ become the coefficients $g_{10}, \ldots, g_{s0}$ of $F_1(y)$ under the substitution $x = a$, i.e., become rational integers. Not all the coefficients $g_1, \ldots, g_s$ can be rational functions of x, since otherwise $F_1(x, y)$ would then belong to $\mathbf{Q}[x, y]$ and the polynomial $F(x, y)$ would be reducible. Let, say, g_1 be irrational, and let $G(x, y) = 0$ be its defining equation, $G(x, y) \in \mathbf{Z}[x, y]$. Then

$$G(a, g_{10}) = 0, \qquad g_{10} \in \mathbf{Z}.$$

Thus, considering all possible divisions of the roots $y_1, \ldots, y_n$ into two groups, we obtain integral polynomials $G_1(x, y), \ldots, G_r(x, y)$ $(r = 2^n - 2)$ of degrees at least 2 with respect to y, irreducible in $\mathbf{Q}[x, y]$, and having the property that the reducibility of $F(a, y)$ implies the equality

$$G_i(a, b_i) = 0, \qquad b_i \in \mathbf{Z},$$

for at least one i ($1 \leqslant i \leqslant r$). Lemma (6.3) shows that the degrees and heights of the polynomials $G_i(x, y)$ are bounded by the quantities $-n_1 n^2$ and $c_{30}(H_F + 1)^{n^3}$, respectively, where $c_{30} = c_{30}(n, n_1)$. Again excluding a finite number of singular numbers a, we can assume that $G_i(a, y)$ has no multiple roots. Application of Theorem 4 to the polynomials $G_i(a + x, y)$ ($i = 1, \ldots, r$) completes the proof of the theorem.

Theorem 1 is one of the simplest examples of the application of the foregoing concepts to specific sequences of the type of the shifted sequences A in Theorem 4, and the proof of Theorem 1 uses the explicit estimates of Lemmas 6.1 and 6.3, inequality (6.9), and a specialization of the function $\chi(m)$ in (6.1). We shall not present the calculations in view of their simplicity.

BIBLIOGRAPHY

1. David Hilbert, *Über die Irreduzibilität ganzer rationaler Functionen mit ganz-zahligen coefficienten*, J. Reine Angew. Math. **110** (1892), 104–129.

2. Emmy Noether, *Gleichungen mit vorgeschriebener Gruppe*, Math. Ann. **78** (1917/18), 221–229.

3. Serge Lang, *Diophantine geometry*, Interscience, 1962.

4. Carl Ludwig Siegel, *Über einige Anwendungen diophantischer Approximationen*, Abh. Preuss. Akad. Wiss. Berlin Phys.-Math. Kl. **1929**, no. 1 (1930).

5. A. Schinzel, *On Hilbert's irreducibility theorem*, Ann. Polon. Math. **16** (1965), 333–340.

6. Michael Fried, *On Hilbert's irreducibility theorem*, J. Number Theory **6** (1974), 211–231.

7. S. D. Cohen, *The distribution of the Galois groups of integral polynomials*, Illinois, J. Math. **23** (1979), 135–152.

8. V. G. Sprindzhuk, *Hilbert's irreducibility theorem and rational points on algebraic curves*, Dokl. Akad. Nauk SSSR **247** (1979), 285–288; English transl. in Soviet Math. Dokl. **20** (1979).

9. ______, *Reducibility of polynomials and rational points on algebraic curves*, Dokl. Akad. Nauk SSSR **250** (1980), 1327–1330; English transl. in Soviet Math. Dokl. **21** (1980).

10. E. C. Titchmarsh, *The theory of functions*, Oxford Univ. Press, 193.

11. Wolfgang M. Schmidt, *Equations over finite fields. An elementary approach*, Lecture Notes in Math., vol. 536, Springer-Verlag, 1976.

12. Michel Waldschmidt, *Transcendence methods*, Queen's Papers in Pure and Appl. Math., no. 52, Queen's Univ., Kingston, Ont., 1979.

13. K. Wiertelak, *On the density of some sets of primes*.II, Acta Arith. **34** (1977/78), 197–210.

Translated by R. LENET

Труды Матем. Инст. Стеклов.
158(1981)

Proceedings of the Steklov
Institute of Mathematics
1983, Issue 4

ON APPROXIMATE FUNCTIONAL EQUATIONS OF THE FIRST KIND

UDC 517.948

A. N. TIKHONOV

ABSTRACT. A new algorithm, based on the method of regularization, is presented for constructing approximate solutions of functional equations $Az = u$ of the first kind that are stable under small changes in the initial information about the equation (A, u).

Bibliography: 6 titles.

1. Let us consider the inverse problem for a linear functional equation $Az = u$ of the first kind [1], where z and u are elements of the Hilbert spaces H_z and H_u, and A $(H_z \to H_u)$ is a linear operator in the Hilbert space of operators with absolute norm

$$\|A\| = N(A) = \left[\sum_{i\,j} (\psi_i, A\varphi_j)^2\right]^{1/2},$$

$\{\varphi_i\}$ and $\{\psi_j\}$ being orthogonal bases in H_u and H_z. It is assumed that A and u are given and z is sought. This problem is determined by the pair (A, u). In the solution of many practical problems an individual approximate equation $\tilde{A}z = \tilde{u}$ is usually given. However, this equation may not have a solution, and even if there is a solution $\tilde{z}$, it may be unstable, i.e, for arbitrarily small perturbations of $\tilde{u}$ and $\tilde{A}$ the solution $\tilde{z}$ (even if it exists) may undergo arbitrarily large changes (the problem may be ill posed) [2], [5]. Thus, the information given by $(\tilde{A}, \tilde{u})$ about the original problem (A, u) is insufficient.

We consider the approximate specification of a functional equation of the first kind as a specification of an individual approximate equation $(\tilde{A}, \tilde{u})$ along with the "accuracy" of its specification. This "accuracy" can be given in various forms, for example as classes $\tilde{\mathfrak{A}} = \{A\}$ and $\tilde{U} = \{u\}$ equivalent in accuracy to $\tilde{A}$ and $\tilde{u}$.

In what follows we assume that the concept of "equivalence in accuracy" is defined with the help of the norms in the Hilbert spaces H_u and H_A:

$$\mathfrak{A}_\mu = \{A \colon \|A - \tilde{A}\| \leqslant \mu\}, \qquad U_\delta = \{u \colon \|u - \tilde{u}\| \leqslant \delta\}.$$

Thus, we consider the approximate specification of the equation $Az = u$ as the specification of the collection of data $\{(\tilde{A}, \tilde{u}), \mu, \delta\}$, where $\tilde{A}z = \tilde{u}$ is an individual approximate problem, and the numbers μ and δ are the accuracy parameters. This collection of data determines a system of equations equivalent to the pair $(\tilde{A}, \tilde{u})$:

$$\tilde{\Sigma}(\delta, \mu) = \{(A, u) \colon \|A - \tilde{A}\| \leqslant \mu, \|u - \tilde{u}\| < \delta\}.$$

1980 *Mathematics Subject Classification.* Primary 47A50, 41A65.

An element z is said to be an *admissible solution* of the system $\tilde{\Sigma}(\delta, \mu)$ if there exists a pair $(A, u) \in \tilde{\Sigma}(\delta, \mu)$ such that $Az = u$. Obviously, the existence of an admissible solution z of the problem $\tilde{\Sigma}(\delta, \mu)$ is equivalent to the solvability of the inequality

$$\|\tilde{u} - Az\| \leqslant \delta, \qquad A \in \tilde{\mathfrak{A}}_\mu = \{A : \|A - \tilde{A}\| \leqslant \mu\}.$$

The collection of all admissible solutions of $\tilde{\Sigma}(\delta, \mu)$ is denoted by

$$\tilde{Z}(\delta, \mu) = \{z : Az = u, (A, u) \in \tilde{\Sigma}(\delta, \mu)\} = \{z : \|\tilde{u} - Az\| \leqslant \delta; \|A - \tilde{A}\| \leqslant \mu\}.$$

An approximate system $\tilde{\Sigma}(\delta, \mu)$ is said to be *comparable* if $\tilde{Z}(\delta, \mu) \neq \varnothing$. If $\tilde{Z}(\delta, \mu)$ contains several admissible solutions, then they are all equivalent in accuracy, by the meaning of the problem.

We consider the functional $\Omega(z) = \|z\|$, which we call the *complexity functional.* A *normal solution* of the approximate system $\tilde{\Sigma}(\delta, \mu)$ with respect to the functional $\Omega(z)$ is defined to be a solution $\tilde{Z}(\delta, \mu)$ of the extremal problem

$$\Omega(\tilde{z}_{\delta\mu}) = \inf \Omega(z) \quad \text{on } \tilde{Z}(\delta, \mu).$$

The problem of finding an normal solution of an approximately specified functional equation of the first kind $\tilde{\Sigma}(\delta, \mu)$ is defined as a variational problem $\tilde{N}^{(0)}(\delta, \mu)$ with two constraints of inequality type: find an element $\tilde{z}_{\delta\mu} \in H_z$ such that

$$\|\tilde{z}_{\delta\mu}\| = \inf\|z\| = \tilde{\eta}(\delta, \mu) \quad \text{on } \tilde{Z}(\delta, \mu) = \{z : \|\tilde{u} - Az\| \leqslant \delta, \|A - \tilde{A}\| \leqslant \mu\}.$$

The quantity $\tilde{\eta}(\delta, \mu)$ is called the *complexity threshold* of the problem $\tilde{Z}(\delta, \mu)$, since there do not exist any admissible solutions $z \in \tilde{Z}(\delta, \mu)$ with $\|z\| < \tilde{\eta}(\delta, \mu)$.

2. Consider the system of equations $Az = u$ for given $z \in H_z$ ($z \neq 0$) and $u \in H_u$ with respect to the operator A.

MAIN LEMMA. *Among the operators having a finite absolute norm there exists a unique operator $\hat{A}$ which is minimal in norm and solves the equation $Az = u$ ($z \neq 0$), and this operator is given by*

$$\hat{A} = \frac{u(z, \cdot)}{\|z\|^2}, \quad or \quad a_{ij} = (\psi_i \hat{A} \varphi_j) = \frac{(\psi_i \cdot u)(z \cdot \varphi_i)}{\|z\|^2} = \frac{u_i z_j}{\|z\|^2}.$$

Recall that the absolute norm is a convex norm:

$$N(A'') \leqslant N(A') + N(\hat{A}) \quad \text{if } A'' = A' + \hat{A}.$$

A direct verification shows that $\hat{A}$ satisfies the condition $\hat{A}z = u$, and

$$\|\hat{A}\|^2 = N^2(\hat{A}) = \sum_{ij}^{\infty} (a_{ij})^2 = \frac{\|u\|^2}{\|z\|^2}, \qquad \|\hat{A}\| = \frac{\|u\|}{\|z\|}.$$

If A' is another solution of the problem that is minimal in norm, then $\|u\| \leqslant \|A'\| \cdot \|z\|$, i.e., $N(A') \geqslant \|u\|/\|z\| = N(\hat{A})$, so that $N(A') = N(\hat{A})$.

We show that if $N(A') = N(\hat{A})$, then $A' = \hat{A}$. Consider the operator

$$A'' = \tfrac{1}{2}(A' + \hat{A}),$$

which also is a solution of the problem. If $A' \not\equiv \hat{A}$, then

$$N(A'') < \tfrac{1}{2}(N(A') + N(\hat{A})) = \|u\|/\|z\|,$$

as follows from the convexity of the absolute norm. This proves the lemma.

3. A scheme will be given later for proving the existence and uniqueness of a solution. We note first that if $\tilde{z}_{\delta\mu}$ exists and $(\overline{A}, \bar{u})$ is an individual system realizing it, $\overline{Z}z_{\delta\mu} = \bar{u}$, then $\overline{A}$ and $\bar{u}$ lie on the boundaries of $\tilde{\mathfrak{A}}_\mu$ and $\tilde{U}_\delta$.

THEOREM 3.1. *If $(\overline{A}, \bar{u})$ realizes a normal solution $\tilde{z}_{\delta\mu}$ of $\tilde{\Sigma}(\delta, \mu)$, then $\|\overline{A} - \tilde{A}\| = \mu$.*

THEOREM 3.2. *If $(\overline{A}, \bar{u})$ realizes a normal solution $\tilde{z}_{\delta\mu}$ of $\tilde{\Sigma}(\delta, \mu)$, then $\|\bar{u} - \tilde{u}\| = \delta$.*

Thus, the problem of determining $\tilde{z}_{\delta\mu}$ reduces to problem $\tilde{N}^{(1)}(\delta, \mu)$: find $\tilde{z}_{\delta\mu}$ such that

$$\|\tilde{z}_{\delta\mu}\| = \inf\|z\| = \tilde{\eta}(\delta, \mu) \quad \text{on } Z^{(1)}(\delta, \mu) = \{z : \|\tilde{u} - Az\| = \delta, \|A - \tilde{A}\| = \mu\}.$$

4. Variational problem: find an operator $\hat{A}$ realizing $\inf\|\tilde{u} - Az\| = \|\tilde{u} - \hat{A}z\|$ on $\tilde{\mathfrak{A}}_\mu = \{A : \|A - \tilde{A}\| \leqslant \mu\}$ for given $z \in H_z$ and $u \in H_u$.

Let $\rho(z) = \tilde{u} - Az$, so that $\tilde{\rho}(z) = \tilde{u} - \tilde{A}z$ and

$$\inf\|\tilde{u} - Az\| = \inf\|\rho(z)\| = \|\hat{\rho}(z)\|, \quad \hat{\rho}(z) = \tilde{\rho}(z) - \Delta z \qquad (\hat{A} = \tilde{A} + \Delta).$$

We also introduce the notation

$$\nu(z) = \|\rho(z)\| - \mu\|z\| = \|\tilde{u} - Az\| - \mu\|z\|,$$
$$\tilde{\nu}(z) = \|\tilde{\rho}(z)\| - \mu\|z\| = \|\tilde{u} - \tilde{A}z\| - \mu\|z\|.$$

THEOREM 4.1. *If $\tilde{\nu}(z) \leqslant 0$, or $\|\tilde{\rho}(z)\| = \mu\theta\|z\|$, $0 < \theta \leqslant 1$, then the equation $Az = \tilde{u}$ is solvable, and the unique solution $\hat{A}$ deviating least in norm from $\tilde{A}$ is equal to $\tilde{A} + \hat{\Delta}$, where*

$$\hat{\Delta} = \frac{\tilde{\rho}(z)(z, \cdot)}{\|z\|^2} = \theta\mu\frac{\tilde{\rho}(z)}{\|\tilde{\rho}(z)\|}\frac{(z, \cdot)}{\|z\|}, \qquad \|\hat{\Delta}\| = \theta\mu.$$

We note that $z \neq 0$ if $\theta > 0$, since otherwise $\|\tilde{\rho}(0)\| = \|\tilde{u}\| = 0$.

But if $\theta = 0$, then $\|\tilde{\rho}(z)\| = 0$, i.e., the equation $\tilde{A}z = \tilde{u}$ is solvable.

COROLLARY. *If $\tilde{z}_{\delta\mu}$ is a normal solution of the system $\tilde{\Sigma}(\delta, \mu)$, then $\tilde{\nu}(\tilde{z}_{\delta\mu}) > 0$.*

Indeed, otherwise $\tilde{z}_{\delta\mu}$ would satisfy the equation $\tilde{u} - \hat{A}z_{\delta\mu} = 0$ ($\hat{A} \in \tilde{\mathfrak{A}}_\mu$), but for a normal solution we have $\|\tilde{u} - \hat{A}\tilde{z}_{\delta\mu}\| = \delta$.

THEOREM 4.2. *If $\tilde{\nu}(z) > 0$, i.e., $\|\tilde{\rho}(z)\| > \mu\|z\|$, then* $\inf\|\tilde{u} - Az\|$ *on $\tilde{\mathfrak{A}}_\mu$ is realized for the unique matrix $\hat{A} + \tilde{A} + \hat{\Delta}$, where*

$$\hat{\Delta} = \mu\frac{\tilde{\rho}(z)}{\|\tilde{\rho}(z)\|}\frac{(z, \cdot)}{\|z\|}, \qquad \|\hat{\Delta}\| = \mu,$$

and, moreover,

$$\|\hat{\rho}(z)\| = \|\tilde{\rho}(z)\| - \mu\|z\|.$$

5. Problem $N^{(2)}(\delta, \mu)$ [3], [4]. Find an element $\tilde{z}_2$ such that

$$\|\tilde{z}_2\| = \inf\|z\| = \eta(\delta, \mu) \quad \text{on } Z^{(2)}(\delta, \mu) = \{z : \tilde{\nu}(z) \leqslant \delta\}.$$

REMARK. For $\tilde{z}_2$ we have $\tilde{\nu}(\tilde{z}_2) = \delta$, i.e.,

$$\|\tilde{u} - \tilde{A}\tilde{z}_2\| - \mu\|\tilde{z}_2\| = \delta.$$

THEOREM 5.2. *The sets of solutions of the problems $\tilde{N}^{(0)}(\delta, \mu)$ and $\tilde{N}^{(2)}(\delta, \mu)$ coincide.*

Thus, there are two possibilities:

1) If the set of solutions of one of the problems $\tilde{N}^{(0)}(\delta, \mu)$ or $\tilde{N}^{(2)}(\delta, \mu)$ is nonempty, then so is the set of solutions of the other problem and they coincide.

2) The sets of solutions of the two problems are empty.

6. Let us turn to the proof of existence and uniqueness theorems for these problems.

The problems

$$\tilde{n}(\delta)\colon \text{find } \tilde{z}_\delta\colon \|\tilde{z}_\delta\| = \inf\|z\| = \tilde{\eta}(\delta) \quad \text{on } \tilde{Z}(\delta) = \{z\colon \|\tilde{\rho}(z)\| \leqslant \delta\},$$

$$[\tilde{n}](\eta)\colon \text{find } \tilde{z}_\eta\colon \|\tilde{\rho}(z)\| = \inf\|\tilde{\rho}(z)\| = \tilde{\delta}(n) \quad \text{on } \tilde{Z}(\eta) = \{z\colon \|z\| \leqslant \eta\}$$

with a fixed operator $\tilde{A}$ have coinciding solutions [6]

$$\tilde{z}_\delta = \tilde{z}_\eta \quad \text{for } \delta = \tilde{\delta}(\eta) \text{ or } \eta = \tilde{\eta}(\delta),$$

whose existence and uniqueness follows from the linearity of $\tilde{A}$.

The solution of the problem $[\tilde{N}^{(2)}](\delta, \mu)$,

$$\text{find } \tilde{z}^{(2)}_{\eta\mu}\colon \tilde{\nu}\left(\tilde{z}^{(2)}_{\eta\mu}\right) = \inf \tilde{\nu}(z) = \tilde{\delta}(\eta, \mu) \quad \text{on } Z_\eta(\tilde{\nu}(z) = \|\tilde{\rho}(z)\| - \mu\|z\|),$$

obviously coincides with the solution of the problem $[\tilde{\eta}](\eta)$. Thus, $\tilde{z}^{(2)}_{\eta\mu} = \tilde{z}_\eta$, and $\tilde{\delta}(\eta, \mu) = \delta(\eta) - \mu\eta$, which implies the existence and uniqueness of $\tilde{z}^{(2)}(\delta, \mu)$.

The problem $\tilde{N}^{(2)}(\delta, \mu)$,

$$\text{find } \tilde{z}_{\delta\mu}\colon \|\tilde{z}_{\delta\mu}\| = \inf\|z\| = \tilde{\eta}(\delta, \mu) \quad \text{on } Z^{(2)}_{\delta\mu} = \{z\colon \|\tilde{\rho}(z)\| - \mu\|z\| = \delta\},$$

has a solution which coincides with $\tilde{z}(\eta, \mu)$ if $\delta = \tilde{\delta}(\eta, \mu)$, so that $\tilde{z}_{\delta\mu} = \tilde{z}_{\hat{\eta}\mu}$ for $\delta = \delta(\hat{\eta}, \mu)$.

Thus, to establish conditions under which there exists a unique $\tilde{z}_{\delta\mu}$ it suffices to establish conditions for the equation $\delta = \delta(\hat{\eta}, \mu)$ to be solvable with respect to η; we denote such conditions by Δ in what follows.

The problem $\tilde{N}^{(0)}(\delta, \mu)$,

$$\text{find } \tilde{z}(\delta, \mu)\colon \|\tilde{z}(\delta, \mu)\| = \inf\|z\| = \eta(\delta, \mu)$$
$$\text{on } Z^{(0)}(\delta, \mu) = \{z\colon \|\tilde{u} - Az\| \leqslant \delta, \|A - \tilde{A}\| \leqslant \mu\},$$

has solutions that coincide with the solutions of $\tilde{N}^{(2)}(\delta, \mu)$, so that the existence and uniqueness conditions for the solution of $\tilde{N}^{(0)}(\delta, \mu)$ coincide with the solvability conditions for $\tilde{N}^{(2)}(\delta, \mu)$ (the conditions Δ).

7. We turn to the condition Δ and an algorithm for computing $\tilde{z}(\delta, \mu)$. The condition Δ amounts to the solvability of the equation

$$\mathring{\tilde{\delta}}(\hat{\eta}, \mu) = \mathring{\tilde{\delta}}(\hat{\eta}) - \mu\hat{\eta} = \delta$$

with respect to $\hat{\eta}$. This is a transcendental equation containing the function $\mathring{\tilde{\delta}}(\eta)$, which is determined by the problem $[\tilde{n}](\delta)$ for the linear functional equation of the first kind $\tilde{A}z = u$, $\|u - \tilde{u}\| \leqslant \delta$, with a fixed operator $\tilde{A}$ and an approximately specified right-hand side u with accuracy δ.

This problem has been thoroughly studied, and approximate calculation of the function $\mathring{\tilde{\delta}}(\eta)$ has been made into a routine on a computer.

Let $h^* = \inf\|\tilde{u} - \tilde{A}z\|$ on H_z. Denote by $\tilde{u}^*$ the projection of $\tilde{u}$ on $U_{\tilde{A}} = \{u\colon u = \tilde{A}z, z \subset H_z\}$, and let z^* be a normal solution of the equation $\tilde{u}^* = \tilde{A}z^*$ (denote

$\|z^*\| = \eta^*$), so that $h^* = \|\tilde{\rho}(z^*)\| = \|\tilde{u} - \tilde{A}z^*\|$. The quantity h^* represents the solvability threshold of the equation $\tilde{A}z = u$, $\|u - \tilde{u}\| \leqslant \delta$, since no solutions of this equation exist for $\delta < h^*$. Thus, condition Δ for the solvability of the equation

$$\check{\delta}(\hat{\eta}, \mu) = \check{\delta}(\hat{\eta}) - \mu\hat{\eta} = \delta$$

is as follows:

$$\Delta \begin{cases} \text{for every } \delta \text{ we have } 0 \leqslant \delta \leqslant \|\tilde{u}\| & \text{if } h^* \leqslant \mu\eta^*\big(\check{\delta}(\eta^*, \mu) \leqslant 0\big), \\ \text{for } \delta \text{ we have } h^* - \mu\eta^* \leqslant \delta \leqslant \|\tilde{u}\| & \text{if } h^* > \mu\eta^*\big(\check{\delta}(\eta^*, \mu) > 0\big), \end{cases}$$

and $0 \leqslant \hat{\eta} \leqslant \eta^*$.

To find $\tilde{z}_{\hat{\eta}\mu} = \tilde{z}(\hat{\eta})$ it now suffices to solve problem $[\tilde{n}](\hat{\eta})$ or problem $n(\delta)$ for $\check{\delta}(\hat{\eta}) = \delta + \mu\hat{\eta}$, and this, as mentioned above, has been made into a routine on a computer.

Thus, we have

THEOREM 7.1. *If the approximated system $\tilde{\Sigma}(\delta, \mu)$ is comparable, then it has a unique normal solution. A necessary and sufficient condition for comparability is the relation Δ.*

8. Let us consider the linear functional equation $\bar{A}z = \bar{u}$ with exactly specified $\bar{A}$ and $\bar{u}$. Denote by $\bar{z}$ a normal solution of this equation.

The system $\tilde{\Sigma}(\delta, \mu)$ will be called an *approximate system* for $(\bar{A}, \bar{u})$ if $(\bar{A}, \bar{u}) \in \tilde{\Sigma}(\delta, \mu)$. We have

THEOREM 8.1. *For any degree of accuracy $\varepsilon > 0$ there exist values $\delta(\varepsilon) > 0$ and $\mu(\varepsilon) > 0$ such that for $\delta < \delta(\varepsilon)$ and $\mu < \mu(\varepsilon)$ a normal solution $\tilde{z}_{\delta\mu}$ of the approximate system $\tilde{\Sigma}(\delta, \mu)$ deviates from the normal solution $\bar{z}$ of $(\bar{A}, \bar{u})$ by not more than ε:*

$$\|\tilde{z}_{\delta\mu} - \bar{z}\| \leqslant \varepsilon \qquad (\delta < \delta(\varepsilon), \mu < \mu(\varepsilon)).$$

This theorem shows the stability of $\tilde{z}_{\delta\mu}$ for small values of δ and μ.

BIBLIOGRAPHY

1. A. N. Tikhonov, *Nonlinear equations of the first kind*, Dokl. Akad. Nauk SSSR **161** (1965), 1023–1026; English transl. in Soviet Math. Dokl. **6** (1965).

2. A. N. Tikhonov and V. Ya. Arsenin, *Methods of solving ill-posed problems*, 2nd ed., "Nauka", Moscow, 1979; English transl. of 1st ed., Wiley, 1977.

3. V. A. Morozov, *The residual principle in solving operator equations by the method of regularization*, Zh. Vychisl. Mat. i Mat. Fiz. **8** (1968), 295–309; English transl. in USSR Comput. Math. and Math. Phys. **8** (1968).

4. A. V. Goncharskiĭ, A. S. Leonov and A. G. Yagola, *A generalization of the residual principle for the case of an operator specified with an error*, Dokl. Akad. Nauk SSSR **203** (1972), 1238–1239; English transl. in Soviet Math. Dokl. **13** (1972).

5. V. K. Ivanov, V. V. Vasin, and V. P. Tanana, *Theory of linear ill-posed problems and its applications*, "Nauka", Moscow, 1978. (Russian)

6. A. N. Tikhonov, *On a reciprocity principle*, Dokl. Akad. Nauk SSSR **253** (1980), 305–308; English transl. in Soviet Math. Dokl. **22** (1980).

Translated by H. H. McFADEN

Труды Матем. Инст. Стеклов.
158(1981)

Proceedings of the Steklov
Institute of Mathematics
1983, Issue 4

AN ASYMPTOTIC FORMULA FOR THE MEAN VALUES OF A FRACTIONAL POWER OF THE RIEMANN ZETA FUNCTION

UDC 511

R. T. TURGANALIEV

ABSTRACT. In this article the following theorem is proved.

THEOREM. *If* $0 < \lambda < 2$ *and* $s = \sigma + ti$, $\frac{1}{2} < \sigma < 1$, *then the following asymptotic formula is valid*:

$$\int_1^T |\zeta(s)|^{2\lambda}\, dt = C(\sigma, \lambda)T + O(T^{1-\kappa}),$$

where $C(\sigma, \lambda)$ *is the sum of the convergent series* $\sum_{n=1}^{\infty} \alpha_\lambda^2(n)/n^{2\sigma}$, *the* $\alpha_\lambda(n)$ *are the coefficients of the Dirichlet series expansion of* $\zeta^\lambda(s)$ *for* $\operatorname{Re} s > 1$, *and* $0 < \kappa = \kappa(\sigma, \lambda)$.

Figures: 2. Bibliography: 10 titles.

Notation

$s = \sigma + it$ is a complex number; $\sigma = \operatorname{Re} s$ and $t = \operatorname{Im} s$.

$\zeta(s)$ is the Riemann zeta function.

$\Gamma(s)$ is the Euler gamma function.

$A \ll B$ or $A = O(B)$ indicates that $|A| \leqslant cB$, where c is a positive constant.

$A \asymp B$ signifies that $c_1 A \leqslant B \leqslant c_2 A$, where c_1 and c_2 are absolute positive constants.

$\prod_p$ is the product over prime numbers.

$[\alpha]$ is the integral part of a real number α.

Introduction

Ingham and Davenport proved the following theorem ([8], Theorem 7.11).

For $0 < \lambda \leqslant 2$

$$\lim_{T\to\infty} \frac{1}{T}\int_1^T |\zeta(\sigma + it)|^{2\lambda}\, dt = \sum_{n=1}^{\infty} \frac{\alpha_\lambda^2(n)}{n^{2\sigma}} \qquad \left(\sigma > \frac{1}{2}\right),$$

where the $\alpha_\lambda(n)$ *are the coefficients of the Dirichlet series expansion of* $\zeta^\lambda(s)$ *for* $\operatorname{Re} s > 1$.

In the course of the proof, by using the finite Euler product

$$\zeta_N^\lambda(s) = \prod_{p<N} (1 - p^{-s})^{-1},$$

it is proved that

$$\lim_{N\to\infty} \lim_{T\to\infty} \frac{1}{T}\int_1^T |\zeta(\sigma + it) - \zeta_N(\sigma + it)|^{2\lambda}\, dt = 0 \qquad \left(\sigma > \frac{1}{2}\right).$$

1980 *Mathematics Subject Classification.* Primary 10H05, 10H25.

0081-5438/84 $1.00 + $.25 per page

Here, as we see, no effective estimate is given. However, an effective estimate with the help of a finite Euler product yields, for example, only a logarithmic reduction of the remainder for the case $\lambda = 1/2$, i.e., the following formula is valid:

$$\int_1^T |\zeta(s)|\, dt = C(\sigma)T + O\left(\frac{T}{\ln T}\right),$$

where $C(\sigma) = \sum_1^\infty \alpha_n^2/n^{2\sigma}$ and the α_n are the coefficients of the Dirichlet series expansion of the function $\zeta_N^{1/2}(s) = \prod_{p<N}(1 - 1/p^s)^{-\lambda}$ for $\operatorname{Re} s > 0$.

In this paper an idea of S. M. Voronin is used: instead of the Euler product, consider the finite partial sum of the first N terms of the Dirichlet series expansion of $\zeta^\lambda(s)$ for $\operatorname{Re} s > 1$. This makes it possible to obtain a remainder in the asymptotic formula having a power reduction.

Let us state the main result of the article.

THEOREM. *If* $0 < \lambda < 2$ *and* $s = \sigma + ti$, $1/2 < \sigma < 1$, *then the following asymptotic formula is valid*:

$$\int_1^T |\zeta(s)|^{2\lambda} dt = C(\sigma, \lambda)T + O(T^{1-\kappa}),$$

where $C(\sigma, \lambda)$ *is the sum of the convergent series* $\sum_{n=1}^\infty \alpha_\lambda^2(n)/n^{2\sigma}$ *and the* $\alpha_\lambda(n)$ *are the coefficients of the Dirichlet series expansion of* $\zeta^\lambda(s)$ *for* $\operatorname{Re} s > 1$ *and* $0 < \kappa = \kappa(\sigma, \lambda)$.

The theorem is first proved for $\lambda = 1/2$, then for $0 < \lambda < 2$ under the assumption that the Riemann hypothesis is valid, and finally for $0 < \lambda < 2$ without assuming any conjectures.

In the course of the proof the lemmas stated below are used, as well as the convexity theorem, Hadamard's three circles inequality, and the density theorem for the number of nontrivial zeros of the Riemann zeta function. The main result can be improved if new, more precise density theorems are used.

This paper was presented at the analytic number theory seminar of Professor A. A. Karacuba, which the author attended during the last two years.

The author is deeply grateful to Professor A. A. Karacuba and Doctor of Physical and Mathematical Sciences, S. M. Voronin, who assisted in the preparation of this paper.

1. Lemmas

LEMMA 1. *Let* $0 < \sigma < 2$. *Then*

$$\sum_{0<m<n<T}\sum \frac{1}{m^\sigma n^\sigma \ln(n/m)} \ll \frac{T^\varepsilon}{(1-\sigma)^2}(T^{2-2\sigma} + 1) \qquad (\sigma \neq 1);$$

$$\sum_{0<m<n<T}\sum \frac{1}{mn \ln(n/m)} \ll T^\varepsilon,$$

where the constants in the $\ll$ *signs depend on* ε.

The proof is similar to the proof of Lemma 1 in [8], Chapter VII, §2.

LEMMA 2. *Let $k \geqslant 2$ be an integer. Then*

$$\sum\sum_{0<m<n<T} \frac{\tau_k(m)\tau_k(n)}{m^\sigma n^\sigma \ln(n/m)} \ll \frac{T^{2-2\sigma+\varepsilon}}{(1-\sigma)^2} \qquad \left(\frac{1}{2} \leqslant \sigma < 1\right);$$

$$\sum\sum_{0<m<n<T} \frac{\tau_k(m)\tau_k(n)}{mn\ln(n/m)} \ll T^\varepsilon,$$

where the constants in the $\ll$ signs depend on ε.

This is a consequence of Lemma 1, since $\tau_k(n) \ll n^\varepsilon$.

LEMMA 3. *For $\sigma > 1$*

$$\sum\sum_{N<m<n} \frac{\tau_k(m)\tau_k(n)}{m^\sigma n^\sigma \ln(n/m)} \ll \frac{[(\sigma-1)\ln N + 1]^{k^2}}{N^{\sigma-1}(\sigma-1)^{k^2+1}},$$

and for $\sigma > 1/2$

$$\sum_{N<n} \frac{\tau_k^2(n)}{n^{2\sigma}} \ll \frac{[(2\sigma-1)\ln N + 1]^{k^2-1}}{N^{2\sigma-1}(2\sigma-1)^{k^2}},$$

where the constants in the $\ll$ signs are absolute.

PROOF (cf. [6], Chapter IX, §1, Proof of Lemma 1.10, and [8]. §9.18). For $n/m > 1$ we have $\ln(n/m) > 1 - 1/\sqrt{n/m}$, whence

$$\frac{1}{\ln(n/m)} < 1 + \frac{1}{(n/m)^{1/2}\ln(n/m)}.$$

Let $\sigma = 1 + \delta$, $\delta > 0$. Then from the preceding inequality we obtain

$$\begin{aligned}\sum &= \sum\sum_{N<m<n} \frac{\tau_k(m)\tau_k(n)}{m^\sigma n^\sigma \ln(n/m)} \\ &< \sum\sum_{N<m<n} \frac{\tau_k(m)\tau_k(n)}{m^{1+\delta}n^{1+\delta}} + \sum\sum_{N<m<\infty} \frac{\tau_k(m)\tau_k(n)}{m^{1/2}n^{1/2}m^\delta n^{1+\delta}\ln(n/m)} \\ &< \left(\sum_{n=1}^{\infty} \frac{\tau_k(n)}{n^{1+\delta}}\right)^2 + \sum\sum_{N<m<n} \frac{\tau_k(m)\tau_k(n)}{(mn)^{1/2}\ln(n/m)} \int_n^\infty \frac{1+\delta}{x^{2+\delta}}\,dx \\ &= \{\zeta(1+\delta)\}^{2k} + \Sigma_1.\end{aligned}$$

We apply Abel's transformation ([4], Chapter III, §5, Lemma 3) to the sum Σ_1:

$$\sum_{a<n\leqslant b} c_n f(n) = -\int_a^b \mathbf{C}(x)f'(x)\,dx + \mathbf{C}(b)f(b), \qquad \mathbf{C}(x) = \sum_{a<n\leqslant x} c_n.$$

We set

$$\mathbf{C}(y) = \sum\sum_{N<m<n<y} \frac{\tau_k(m)\tau_k(n)}{(mn)^{1/2}\ln(n/m)}, \qquad f(x) = \int_X^Y \frac{1+\delta}{t^{2+\delta}}\,dt.$$

We then have

$$\Sigma_1 = \int_N^Y \left(\sum\sum_{N<m<n\leqslant X} \frac{\tau_k(m)\tau_k(n)}{(mn)^{1/2}\ln(n/m)}\right) \frac{1+\delta}{x^{2+\delta}}\,dx. \tag{1}$$

To estimate the double sum under the integral, we shall prove that for $k \geqslant 2$ and $1/2 \leqslant \sigma < 1$

$$\Sigma_2 = \sum\sum_{0<m<n\leqslant X} \frac{\tau_k(m)\tau_k(n)}{m^\sigma n^\sigma \ln(n/m)} \ll X(\ln X)^{k^2}. \tag{2}$$

Indeed, since (cf. [9])

$$\frac{1}{\ln(n/m)} = \frac{1}{-\ln(1-(n-m)/n)} < 1 + \frac{m}{n-m}$$
$$= 1 + \frac{\sqrt{mn}}{m-n} \leqslant 1 + \frac{m^\sigma n^\sigma}{n-m},$$

we have

$$\Sigma_2 < \sum\sum_{0<m<n\leqslant X} \frac{\tau_k(m)\tau_k(n)}{m^\sigma n^\sigma} + \sum\sum_{0<m<n\leqslant X} \frac{\tau_k(m)\tau_k(n)}{n-m} = \Sigma_3 + \Sigma_4,$$

$$\Sigma_3 < \left(\sum_{n=1}^{X} \frac{\tau_k(n)}{n^\sigma}\right)^2 = \left(1 + \sum_{1<n\leqslant X} \frac{\tau_k(n)}{n^\sigma}\right)^2.$$

We apply Abel's transformation to the last sum. Then on the basis of the formula (see [8], §12.1)

$$\sum_{0<n\leqslant X} \tau_k(n) = X\mathcal{P}_k(\ln X) + O(X^{(k-1)/(k+1)} \ln^{k-2} X),$$

where $\mathcal{P}_k$ is a polynomial of degree $k-1$, $k = 2, 3, \ldots$, we have

$$\Sigma_3 \ll X^{2-2\sigma} \mathcal{P}_k^2(\ln X). \tag{3}$$

To estimate Σ_4 we introduce the notation $n - m = r$. We have

$$\Sigma_4 = \sum_{r=1}^{X-1} \frac{1}{r} \sum_{m=1}^{X-r} \tau_k(m)\tau_k(m+r).$$

Applying Cauchy's inequality to the inner sum, on the basis of the inequality (see [3], Chapter I, Lemma 17)

$$\sum_{0<n\leqslant X} (\tau_k(n))^2 \ll X(\ln X + 1)^{k^2-1} \tag{4}$$

we obtain

$$\Sigma_4 \ll X(\ln X)^{k^2-1} \sum_{r=1}^{X-1} \frac{1}{r} \ll X(\ln X)^{k^2}.$$

Therefore, by (3) and (4) inequality (2) is proved, and from (1) we obtain

$$\Sigma_1 \ll \int_N^\infty \frac{(\ln x)^{k^2}}{x^{1+\delta}} dx.$$

Furthermore, setting $x = Nu^{1/\delta}$, we have

$$\int_N^\infty \frac{(\ln x)^{k^2}}{x^{2\sigma}} dx = \frac{1}{\delta N^\delta} \int_1^\infty \frac{(\ln N + \delta^{-1} \ln u)^{k^2}}{u^2} du.$$

Expanding the binomial by Newton's formula and noting that

$$\int_1^\infty \frac{du}{u^2}=1,\ \int_1^\infty \frac{\ln u}{u^2}du=1,\ \int_1^\infty \frac{(\ln u)^2}{u^2}du=2!,\dots,\int_1^\infty \frac{(\ln u)^{k^2}}{u^2}=(k^2)!,$$

we obtain

$$\Sigma_1 \ll \frac{1}{\delta N^\delta}\left(\ln N+\frac{1}{\delta}\right)^{k^2},\quad \text{where } \delta=\sigma-1.$$

Therefore, the first part of the lemma is proved. The second part of the lemma is obtained by using Abel's transformation and (3).

LEMMA 4. *Let φ be a finite sequence of numbers $s=\sigma+ti$, and let T_0, T, σ_0, and δ be real numbers such that*

$$T_0+\delta/2\leqslant t\leqslant T_0+T-\delta/2,\qquad \sigma>\sigma_0,$$
$$|t-t'|\geqslant\delta\quad \text{for } s\in\varphi,\ s'\in\varphi,\ s\neq s'.$$

Then

$$\sum_{s\in\varphi}\left|\sum_{n=1}^N \frac{a_n}{n^s}\right|^2 \ll (T+N)(\delta^{-1}+\ln N)\sum_{n=1}^N \frac{|a_n|^2}{n^{2\sigma_0}}\left(1+\log\frac{\log 2N}{\log 2n}\right).$$

The proof follows from [5] (Theorem 7.5).

LEMMA 5. *The following formulas are valid*:

$$\lim_{T\to\infty}\frac{1}{T}\int_1^T|\zeta(\sigma+it)|^4\,dt=\frac{\zeta^4(2\sigma)}{\zeta(4\sigma)}\qquad\left(\sigma>\frac{1}{2}\right),$$

$$\int_0^T\left|\zeta\left(\frac{1}{2}+it\right)\right|^4 dt=a_4T(\log T)^4+a_3T(\log T)^3+a_2T(\log T)^2$$
$$+a_1T\log T+a_0T+O(T^{7/8+\varepsilon}),$$

where $T\geqslant 2$ and $a_0,\dots,a_4$ are constants.

The first assertion is the theorem in [8], Chapter VII, §5, and the second is Theorem 1 in [10].

§2. Asymptotic formula for the integral of the modulus of the Riemann zeta function

In this section we prove the formula

$$\int_1^T|\zeta(s)|\,dt=C(\sigma)T+O(T^{5/4-\sigma/2+\varepsilon})\qquad\left(\frac{1}{2}<\sigma<1\right),\tag{1}$$

where $C(\sigma)=\sum_1^\infty \alpha_n^2/n^{2\sigma}$ and α_n are the coefficients of the Dirichlet series expansion of the function $\zeta^{1/2}(s)$ for $\operatorname{Re}s>1$.

PROOF. It is known (see [4], Chapter III, §5, Theorem 6) that for $0<\sigma_0\leqslant\sigma\leqslant 2$ and $X\geqslant|t|/\pi$

$$\zeta(s)=\sum_{n\leqslant X}\frac{1}{n^s}+\frac{X^{1-s}}{s-1}+O(X^{-\sigma}),$$

where $s = \sigma + ti$ and the constant in the O sign depends only on σ_0. Setting $x = T$, we obtain

$$\zeta(s) = \sum_{n \leqslant T} \frac{1}{n^s} + R_2, \quad \text{where } R_2 = O(T^{-\sigma}), \quad \sigma \geqslant \sigma_0 > 0. \tag{2}$$

We raise

$$\zeta(s) = \prod_p \left(1 - \frac{1}{p^s}\right)^{-1} = \sum_{n=1}^{\infty} \frac{1}{n^s} \qquad (\sigma > 1) \tag{3}$$

to the $1/2$ power and expand $(1 - p^{-s})^{-1/2}$ into a binomial series (the latter converges absolutely as a power of an absolutely convergent geometric progression):

$$\prod_p \left(1 - \frac{1}{p^s}\right)^{-1/2} = \prod_p \left(1 + \frac{c_1}{p^s} + \frac{c_2}{p^{2s}} + \cdots + \frac{c_r}{p^{rs}} + \cdots\right),$$

where

$$c_r = \frac{3 \cdot 5 \cdot \cdots \cdot (2r-1)}{r!2^r} < 1.$$

Expanding the product, we obtain

$$\zeta^{1/2}(s) = \sum \frac{c_{r_1} c_{r_2} \cdots c_{r_k}}{(p_1^{r_1} p_2^{r_2} \cdots p^r k)^s} = \sum_{n=1}^{\infty} \frac{\alpha_n}{n^s}, \tag{4}$$

where $\alpha_1 = 1$ and $0 < \alpha_n = c_{r_1} \cdots c_{r_k} < 1$. We square (4) for $\sigma > 1$. Then according to (3) we obtain

$$\zeta(s) = \left(1 + \frac{\alpha_2}{2^s} + \frac{\alpha_2}{3^s} + \cdots\right)^2 = 1 + \frac{2\alpha_2}{2^s} + \frac{2\alpha_3}{3^s} + \frac{4\alpha_4 + \alpha_2^2}{4^s} + \cdots$$
$$= 1 + \frac{1}{2^s} + \frac{1}{3^s} + \cdots,$$

whence

$$\sum_{\substack{n = n_1 n_2, \\ 1 \leqslant n_1, n_2 \leqslant n}} \alpha_{n_1} \alpha_{n_2} = 1.$$

Here the "complete" sum is taken over all divisors of n, and so the number of terms is equal to $\tau(n)$.

We select from (4) the first $[\sqrt{T}]$ coefficients, and for $\operatorname{Re} s > 0$ we compose the finite sum

$$\zeta_{\sqrt{T}}(s) = \sum_{n \leqslant \sqrt{T}} \frac{\alpha_n}{n^s}.$$

Then

$$\zeta_{\sqrt{T}}^2(s) = \left(\sum_{n \leqslant \sqrt{T}} \frac{\alpha_n}{n^s}\right)^2 = 1 + \frac{1}{2^s} + \cdots + \frac{1}{\left[\sqrt{T}\right]^s} + \sum_{[\sqrt{T}] < n \leqslant [\sqrt{T}]^2} \frac{\beta_n}{n^s},$$

where $0 \leqslant \beta_n < 1$. We have

$$\sum_{n \leqslant T} \frac{1}{n^s} = \zeta_{\sqrt{T}}^2(s) + \sum_{n \leqslant T} \frac{1}{n^s} - \zeta_{\sqrt{T}}^2(s) = \zeta_{\sqrt{T}}^2(s) + \sum_{\sqrt{T} < n \leqslant T} \frac{\gamma_n}{n^s},$$

where $\gamma_n = 1 - \beta_n$ for $\sqrt{T} < n \leqslant [\sqrt{T}]^2$ and $\gamma_n = 1$ for $[\sqrt{T}]^2 < n \leqslant T$. Equality (2) assumes the form

$$\zeta(s) = \zeta^2_{\sqrt{T}}(s) + R_1 + R_2,$$

where

$$R_1 = \sum_{\sqrt{T}<n\leqslant T} \frac{\gamma_n}{n^s}, \quad R_2 = O(T^{-\sigma}), \qquad 0 < \gamma_n \leqslant 1, \operatorname{Re} s > 0.$$

By the triangle inequality we have

$$\int_{T/2}^{T} |\zeta_{\sqrt{T}}(s)|^2\,dt - \int_{T/2}^{T} |R_1|\,dt - \int_{T/2}^{T} |R_2|\,dt$$
$$\leqslant \int_{T/2}^{T} |\zeta(s)|^2\,dt \leqslant \int_{T/2}^{T} |\zeta_{\sqrt{T}}(s)|^2 + \int_{T/2}^{T} |R_1|\,dt + \int_{T/2}^{T} |R_2|\,dt. \tag{5}$$

In what follows we shall select the principal term from the first integral I on the right side of (5), and all the remaining ones will yield the remainder:

$$I = \int_{T/2}^{T} \left| \sum_{n\leqslant\sqrt{T}} \frac{\alpha_n}{n^s} \right|^2 dt = \int_{T/2}^{T} \left(\sum_{m\leqslant\sqrt{T}} \frac{\alpha_m}{m^{\sigma+it}} \right)\left(\sum_{n\leqslant\sqrt{T}} \frac{\alpha_n}{n^{\sigma-it}} \right) dt$$
$$= \sum_{m\leqslant\sqrt{T}} \sum_{n\leqslant\sqrt{T}} \frac{\alpha_m\alpha_n}{m^\sigma n^\sigma} \int_{T/2}^{T} \left(\frac{n}{m}\right)^{it} dt$$
$$= \frac{T}{2} \sum_{n=1}^{\infty} \frac{\alpha_n^2}{n^{2\sigma}} - \frac{T}{2} \sum_{n>\sqrt{T}} \frac{\alpha_n^2}{n^{2\sigma}} + O\left(\sum\sum_{0<m<n\leqslant\sqrt{T}} \frac{1}{m^\sigma n^\sigma \ln(n/m)} \right).$$

Since $\sigma > 1/2$, the first sum converges, and we shall denote it by $C(\sigma)$. The second sum does not exceed

$$\sum_{n>\sqrt{T}} \frac{\alpha_n^2}{n^{2\sigma}} \leqslant \sum_{n>\sqrt{T}} \frac{1}{n^{2\sigma}} \leqslant \int_{\sqrt{T}}^{\infty} \frac{dx}{x^{2\sigma}} \ll T^{1/2-\sigma+\varepsilon},$$

and by Lemma 1 the remainder is equal to $O(T^{1-\sigma+\varepsilon})$. Therefore, $I = (T/2)C(\sigma) + O(T^{1-\sigma+\varepsilon})$.

Let us estimate $I_1 = \int_{T/2}^{T} |R_1|\,dt$ by the Cauchy-Schwarz inequality. We have

$$I_1 \ll \sqrt{T} \left(\int_{T/2}^{T} |R_1|^2\,dt \right)^{1/2}.$$

Here the integral in parentheses can be estimated in the same manner in which we estimated I. We obtain

$$\int_{T/2}^{T} \left| \sum_{\sqrt{T}<n\leqslant T} \frac{\gamma_n}{n^s} \right|^2 dt = O(T^{3/2-\sigma+\varepsilon}), \qquad I_1 = O(T^{5/4--\sigma/2+\varepsilon}).$$

Finally,

$$I_2 = \int_{T/2}^{T} |R_2|\,dt = O(T^{1-\sigma}).$$

Therefore, (5) can be written in the following form:

$$\int_{T/2}^{T} |\zeta(s)|\,dt = \frac{T}{2} C(\sigma) + O(T^{5/4-\sigma/2+\varepsilon}). \tag{6}$$

We divide $(1, T)$ into intervals of the form $[T/2^i, T/2^{i-1}]$, where $i = 1,\dots,r$, and r is such that $r \leqslant \ln T/\ln 2 < r+1$. By (6) we have

$$\int_{T\cdot 2^{-i}}^{T\cdot 2^{-i+1}} |\zeta(s)|\,dt = \frac{1}{2^i} C(\sigma) + O(T^{5/4-\sigma/2+\varepsilon}).$$

Summing over $i = 1,\dots,r$, we obtain

$$\int_1^T |\zeta(s)|\,dt = C(\delta)T + O(T^{5/4-\sigma/2+\varepsilon}).$$

This proves (1).

§3. Asymptotic formula for the mean value of a fractional power of $\zeta(s)$ under the Riemann hypothesis

Let us prove the theorem, assuming that the Riemann conjecture is valid.

Let $0 < \lambda < 2$. For $\operatorname{Re} s > 1$ we have

$$\zeta^\lambda(s) = \prod_p \left(1 - \frac{1}{p^s}\right)^{-1}$$
$$= \prod_p \left(1 + \frac{\lambda}{p^s} + \frac{\lambda(\lambda+1)}{2!p^{2s}} + \cdots + \frac{\lambda(\lambda+1)\cdots(\lambda+r-1)}{r!p^{rs}} + \cdots\right).$$

The binomial series in parentheses converges absolutely for $\operatorname{Re} s > 0$ as the λth power of an absolutely convergent geometric progression. We introduce the notation

$$c_2 = \frac{\lambda(\lambda+1)\cdots(\lambda+r+1)}{r!}. \tag{1}$$

We have

$$\zeta^\lambda(s) = \prod_p \left(1 + \frac{c_1}{p^s} + \frac{c_2}{p^{2s}} + \cdots\right) = \sum \frac{c_{r_1}\cdots c_{r_k}}{(p_1^{r_1}p_2^{r_2}\cdots p_r^{r_k})^s} = \sum_{n=1}^{\infty} \frac{\alpha_\lambda(n)}{n^s}, \tag{2}$$

where $\alpha_\lambda(n) = c_{r_1}\cdots c_{r_k}$. We square (2) for $\sigma > 1$:

$$\zeta^{2\lambda}(s) = 1 + \frac{2c_1}{2^s} + \frac{2c_1}{3^s} + \frac{c_1^2 + 2c_2}{4^s} + \cdots$$
$$= 1 + \frac{2\alpha_\lambda(2)}{2^s} + \cdots + \frac{2\alpha_\lambda(6) + 2\alpha_\lambda(2)\alpha_\lambda(3)}{6^s} + \cdots,$$

i.e.,

$$\zeta^{2\lambda}(s) = \sum_{n=1}^{\infty} \frac{\delta_\lambda(n)}{n^s}, \tag{3}$$

where

$$\delta_\lambda(n) = \sum_{\substack{n=n_1n_2\\ 1\leqslant n_1,n_2\leqslant n}} \alpha_\lambda(n_1)\alpha_\lambda(n_2), \qquad \delta_\lambda(1) = 1. \tag{4}$$

The number of terms in the last sum is equal to $\tau(n)$.

It is evident from the binomial coefficients (1) that to a larger corresponds λ a larger c_r, and so the same holds for $\alpha_\lambda(n)$ and $\delta_\lambda(n)$. Since $\lambda < 2$ and it is known (see [8], §1.2) that

$$\zeta^2(s) = \sum_{n=1}^{\infty} \frac{\tau(n)}{n^s} \quad (\sigma > 1), \qquad \zeta^4(s) = \sum_{n=1}^{\infty} \frac{\tau_4(n)}{n^s} \quad (\sigma > 1),$$

we have

$$\alpha_\lambda(n) \leqslant \alpha_2(n) = \tau(n), \qquad \delta_\lambda(n) \leqslant \delta_2(n) = \tau_4(n). \tag{5}$$

Moreover,

$$\begin{aligned} &\text{for } 0 < \lambda < 1\text{:} \quad 0 < c_r < 1, 0 < \alpha_\lambda(n) < 1, \delta_\lambda(n) < \tau(n); \\ &\text{for } \lambda = 1\text{:} \quad c_r = \alpha_\lambda(n) = \alpha_1(n) = 1, \delta_\lambda(n) = \tau(n); \\ &\text{for } 1 < \lambda \leqslant 2\text{:} \quad c_r > 1, \alpha_\lambda(n) > 1, \delta_\lambda(n) > \tau(n). \end{aligned} \tag{6}$$

We select the first n coefficients from (2), and for $\operatorname{Re} s > 0$ we compose the finite sum

$$\zeta_N(s) = \sum_{n \leqslant N} \frac{\alpha_\lambda(n)}{n^s}. \tag{7}$$

We square it. Let

$$\zeta_N^2(s) = \sum_{n \leqslant N^2} \frac{\beta_\lambda(n)}{n^s}. \tag{8}$$

It then follows from (4) that $\beta_\lambda(n) = \delta_\lambda(n)$ for $n = 1, \dots, N$ and that $\beta_\lambda(n) \leqslant \delta_\lambda(n)$ for any $n \leqslant N^2$, i.e.,

$$\zeta_N^2(s) = \sum_{n=1}^{N} \frac{\delta_\lambda(n)}{n^s} + \sum_{N < n \leqslant N^2} \frac{\beta_\lambda(n)}{n^s}. \tag{9}$$

Suppose that the Riemann hypothesis is valid, i.e., that $\zeta(s)$ has no nontrivial zeros outside the line $\operatorname{Re} s = 1/2$. Then $\zeta^\lambda(s)$ has no branch points to the right of the line $\operatorname{Re} s = 1/2$ that are zeros of $\zeta(s)$.

By the triangle inequality we have

$$\begin{aligned} \int_1^T |\zeta_N(s)|^2\, dt - \int_1^T |\zeta^{2\lambda}(s) - \zeta_N^2(s)|\, dt &\leqslant \int_1^T |\zeta(s)|^{2\lambda}\, dt \\ &\leqslant \int_1^T |\zeta_N(s)|^2\, dt + \int_1^T |\zeta^{2\lambda}(s) - \zeta_N^2(s)|\, dt. \end{aligned} \tag{10}$$

In what follows we shall select the principal term from the first integral on the right side of (10), and the remaining terms will occur in the remainder.

The first integral on the right side of (10) equals

$$\int_1^T |\zeta_N(s)|^2\,dt = \int_1^T \left| \sum_{n\le N} \frac{\alpha_\lambda(n)}{n^s} \right|^2 dt$$

$$= \int_1^T \left(\sum_{m\le N} \frac{\alpha_\lambda(m)}{m^{\sigma+it}} \right) \left(\sum_{n\le N} \frac{\alpha_\lambda(n)}{n^{\sigma-it}} \right) dt$$

$$= \sum_{n\le N} \frac{\alpha_\lambda(n)}{n^{2\sigma}}(T-1) + \sum_{m\le N} \sum_{n\le N} \frac{\alpha_\lambda(m)\alpha_\lambda(n)}{m^\sigma n^\sigma} \int_1^T \left(\frac{n}{m}\right)^{it} dt$$

$$= T \sum_{n\le N} \frac{\alpha_\lambda^2(n)}{n^{2\sigma}} - \sum_{n\le N} \frac{\alpha_\lambda^2(n)}{n^{2\sigma}} + \sum\sum_{m\ne n} \frac{\alpha_\lambda(m)\alpha_\lambda(n)}{m^\sigma n^\sigma} \frac{(n/m)^{iT} - (n/m)}{i\ln(n/m)}.$$

Since, by (5), $\alpha_\lambda(n) \le \tau(n)$ and $\tau(n) \ll n^\varepsilon$ ([2], Chapter 1, Lemma 5), by Lemma 2 we have

$$\int_1^T |\zeta_N(s)|^2\,dt = T\sum_{n=1}^\infty \frac{\alpha_\lambda^2(n)}{n^{2\sigma}} - T\sum_{n<N} \frac{\alpha_\lambda^2(n)}{n^{2\sigma}} - \sum_{n\le N} \frac{\alpha_\lambda^2(n)}{n^{2\sigma}}$$

$$+ O\left(\sum\sum_{0<m<n\le N} \frac{\alpha_\lambda(m)\alpha_\lambda(n)}{m^\sigma n^\sigma \ln n/m} \right)$$

$$= C(\sigma,\lambda)T + O\left(T\int_N^\infty \frac{dx}{x^{2\sigma-2\varepsilon}} \right) + O\left(\int_1^N \frac{dx}{x^{2\sigma-2\varepsilon}} \right) + O(N^{2-2\sigma+2\varepsilon}),$$

i.e.,

$$\int_1^T |\zeta_N(s)|^2\,dt = C(\sigma,\lambda)T + O(TN^{1-2\sigma+\varepsilon}) + O(N^{2-2\sigma+\varepsilon}), \tag{11}$$

where $C(\sigma,\lambda)$ is the sum of the absolutely convergent series

$$\sum_{n=1}^\infty \frac{\alpha_\lambda^2(n)}{n^{2\sigma}}.$$

Next, consider the second integral on the right side of (10). It is known ([8], p. 203) that if $\alpha < \sigma < \beta$, then

$$\left\{ \int_1^T |f(\sigma+ti)|^{1/(p\nu+q\mu)}\,dt \right\}^{p\nu+q\mu}$$

$$\ll \left\{ \left[\int_1^T |f(\alpha+it)|^{1/\nu}\,dt \right]^\nu \right\}^p \left\{ \left[\int_1^T |f(\beta+it)|^{1/\mu}\,dt \right]^\mu \right\}^q, \tag{12}$$

where $p = (\beta-\sigma)/(\beta-\alpha)$, $q = (\sigma-\alpha)/(\beta-\alpha)$, and $p+q=1$. Under the Riemann hypothesis, for the function $f(s) = \zeta^{2\lambda}(s) - \zeta_N^2(s)$ inequality (12) is applicable if we set $\alpha = 1/2+\delta_1$ and $\beta = 1+\delta$; $\mu=\nu=1$. We have

$$\int_1^T |f(\sigma+ti)|\,dt \ll \left\{ \int_1^T |f(\alpha+it)|\,dt \right\}^p \left\{ \int_1^T |f(\beta+it)|\,dt \right\}^q, \tag{13}$$

where

$$p = \frac{1+\delta-\sigma}{1/2+(\delta-\delta_1)}, \qquad q = \frac{\sigma-1/2-\delta_1}{1/2+(\delta-\delta_1)}.$$

The first integral on the right side of (13) equals

$$\int_1^T |\zeta^{2\lambda}\left(\frac{1}{2}+\delta_1+it\right) - \zeta_N^2\left(\frac{1}{2}+\delta_1+it\right)|\,dt$$
$$\leqslant \int_1^T |\zeta^{2\lambda}\left(\frac{1}{2}+\delta_1+it\right)|\,dt + \int_1^T |\zeta_N^2\left(\frac{1}{2}+\delta_1+it\right)|\,dt.$$

Here, by Lemma 5, the first term yields $O(T^{1+\varepsilon})$ and the second term, on the basis of (5) and Lemma 2, equals

$$\int_1^T \left|\zeta_N\left(\frac{1}{2}+\delta_1+it\right)\right|^2 dt = \int_1^T \left|\sum_{n\leqslant N}\frac{\alpha_\lambda(n)}{n^{1/2+\delta_1+it}}\right|^2 dt$$
$$= \int_1^T \left(\sum_{m\leqslant N}\frac{\alpha_\lambda(m)}{m^{1/2+\delta_1+it}}\right)\left(\sum_{n\leqslant N}\frac{\alpha_\lambda(m)}{n^{1/2+\delta_1+it}}\right) dt$$
$$= T\sum_{n\leqslant N}\frac{\alpha_\lambda^2(n)}{n^{1+2\delta_1}} + \sum\sum_{m\leqslant N, m\neq n, n\leqslant N}\frac{\alpha_\lambda(m)\alpha_\lambda(n)}{m^{1/2+\delta_1}n^{1/2+\delta_1}}\frac{(n/m)^{iT}-(n/m)^i}{i\ln(n/m)}$$
$$= O(TN^\varepsilon) + O(N^{2-2(1/2+\delta_1-\varepsilon_1)-\varepsilon}) = O(TN^\varepsilon) + O(N^{1-2\delta_1+\varepsilon}).$$

Thus, in (13)

$$\int_1^T |f(\alpha+it)|\,dt = O(T^{1+\varepsilon}) + O(TN^\varepsilon) + O(N^{1-2\delta_1+\varepsilon}). \tag{14}$$

Since $\operatorname{Re} s = \beta = 1+\delta > 1$ in the second integral in (13), $\zeta^{2\lambda}(1+\delta+it)$ has an absolutely convergent Dirichlet series expansion (3), and $\zeta_N^2(1+\delta+it)$ can be represented in the form of the sum (8), where their first n terms are identical. Therefore, on the basis of (5) we have

$$\int_1^T |\zeta^{2\lambda}(1+\delta+it) - \zeta_N^2(1+\delta+it)|\,dt = \int_1^T \left|\sum_{N<n}\frac{b_\lambda(n)}{n^{1+\delta+it}}\right| dt,$$

where $b_\lambda(n) \leqslant \delta_\lambda(n) < \tau_4(n)$.

Furthermore, from the Cauchy-Schwarz inequality we obtain

$$\int_1^T \left|\sum_{N<n}\frac{b_\lambda(n)}{n^s}\right| dt \leqslant T^{1/2}\left(\int_1^T \left|\sum_{N<n}\frac{b_\lambda(n)}{n^s}\right|^2 dt\right)^{1/2}. \tag{15}$$

But the integral in parentheses on the right side of the inequality equals

$$\int_1^T \left|\sum_{N<n}\frac{b_\lambda(n)}{n^s}\right|^2 dt = \int_1^T \sum_{N<n}\sum_{N<m}\frac{b_\lambda(m)b_\lambda(n)}{m^{1+\delta}n^{1+\delta}}\left(\frac{n}{m}\right)^{it} dt$$
$$= T\sum_{N<n}\frac{b_\lambda^2(n)}{n^{2+2\delta}} + 2\sum\sum_{N<m<n}\frac{b_\lambda(m)b_\lambda(n)}{(mn)^{1+\delta}}\frac{(n/m)^{iT}-(n/m)^i}{i\ln(n/m)}.$$

Since $b_\lambda(n) < \tau_4(n)$, by Lemma 3 we have

$$\int_1^T \left|\sum_{N<n}\frac{b_\lambda(n)}{n^s}\right|^2 dt \ll TN^{-1-2\delta+\varepsilon} + \frac{1}{\delta}N^{-\delta+\varepsilon}.$$

Substituting this into (15), we find that

$$\int_1^T \left| \sum_{N<n} \frac{b_\lambda(n)}{n^s} \right| dt \ll TN^{-1/2-\delta} + \frac{1}{\sqrt{\delta}} T^{1/2} N^{-\delta/2+\varepsilon}. \tag{16}$$

Substituting (14) and (16) into (13), we obtain

$$\int_1^T |\zeta^{2\lambda}(\sigma+ti) - \zeta_N(\sigma+ti)|^2\, dt$$
$$\ll \{T^{1+\varepsilon} + TN^\varepsilon + N^{1-2\delta_1+\varepsilon}\}^p \left\{ TN^{-1/2-\delta+\varepsilon} + \frac{1}{\sqrt{\delta}} T^{1/2} N^{-\delta/2+\varepsilon} \right\}^q.$$

Setting $\delta_1 = \delta = 1/\ln N$ and $N = T$, we obtain $T^{1+\varepsilon}$ and $T^{1/2+\varepsilon}$, respectively, for the expressions in braces. We have

$$\begin{aligned} p(1+\varepsilon) + q(1/2+\varepsilon) &= p + q/2 + (p+q)\varepsilon \\ &= p + q/2 + \varepsilon = 3/2 - \sigma + 1/\ln T + \varepsilon, \end{aligned}$$

$$\int_1^T |\zeta^{2\lambda}(\sigma+ti) - \zeta_N^2(\sigma+ti)|\, dt = O(T^{3/2-\sigma+\varepsilon}). \tag{17}$$

In addition, from (11) we obtain

$$\int_1^T |\zeta_N(s)|^2\, dt = C(\sigma, \lambda)T + O(T^{2-2\sigma+\varepsilon}). \tag{18}$$

Thus, the principal term has been selected from the first integral on the right side of (10), and all remaining terms have occurred in the remainder. Since $3/2 - \sigma > 2 - 2\sigma$ for $\sigma > 1/2$, from (17), (18), and (10) we obtain

$$\int_1^T |\zeta(s)|^{2\lambda}\, dt = C(\sigma, \lambda)T + O(T^{3/2-\sigma+\varepsilon}), \tag{19}$$

where

$$C(\sigma, \lambda) = \sum_{n=1}^\infty \frac{\alpha_\lambda^2(n)}{n^{2\sigma}},$$

the $\alpha_\lambda(n)$ are the coefficients of the Dirichlet series expansion of $\zeta^\lambda(s)$ for $\operatorname{Re} s > 1$, and $1/2 < \sigma < 1$.

REMARK. If we had taken $N = \sqrt{T}$ instead of $N = T$, then in (19) we would have obtained $5/4 - \sigma/2$ instead of $3/2 - \sigma$, which corresponds to the result of §2.

§4. Proof of the theorem

In what follows we shall not assume that the Riemann hypothesis is valid. The following density theorem is known ([8], Theorem 9.19(B)): the number of nontrivial zeros $\rho = \beta + \gamma i$ of the function $\zeta(s)$ in the rectangle $\sigma \leqslant \operatorname{Re}\rho < 1$, $|\operatorname{Im}\rho| \leqslant T$ equals

$$N(\sigma, T) = O(T^{3(1-\sigma)/2-\sigma} \ln^5 T), \tag{1}$$

where $1/2 \leqslant \sigma < 1$. Here the quantity in O has an exponent less than one. Therefore, there exist rectangles $\Pi^{(j)}$ with vertices at the points $A^{(j)}$, $B^{(j)}$, $C^{(j)}$, and $D^{(j)}$ that are devoid of zeros of the Riemann zeta function, where

$$A^{(j)}\big(\sigma - 1/\ln T + i(T_1^{(j)} - \ln T)\big), \quad B^{(j)}\big(\sigma - 1/\ln T + i(T_2^{(j)} + \ln T)\big),$$
$$C^{(j)}\big(1 + i(T_2^{(j)} + \ln T)\big), \quad D^{(j)}\big(1 + i(T_1^{(j)} - \ln T)\big).$$

In Figure 1 two such rectangles are drawn.

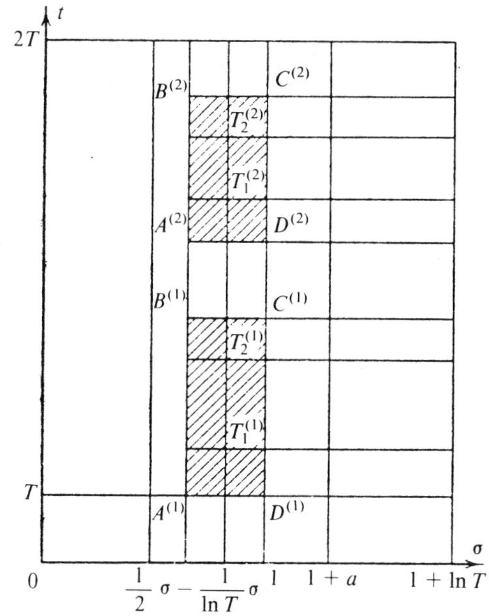

FIGURE 1

We call the intervals $(T_1^{(j)}, T_2^{(j)})$, $j = 1, 2, \ldots$, "basic" (they lie on the line $\operatorname{Re} s = \sigma$), and we also call their corresponding rectangles $A^{(j)}B^{(j)}C^{(j)}D^{(j)}$ "basic" (they are shaded in Figure 1). The coordinates of the points $A^{(j)}$, $B^{(j)}$, $C^{(j)}$, and $D^{(j)}$ are indicated above. A point M on the line $\operatorname{Re} s = \sigma$ must be understood as a point with ordinate iM. For example, the point $T_1^{(j)}$ is the point $\sigma + T_1^{(j)}i$, the point $T_1^{(j)} - \ln T$ is the point $\sigma + (T_1^{(j)} - \ln T)i$, etc.

We denote the union of all basic intervals from T to $2T$ by A, and we denote the union of all their corresponding rectangles by $\overline{A}$.

Let us show that the length of each basic interval can be taken larger than $\ln T$. Indeed, the height of the "large" rectangle equals $2T - T = T$, and the number of zeros of $\zeta(s)$ in it is $N(\sigma, T) < T$. We divide the interval $[T, 2T]$ (i.e., the "large" rectangle $T \leqslant \operatorname{Im} s \leqslant 2T$, $\sigma - 1/\ln T < \operatorname{Re} s > 1$) into $T^{3(1-\sigma)/(2-\sigma)} \ln^5 T$ parts. Since the number $N(\sigma, T)$ of zeros is less than the number of parts, in some parts there are no zeros of $\zeta(s)$. The height of each part equals T: $T^{3(1-\sigma)/(2-\sigma)} \ln^5 T$, and this is obviously larger than $\ln T$.

The number of basic intervals, in general, is "small," i.e., at most $T^{3(1-\sigma)/(2-\sigma)}$, where $1/2 < \sigma < 1$.

We call the intervals (see Figure 1) $(T_2^{(1)}, T_1^{(2)})$, $(T_2^{(2)}, T_1^{(3)}), \ldots, (T_2^{(j)}, T_1^{(j+1)}), \ldots$ "complementary", and we also call their corresponding rectangles with height $T_1^{(j+1)} - T_2^{(j)}$ "complementary". We denote the union of all complementary intervals by B and the union of complementary rectangles by $\overline{B}$.

All nontrivial zeros of $\zeta(s)$ lying in the "large" rectangle are situated in the region B.

Next, consider the intervals of the form

$$(\sigma + i(\gamma - 2\ln T), \sigma + i(\gamma + 2\ln T)), \tag{2}$$

where $\gamma = \operatorname{Im}\rho$. If we denote by C the union of these intervals over all zeros $\rho = \beta + i\gamma$ lying in the "large" rectangle, then C covers B and the union $\bar{C}$ of corresponding rectangles covers the region $\bar{B}$.

We have

$$\int_T^{2T} |\zeta^{2\lambda}(s)|\,dt = \int_T^{2T} |\zeta(s)|^{2\lambda}\,dt = \sum_A \int_{T_1^{(j)}}^{T_2^{(j)}} |\zeta(s)|^{2\lambda}\,dt + \sum_B \int_{T_2^{(j)}}^{T_1^{(j+1)}} |\zeta(s)|^{2\lambda}\,dt. \quad (3)$$

The zeros of $\zeta(s)$ are branch points for $\zeta^\lambda(s)$. We continue $\zeta^\lambda(s)$ analytically to basic rectangles, where $\zeta^\lambda(s)$ has no zeros. Then on the basis of the triangle inequality, for the basic intervals we can write

$$\begin{aligned}\int_{T_1^{(j)}}^{T_2^{(j)}} |\zeta(s)|^{2\lambda}\,dt &= \int_{T_1^{(j)}}^{T_2^{(j)}} |\zeta^{2\lambda}(s)|\,dt \\ &= \int_{T_1^{(j)}}^{T_2^{(j)}} |\zeta_N^2(s)|\,dt + O\left(\int_{T_1^{(j)}}^{T_2^{(j)}} |\zeta^{2\lambda}(s) - \zeta_N^2(s)|\,dt\right). \quad (4)\end{aligned}$$

Summing over all basic intervals and substituting into (3), we obtain

$$\begin{aligned}\int_T^{2T} |\zeta(s)|^{2\lambda}\,dt = \sum_A \int_{T_1^{(j)}}^{T_2^{(j)}} |\zeta_N(s)|^2\,dt + O\left(\sum_A \int_{T_1^{(j)}}^{T_2^{(j)}} |\zeta^{2\lambda}(s) - \zeta_N^2(s)|\,dt\right) \\ + O\left(\sum_B \int_{T_2^{(j)}}^{T_1^{(j+1)}} |\zeta(s)|^{2\lambda}\,dt\right). \quad (5)\end{aligned}$$

In what follows we shall select the principal term from the first sum on the right side of (6), and all remaining terms will occur in the remainder.

We then have

$$\int_{A^{(j)}}^{A^{(j+1)}} |\zeta_N(s)|^2\,dt = \int_{T_1^{(j)}-\ln T}^{T_1^{(j)}} + \int_{T_1^{(j)}}^{T_2^{(j)}} + \int_{T_2^{(j)}}^{T_2^{(j)}+\ln T} + \int_{T_2^{(j)}+\ln T}^{T_1^{(j+1)}-\ln T} |\zeta_N(s)|^2\,dt.$$

Hence

$$\int_{T_1^{(j)}}^{T_2^{(j)}} |\zeta_N(s)|^2\,dt = \int_{T_1^{(j)}-\ln T}^{T_1^{(j+1)}-\ln T} + \int_{T_1^{(j)}-\ln T}^{T_1^{(j)}} - \int_{T_2^{(j)}}^{T_2^{(j)}+\ln T} - \int_{T_2^{(j)}+\ln T}^{T_1^{(j+1)}-\ln T} |\zeta_N(s)|^2\,dt.$$

Summing over all j in the interval from T to $2\,T$, we have

$$\sum_A \int_{T_1^{(j)}}^{T_2^{(j)}} |\zeta_N(s)|^2\,dt = \int_T^{2T} |\zeta_N(s)|^2\,dt + O\left(\sum_B \int_{T_2^{(j)}}^{T_1^{(j+1)}} |\zeta_N(s)|^2\,dt\right). \quad (6)$$

We introduce the characteristic function

$$\chi(t) = \begin{cases} 1, & \text{if } t \in \left(T_2^{(j)}, T_1^{(j+1)}\right), \text{ i.e. } t \in B; \\ 0, & \text{if } t \in \left(T_1^{(j)}, T_2^{(j)}\right), \text{ i.e. } t \in A. \end{cases}$$

It is clear that $\chi^k(t) = \chi(t)$. Therefore, the sum in O in (6) can be represented in the form

$$\sum_B \int_{T_2^{(j)}}^{T_1^{(j+1)}} |\zeta_N(s)|^2\,dt = \sum_{A\cup B} \int_{T_1^{(j)}-\ln T}^{T_1^{(j+1)}-\ln T} \chi(t)|\zeta_N(s)|^2\,dt = \int_T^{2T} \chi(t)|\zeta_N(s)|^2\,dt.$$

We obtain

$$\sum_A \int_{T_1^{(j)}}^{T_2^{(j)}} |\zeta_N(s)|^2\,dt = \int_T^{2T} |\zeta_N(s)|^2\,dt + O\left(\int_T^{2T} \chi(t)|\zeta_N(s)|^2\,dt\right). \quad (7)$$

The first integral on the right side of (7) is computed in the same way as (11) in §3. It equals

$$\int_T^{2T} |\zeta_N(s)|^2\,dt = \int_T^{2T}\left|\sum_{n\leqslant N}\frac{\alpha_\lambda(n)}{n^s}\right|^2 dt$$
$$= C(\sigma,\lambda)T + O(TN^{1-2\sigma+\varepsilon}) + O(N^{2-2\sigma+\varepsilon}), \tag{8}$$

where $C(\sigma,\lambda)$ is the sum of the series

$$\sum_{n=1}^{\infty}\frac{\alpha_\lambda^2(n)}{n^{2\sigma}}$$

which is absolutely convergent for $\sigma > 1/2$.

For the integral in O in (7) we apply the Cauchy-Schwarz inequality. We have

$$\int_T^{2T}\chi(t)|\zeta_N(s)|^2\,dt \leqslant \left(\int_T^{2T}\chi^2(t)\,dt\right)^{1/2}\left(\int_T^{2T}|\zeta_N(s)|^4\,dt\right)^{1/2}. \tag{9}$$

We saw above that the union C covers the set B. Therefore, the sum of the lengths of the intervals of the form (2) over all zeros is not less than the sum of the lengths of all complementary intervals (the latter are disjoint), which equals the integral $\int_T^{2T}\chi(t)\,dt$. But the length of each interval of the form (2) equals $4\ln T$, the number of them equals the number of zeros in the "large" rectangle, and the number of zeros, according to (1), equals

$$O(T^{3(1-\sigma)/(2-\sigma)}\ln^5 T).$$

Therefore, the sum of the lengths of the intervals of the form (2) over all zeros equals

$$O(T^{3(1-\sigma)/(2-\sigma)}\ln^6 T),$$

and so

$$\int_T^{2T}\chi(t)\,dt = O(T^{3(1-\sigma)/(2-\sigma)}\ln^6 T). \tag{10}$$

Furthermore, by Lemmas 3 and 2 we have

$$\int_T^{2T}|\zeta_N(s)|^4\,dt = \int_T^{2T}|\zeta_N^2(s)|^2\,dt = \int_T^{2T}\left|\sum_{n\leqslant N^2}\frac{\beta_\lambda(n)}{n^s}\right|^2 dt$$
$$= \int_T^{2T}\left(\sum_{m\leqslant N^2}\frac{\beta_\lambda(m)}{m^{\sigma-it}}\right)\left(\sum_{n\leqslant N^2}\frac{\beta_\lambda(n)}{n^{\sigma-it}}\right)dt$$
$$= T\sum_{n\leqslant N^2}\frac{\beta_\lambda^2(n)}{n^{2\sigma}} + O\left(\sum\sum_{m<n\leqslant N^2}\frac{\beta_\lambda(m)\beta_\lambda(n)}{m^\sigma n^\sigma\ln(n/m)}\right)$$
$$\leqslant T\sum_{n=1}^{\infty}\frac{\tau_4^2(n)}{n^{2\sigma}} - T\sum_{N^2<n}\frac{\tau_4^2(n)}{n^{2\sigma}} + O(N^{2(2-2\sigma+\varepsilon)})$$
$$= O(T) + O(TN^{2-4\sigma+\varepsilon}) + O(N^{4-4\sigma+\varepsilon}).$$

Substituting this and (10) into (9), we obtain

$$\sum_B\int_{T_2^{(j)}}^{T_1^{(j+1)}}|\zeta_N(s)|^2\,dt = \int_T^{2T}\chi(t)|\zeta_N(s)|^2\,dt$$
$$\ll T^{3(1-\sigma)/2(2-\sigma)}\{T^{1/2} + T^{1/2}N^{1-2\sigma+\varepsilon} + N^{2-2\alpha+\varepsilon}\}T^\varepsilon,$$

and substituting the latter and (8) into (7), we obtain

$$\sum_A \int_{T_1^{(j)}}^{T_2^{(j)}} |\zeta_N(s)|^2\,dt = C(\sigma,\lambda)T + O(TN^{1-2\sigma+\varepsilon}) + O(N^{2-2\sigma+\varepsilon})$$
$$+O\{T^{3(1-\sigma)/2(s-\sigma)+\varepsilon}(T^{1/2} + T^{1/2}N^{1-2\sigma+\varepsilon} + N^{2-2\sigma+\varepsilon})\}. \quad (11)$$

This is the first sum on the right side of (5).

Let us estimate the second sum on the right side of (5). As we noted above, each basic interval $(T_1^{(j)}, T_2^{(j)}) \in A$ has length at least $\ln T$. Being augmented above and below by $\ln T$ and on the left by $1/\ln T$, the rectangles corresponding to these intervals to the right of the line $\operatorname{Re} s = \sigma - 1/\ln T$ do not contain zeros of $\zeta(s)$ (the shaded part of Figure 1).

Let us now divide A into subintervals of length of order $\ln T$, i.e., into intervals $I_1, I_2, \dots$ such that 1) they are disjoint, 2) their union yields A, and 3) $\frac{1}{2}\ln T <$ length of $I_\nu < 2\ln T$. We take the intervals I_ν on the line $\operatorname{Re} s = \sigma$. Being augmented above and below by $\ln T$ and on the left by $1/\ln T$, the rectangles Π_ν to the right of the line $\operatorname{Re} s = \sigma - 1/\ln T$ corresponding to these intervals do not contain zeros of $\zeta(s)$, and these rectangles lie inside the rectangles $\Pi^{(j)}$ shaded in Figure 1.

Consider the values of the function $f(s) = \zeta^{2\lambda}(s) - \zeta_N^2(s)$ in I_ν. Since $f(s)$ is continuous, there exists a point s_ν such that

$$\sup_{s\in I_\nu} |f(s)| \ll |f(s_\nu)|. \quad (12)$$

If s_ν coincides with the common upper (lower) endpoint of I_ν and $[T_1^{(j)}, T_2^{(j)}]$, then above (below) the point s_ν there still remains a band of width $\ln T$ inside of which $\zeta(s)$ has no zeros to the right of the line $\operatorname{Re} s = \sigma - 1/\ln T$.

In Figure 2 the part of Figure 1 corresponding to the interval I_ν has been singled out. On the line $\operatorname{Re} s = \sigma$ we mark the point $M_2(s_\nu)$ with ordinate equal to the imaginary part of s_ν, and we find the point O_ν as the intersection of the lines $\operatorname{Im} s = \operatorname{Im} s_\nu$ and $\operatorname{Re} s = 1 + \ln T$. Next, we draw two circles $\Gamma_{1\nu}$ and $\Gamma_{2\nu}$ with center

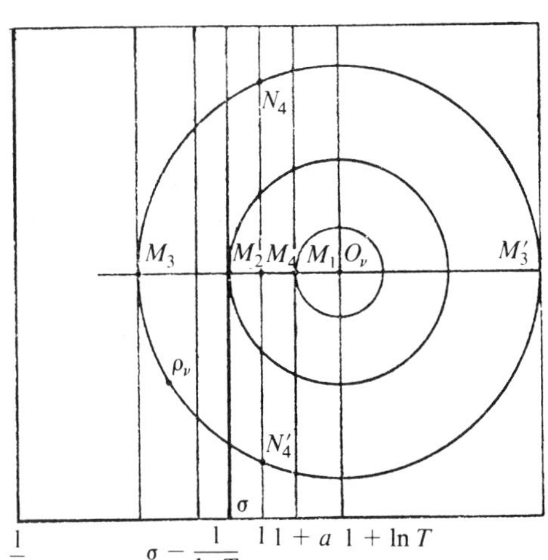

FIGURE 2

at O_ν and radii $r_1 = \ln T - a$ and $r_2 = 1 + \ln T - \sigma$. Consider the interval I_ν containing s_ν. Let ρ_ν be the first zero of $\zeta(s)$ lying to the left of the rectangle Π_ν (this zero lies to the left of the line $\operatorname{Re} s = \sigma - 1/\ln T$) if we go left to the line $\operatorname{Re} s = 1/2$. Through the point ρ_ν we draw a third circle $\Gamma_{3\nu}$ with center at O_ν and radius $O_\nu M_3 = r_3(\rho_\nu) = r_3$.

Let us show that $\zeta(s)$ has no zeros inside the circle $\Gamma_{3\nu}$. Indeed, as we know, $\zeta(s)$ has no zeros to the right of or on the line $\operatorname{Re} s = 1$. From elementary geometry it is known that (see Figure 2)

$$|M_4 M_4|^2 = |M_3 N_4||M_4 M_3'| = |M_3 M_4|(2r_3 - |M_3 M_4|),$$

where

$$1 - (\sigma - 1/\ln T) < |M_3 M_4| < 1 - 1/2 = 1/2$$

and

$$(1 + \ln T) - (\sigma - 1/\ln T) < r_3 < (1 + \ln T) - 1/2,$$

i.e.,

$$1 - \sigma + \frac{1}{\ln T} < |M_3 M_4| < \frac{1}{2}, \qquad (1 - \sigma) + \ln T + \frac{1}{\ln T} < r_3 < \frac{1}{2} + \ln T.$$

Therefore,

$$|M_4 N_4|^2 < \frac{1}{2}\left[2\left(\frac{1}{2} + \ln T\right) - \left(1 - \sigma + \frac{1}{\ln T}\right)\right] = \ln T + \frac{\sigma}{2} - \frac{1}{2\ln T}.$$

Since $4\ln T + 2 > 1$, we have $\sqrt{4\ln T} < \ln T$ for $t > e^{3.5}$. Thus, for $T > e^{3.5}$ we have $|N_4' N_r| < \ln T$. Therefore, the horizontal band corresponding to the interval $N_4' N_4$ lies entirely inside the band corresponding to the interval $(M_2(s_\nu) + \ln T,\ M_2(s_j) - \ln T)$. In the latter band $\zeta(s)$ has no zeros (see Figure 2) to the right of the line $\operatorname{Re} s = \sigma - 1/\ln T$. Therefore, $\zeta(s)$ also has no zeros to the right of the line $\operatorname{Re} s = \sigma - 1/\ln T$ in the band corresponding to the interval $N_4' N_4$. Then there are also no zeros to the right of this in the circle $\Gamma_{3\nu}$. Since ρ_ν is the first zero (by construction), there are also no zeros inside $\Gamma_{3\nu}$ to the left of the line $\operatorname{Re} s = \sigma - 1/\ln T$. Thus $\zeta(s)$ has no zeros inside $\Gamma_{3\nu}$, and ρ_ν lies on $\Gamma_{3\nu}$ (there can be several such ρ_ν). If there is no "first zero" up to the line $\operatorname{Re} s = 1/2$ to the left of the line $\operatorname{Re} s = \sigma - 1/\ln T$, then $\Gamma_{3\nu}$ is tangent to the line $\operatorname{Re} s = 1/2$. We introduce the notation

$$M_\nu(r_1) = \max_{s \in \Gamma_{1\nu}} |f(s)|, \quad M_\nu(r_2) = \max_{s \in \Gamma_{2\nu}} |f(s)|, \quad M_\nu(r_3) = \max_{s \in \Gamma_{3\nu}} |f(s)|,$$

where r_1, r_2, and r_3 are the radii of the circles $\Gamma_{1\nu}$, $\Gamma_{2\nu}$, and $\Gamma_{3\nu}$. It is clear that for $s \in I_\nu$ we have (see (12))

$$|f(s)| \leqslant \sup_{s \in I_\nu} |f(s)| \ll |f(s_\nu)| \ll M_\nu(r_2).$$

We already know that r_1 is identical for all $\Gamma_{1\nu}$ and equals $\ln T - a$, that r_2 is also identical for all $\Gamma_{2\nu}$ and equals $1 - \sigma + \ln T$, and that r_3 depends on ρ_ν. Hadamard's three-circle inequality ([6], Appendix §9, Theorem 9.2) is well known:

$$M(r_2) \leqslant M(r_1)^{\ln(r_3/r_2)/\ln(r_3/r_1)} \cdot M(r_3)^{\ln(r_2/r_1)/\ln(r_3/r_1)},$$

where $r_1 < r_2 < r_3$.

We have

$$\sum_A \int_{T_1^{(j)}}^{T_2^{(j)}} |f(s)|\,dt = \sum_A \int_{I_\nu} |f(s)|\,dt \ll \sum_{A;\, s_\nu \in I_\nu} |f(s_\nu)| \int_{I_\nu} dt$$

$$\ll \sum_{A;\, s_\nu \in I_\nu} |f(s_\nu)|\,(\text{length of } I_\nu) \ll \ln T \sum_{A;\, s_\nu \in I_\nu} |f(s_\nu)|$$

$$\ll \ln T \sum_{A;\, s_\nu \in I_\nu} |f(s_\nu)| \ll \ln T \sum_\nu M_\nu(r_2)$$

$$\ll \ln T \sum_\nu M_\nu(r_1)^{\ln(r_3/r_2)/\ln(r_3/r_1)} \cdot M_\nu(r_3(\nu))^{\ln(r_2/r_1)/\ln(r_3/r_1)}. \tag{13}$$

Since r_3 depends on ρ_ν, we cannot apply Hölder's inequality to the last sum. Therefore, we partition the interval $(1/2, \sigma - 1/\ln T)$ into $l \asymp \ln T$ intervals of length of order $1/\ln T$; i.e., if $\sigma_1 = \sigma - 1/\ln T > \sigma_2 > \cdots > \sigma_l = 1/2$ are partition points, then by definition the length of the interval $[\sigma_{k+1}, \sigma_k]$ satisfies the condition

$$c_1(1/\ln T) < \sigma_k - \sigma_{k+1} < c_2/\ln T, \quad \text{where } k = 1, 2, \ldots, l-1.$$

The above-mentioned "first zero" ρ_ν lying to the left of the line $\operatorname{Re} s = \sigma - 1/\ln T = \sigma_1$ and corresponding to the interval I_ν can be situated at different distances from this line for different I_ν. Therefore, $r_3(\rho_\nu)$ can be different for different I_ν. In addition, the left ends M_3 of the horizontal diameters of the circles $\Gamma_{3\nu}$ can also be situated in different vertical bands $(\sigma_{k+1}, \sigma_k]$ of width of order $1/\ln T$ for different I_ν. Moreover, for several I_ν their corresponding points M_3 can be located in the same vertical band $(\sigma_{k+1}, \sigma_k]$. The radii $\Gamma_{3\nu}$ corresponding to these I_ν are identical up to $1/\ln T$.

If $\zeta(s)$ has no zeros to the left of the interval I_ν (I_ν lies on the line $\operatorname{Re} s = \sigma$) up to the line $\operatorname{Re} s = 1/2$, then the corresponding circle $\Gamma_{3\nu}$ is tangent to the line $\operatorname{Re} s = 1/2$; i.e., M_3 is located on the line $\operatorname{Re} s = 1/2$, and r_3 equals $1/2 + \ln T$.

Thus, the collection of all $I_1, I_2, \ldots$ of the set A can be divided into classes $(\bar\sigma_1), (\bar\sigma_2), \ldots$ according to the following criterion: we place in the same class $(\bar\sigma_k)$ all the intervals $I_{\nu k}$ for which the left ends (M_3) of the diameters of the corresponding third circles are in the same vertical band $(\sigma_{k+1}, \sigma_k]$. We denote the collection of indices of all intervals of the class $(\bar\sigma_k)$ by (σ_k).

As already noted, the radii of the third circles $\Gamma_{3\nu}$ corresponding to the class $(\bar\sigma_k)$ are identical up to $1/\ln T$ and equal $r_3(k) = 1 + \ln T - \sigma_k$ (we shall also speak of a "circle of the class $(\bar\sigma_k)$").

We note that since the total number of zeros of $\zeta(s)$ in the "large" rectangle equals $N(\sigma, T) \ll T^{3(1-\sigma)/(2-\sigma)} \ln^5 T$, the number of intervals $I_\nu^{(k)}$ of a given class $(\bar\sigma_k)$ is also $\ll N(\sigma, T)$ (for the class $(\bar\sigma_k)$ the index ν also assumes at most $N(\sigma, T)$ values).

Therefore, (13) can be represented in the form

$$\sum_A \int_{I_\nu} |f(s)|\,dt \ll \ln T \sum_{k=1}^{l} \sum_{\nu \in (\sigma_k)} [M_{\nu k}(r_1)]^{\ln(r_3/r_1)/\ln(r_2/r_1)}$$

$$\cdot [M_{\nu k}(r_3)]^{\ln(r_2/r_1)/\ln(r_3/r_1)}, \tag{14}$$

where

$$r_3 = r_3(k) = 1 + \ln T - \sigma_k, \quad r_1 = \ln T - a, \quad r_2 = 1 - \sigma + \ln T, \tag{15}$$

$$M_{\nu k}(r_1) = \max_{s \in \Gamma_{1\nu}^{(k)}} |f(s)|, \qquad M_{\nu k}(r_3) = \max_{s \in \Gamma_{3\nu}^{(k)}} |f(s)|. \tag{16}$$

Hölder's inequality can now be applied to the inner sum in (14).

As a preliminary let us simplify the exponents in (14). By (15) we have

$$\ln \frac{r_3}{r_1} = \ln\left(1 + \frac{r_3 - r_1}{r_1}\right) = \ln\left(1 + \frac{1 - \sigma_k + a}{\ln T - a}\right) = \frac{1 - \sigma_k + a}{\ln T - a} + O\left(\frac{1}{\ln^2 T}\right),$$

since $(1 - \sigma_k + a)/(\ln T - a) < 1$.

Similarly,

$$\ln \frac{r_3}{r_2} = \frac{\sigma - \sigma_k}{1 - \sigma + \ln T} + O\left(\frac{1}{\ln^2 T}\right), \qquad \ln \frac{r_2}{r_1} = \frac{1 - \sigma + a}{\ln T - a} + O\left(\frac{1}{\ln^2 T}\right),$$

$$\frac{\ln(r_3/r_2)}{\ln(r_2/r_1)} = \frac{\sigma - \sigma_k + O(1/\ln T)}{(1 - \sigma_k + a) + O(1/\ln T)} = \frac{\sigma - \sigma_k}{1 - \sigma_k + a} + O\left(\frac{1}{\ln T}\right),$$

$$\ln \frac{r_2}{r_1} = \frac{1 - \sigma + a}{1 - \sigma_k + a} + O\left(\frac{1}{\ln T}\right).$$

Substituting all of this into (14), we obtain

$$\sum_A \int_{I_\nu} |f(s)|\, dt \ll \ln T \sum_{k=1}^{l} \sum_{\nu \in (\sigma_k)} [M_{\nu k}(r_1)]^{(\sigma - \sigma_k)/(1 - \sigma_k + a)} \cdot [M_{\nu k}(r_3)]^{(1 - \sigma + a)/(1 - \sigma_k + a)}.$$

We apply Hölder's inequality:

$$\sum u_\nu v_\nu < \left(\sum u_\nu^p\right)^{1/p} \left(\sum v_\nu^{p/(p-1)}\right)^{(p-1)/p} \qquad (p > 1),$$

taking

$$0 < \alpha = \frac{\sigma - \sigma_k}{1 - \sigma_k + a} < 1, \quad 0 < \beta = \frac{1 - \sigma + a}{1 - \sigma_k + \alpha} < 1, \quad p = \frac{2}{\lambda\beta} > 1; \quad \alpha + \beta = 1;$$

$$u_\nu = M_{\nu k}^{\beta}(r_3), \quad v_\nu = M_{\nu k}^{\alpha}(r_1); \quad \frac{1}{p} = \frac{\lambda\beta}{2}, \quad \frac{p}{p-1} = \frac{2}{2 - \lambda\beta}, \quad \frac{p-1}{p} = \frac{2 - \lambda\beta}{2};$$

$$\sum_A \int_{I_\nu} |f(s)|\, dt \ll \ln T \sum_{k=1}^{l} \left(\sum_{\nu \in (\sigma_k)} M_{\nu k}^{\beta}(r_3) M_{\nu k}^{\alpha}(r_1) \right)$$

$$\ll \ln T \sum_{k=1}^{l} \left\{ \left[\sum_{\nu \in (\sigma_k)} \left(M_{\nu k}^{\beta}(r_3)\right)^{2/\lambda\beta} \right]^{\lambda\beta/2} \left[\sum_{\nu \in (\sigma_k)} \left(M_{\nu k}^{1-\beta}(r_1)\right)^{2/(2-\lambda\beta)} \right]^{(2-\lambda\beta)/2} \right\}$$

$$= \ln T \sum_{k=1}^{l} \left[\sum_{\nu \in (\sigma_k)} \left(M_{\nu k}(r_3)\right)^{2/\lambda} \right]^{\lambda\beta/2} \left[\sum_{\nu \in (\sigma_k)} \left(M_{\nu k}(r_1)\right)^{2(1-\beta)/(2-\lambda\beta)} \right]^{(2-\lambda\beta)/2}, \tag{17}$$

where

$$0 < \frac{2(1 - \beta)}{2 - \lambda\beta} < 1, \quad \frac{\lambda\beta}{2} + \frac{2 - \lambda\beta}{2} = 1, \quad \frac{\lambda\beta}{2} + \frac{2(1 - \beta)}{2 - \lambda\beta} \cdot \frac{2 - \lambda\beta}{2} < 1. \tag{18}$$

By (16),

$$M_{\nu k}(r_3) = \max_{s \in \Gamma_{3\nu}^{(k)}} |f(s)|.$$

Let $\xi_{\nu k}$ be a point on the circle $\Gamma_{3\nu}^{(k)}$ where $|f(s)|$ attains its maximum (since $f(s)$ is continuous, such a point exists). For the first inner sum in (17) we have

$$\begin{aligned}\sum_{\nu\in(\sigma_k)} M_{\nu k}^{2/\lambda}(r_3) &= \sum_{\nu\in(\sigma_k)} |f(\xi_{\nu k})|^{2/\lambda} = \sum_{\nu\in(\sigma_k)} |\zeta^{2\lambda}(\xi_{\nu k}) - \zeta_N^2(\xi_{\nu k})|^{2/\lambda} \\ &\ll \sum_{\nu\in(\sigma_k)} |\zeta^{2\lambda}(\xi_{\nu k})|^{2/\lambda} + \sum_{\nu\in(\sigma_k)} |\zeta^2(\xi_{\nu k})|^{2/\lambda} \\ &= \sum_{\nu\in(\sigma_k)} |\zeta(\xi_{\nu k})|^4 + \sum_{\nu\in(\sigma_k)} |\zeta_N(\xi_{\nu k})|^{4/\lambda}. \qquad (19)\end{aligned}$$

The following approximate functional equation is known ([8], §4.18):

$$\zeta^2(s) = \sum_{n\leqslant X} \frac{\tau(n)}{n^s} + \chi_1^2(s) \sum_{n\leqslant Y} \frac{\tau(n)}{n^{1-s}} + O((X+Y)^{1/2-\sigma}\ln t),$$

where

$$0\leqslant\sigma\leqslant 1, \qquad XY = (t/2\pi)^2, \qquad X\geqslant h>0, \qquad Y>h\geqslant 0,$$

$$\chi_1(s) = \frac{2^{s-1}\pi^s}{\cos(\pi s/2)\Gamma(s)} = \left(\frac{2\pi}{t}\right)^{\sigma+it-1/2} e^{i(t+\pi/4)}\left\{1+O\left(\frac{1}{t}\right)\right\}.$$

Setting $X=Y=T$, we obtain $T^2=(t/2\pi)^2$, $t=2\pi T$, $|\chi_1^2(s)|=O(T^{1-2\sigma})$, and

$$\begin{aligned}\zeta^2(s) &= \sum_{n\leqslant T} \frac{\tau(n)}{n^s} + O\left(T^{1-2\sigma}\left|\sum_{n\leqslant T}\frac{\tau(n)}{n^{1-s}}\right|\right) + O(T^{1/2-\sigma}\ln T) \\ &= S_1(s) + O(T^{1-2\sigma})S_2(s) + O(T^{1/2-\sigma}\ln T),\end{aligned}$$

where

$$S_1(s) = \sum_{n\leqslant T}\frac{\tau(n)}{n^s}, \qquad S_2(s) = \sum_{n\leqslant T}\frac{\tau(n)}{n^{1-s}}.$$

For $s=\xi_{\nu k}$ we have (since $\operatorname{Re}\xi_{\nu k}=\sigma>\sigma_k$, we replace σ by σ_k)

$$\zeta^2(\xi_{\nu k}) = S_1(\xi_{\nu k}) + O(T^{1-2\sigma_k})S_2(\xi_{\nu k}) + O(T^{1/2-\sigma_k}\ln T).$$

By Cauchy's inequality ([4], Chapter V, §1, Corollary 1 of Lemma 3) and by (1), for the first sum in (19) we have

$$\begin{aligned}\sum_{\nu\in(\sigma_k)} |\zeta(\xi_{\nu k})|^4 &= \sum_{\nu\in(\sigma_k)} |\zeta^2(\xi_{\nu k})|^2 \\ &\leqslant \sum_{\nu\in(\sigma_k)} \left(|S_1(\xi_{\nu k})| + O(T^{1-2\sigma_k})|S_2(\xi_{\nu k})| + O(T^{1/2-\sigma_k}\ln T)\right)^2 \\ &\leqslant \sum_{\nu\in(\sigma_k)} 3\left[|S_1(\xi_{\nu k})|^2 + O(T^{2-4\sigma_k})|S_2(\xi_{\nu k})|^2 + O(T^{1-\sigma}\ln^2 T)\right] \\ &\ll \sum_{\nu\in(\sigma_k)} |S_1(\xi_{\nu k})|^2 + O(T^{2-4\sigma_k}) \\ &\quad\times \sum_{\nu\in(\sigma_k)} |S_2(\xi_{\nu k})|^2 + O(T^{3(1-\sigma_k)/(2-\sigma_k)+1-2\sigma_k}\ln^7 T), \qquad (20)\end{aligned}$$

where

$$\frac{3(1-\sigma_k)}{2-\sigma_k}-(2\sigma_k-1)<1, \qquad \sigma_k \geqslant \frac{1}{2}.$$

We now estimate the first and second sum in (20) by Lemma 4. Let φ be the set of points $\xi_{\nu k}$ belonging to the class (σ_k), $\delta = \ln T$, $T_0 = T = N$, and

$$a_n = \tau(n), \quad \sigma_0 = \sigma_k, \quad T + \delta/2 \leqslant \operatorname{Im} \xi_{\nu k} \leqslant 2T - \delta/2; \quad |\operatorname{Im} \xi_{\nu k} - \operatorname{Im} \xi_{\nu' k}| > \delta$$

for $\xi_{\nu k} \neq \xi_{\nu' k}$ (since the $\xi_{\nu k}$ are taken from different circles $\Gamma_{3\nu}^{(k)}$, the last inequality holds).

Since s and φ do not appear on the right side of the inequality in Lemma 4, in place of the condition $s \in \varphi$ we can take the condition $\nu \in (\sigma_k)$. We have

$$\sum_{\nu \in (\sigma_k)} |S_1(\xi_{\nu k})|^2 = \sum_{\nu \in (\sigma_k)} \left| \sum_{n=1}^{T} \frac{\tau(n)}{n^{\xi_{\nu k}}} \right|^2$$

$$\ll T(\delta^{-1} + \ln T) \sum_{n=1}^{T} \frac{|\tau(n)|^2}{n^{2\sigma_k}} \left(1 + \log \frac{\log 2T}{\log 2n}\right).$$

Since

$$\ln T \ll T^{\varepsilon}, \quad \delta = \ln T \ll T^{\varepsilon}, \quad 1 + \log \frac{\log 2T}{\log 2n} = \log \frac{e \log 2T}{\log 2n} \ll T^{\varepsilon},$$

$$\tau(n) \ll n^{\varepsilon} \ll T^{\varepsilon},$$

$$\sum_{n=1}^{T} \frac{|\tau(n)|^2}{n^{2\sigma_k}} \ll \int_1^T \frac{dx}{x^{2\sigma_k - 2\varepsilon}} \ll T^{1-2\sigma_k+\varepsilon} \qquad \left(\sigma_k > \frac{1}{2}\right),$$

we obtain

$$\sum_{\nu \in (\sigma_k)} |S_1(\xi_{\nu k})|^2 \ll T^{2-2\sigma_k+\varepsilon}. \tag{21}$$

As above, the second sum in (20) is

$$\ll O(T^{2-4\sigma_k}) \sum_{\nu \in (\sigma_k)} |S_2(\xi_{\nu k})|^2 = O(T^{2-4\sigma_k}) \sum_{\nu \in (\sigma_k)} \left| \frac{\tau(n)}{n^{1-\xi_{\nu k}}} \right|^2$$

$$\ll T^{2-2\sigma_k} T(\delta^{-1} + \ln T) \sum_{n=1}^{T} \frac{|\tau(n)|^2}{n^{2-2\sigma_k}} \left(1 + \log \frac{\log 2T}{\log 2n}\right)$$

$$\ll T^{3-4\sigma_k+\varepsilon} \int_1^T x^{-2+2\sigma_k+\varepsilon}\, dx \ll T^{2-2\sigma_k+\varepsilon},$$

where $2 - 2\sigma_k + \varepsilon < 1$.

Substituting this and (21) into (20), for the first sum in (19) we have

$$\sum_{\nu \in (\sigma_k)} |\zeta(\xi_{\nu k})|^4 = O(T^{2-2\sigma_k+\varepsilon}) + O(T^{3(1-\sigma_k)/(2-2\sigma_k)-(2\sigma_k-1)})$$

$$= O(T^{2-2\sigma_k+\varepsilon}). \tag{22}$$

We apply Hölder's inequality to the second sum in (19), setting $p = r\lambda/2$, where r is the smallest positive integer such that $r\lambda/2 > 1$, i.e., $r = [2/\lambda] + 1$. For such an r

we have $r > 2/\lambda$, and since $2/\lambda > 1$, we have $r \geqslant 2$. for a fixed λ the integer r is also fixed:

$$\sum_{\nu\in(\sigma_k)} |\zeta_N(\xi_{\nu k})|^{4/\lambda} = \sum_{\nu\in(\sigma_k)} |\zeta_N(\xi_{\nu k})|^{4/\lambda}\cdot 1^{4/\lambda}$$

$$\leqslant\left[\sum_{\nu\in(\sigma_k)} (|\zeta_N(\xi_{\nu k})|^{4/\lambda})^{r\lambda/2}\right]^{2/r\lambda}\left[\sum_{\nu\in(\sigma_k)} (1^{4/\lambda})^{4\lambda/(r\lambda-2)}\right]^{(r\lambda-2)/r\lambda},$$

where

$$\frac{r\lambda-2}{r\lambda}<1, \qquad \frac{r\lambda}{r\lambda-2}+\frac{r\lambda-2}{r\lambda}=1.$$

Since

$$\frac{3(1-\sigma)}{2-\sigma}<\frac{3(1-\sigma_k)}{2-\sigma_k},$$

where $\sigma = \operatorname{Re}\xi_{\nu k}$, by (1) we have

$$\sum_{\nu\in(\sigma_k)} |\zeta_N(\xi_{\nu k})|^{4/\lambda} \ll \left\{\sum_{\nu\in(\sigma_k)} |\zeta_N^r(\xi_{\nu k})|^2\right\}^{2/r\lambda} T^{(3(1-\sigma_k)/(2-\sigma_k))((r\lambda-2)/r\lambda)}, \quad (23)$$

where

$$\frac{2}{r\lambda}+\frac{3(1-\sigma_k)}{2-\sigma_k}\,\frac{r\lambda-2}{r\lambda}<1.$$

Let

$$\zeta_N^r(s)=\left(\sum_{n=1}^{N}\frac{\alpha_\lambda(n)}{n^s}\right)^r=\sum_{n\leqslant N^r}\frac{a_\lambda(n)}{n^s} \qquad (\operatorname{Re}s>0),$$

$$(\zeta^\lambda(s))^r=\left(\sum_{n=1}^{\infty}\frac{\alpha_\lambda(n)}{n^s}\right)^r=\sum_{n=1}^{\infty}\frac{\alpha_{\lambda r}(n)}{n^s} \qquad (\operatorname{Re}s>1).$$

Then $\alpha_{\lambda r}(n)\leqslant\alpha_{[\lambda r]+1}(n)=\tau_{[\lambda r]+1}(n)$ and $\alpha_\lambda(n)\leqslant\alpha_{\lambda r}(n)$.

Therefore, $\alpha_\lambda(n)\leqslant\tau_{[\lambda r]+1}(n)\ll n^\varepsilon$. Again by Lemma 4 we have

$$\sum_{\nu\in(\sigma_k)} |\zeta_N^r(\xi_{\nu k})|^2=\sum_{\nu\in(\sigma_k)}\left|\sum_{n\leqslant N^r}\frac{a_\lambda(n)}{n^{\xi_{\nu k}}}\right|^2$$

$$\ll(T+N^r)(\delta^{-1}+\ln N^r)\sum_{n\leqslant N^r}\frac{|a_\lambda(n)|^2}{n^{2\sigma_k}}\left(1+\log\frac{\log 2N^r}{\log 2n}\right),$$

where $\sigma_k>1/2$, $\delta=\ln T$, and r is defined above.

We set $N=T^{1/r}$. We have

$$\ln T\ll T^\varepsilon, \qquad 1+\log\frac{\log 2T}{\log 2n}\ll T^\varepsilon,$$

and

$$\sum_{\nu\in(\sigma_k)} |\zeta_N^r(\xi_{\nu k})|^2\ll T^{1+\varepsilon}\int_1^T\frac{dx}{x^{2\sigma_k-2\varepsilon}}\ll T^{2-2\sigma_k+\varepsilon},$$

where $2 - 2\sigma_k + \varepsilon < 1$. Substituting this into (23), we obtain

$$\sum_{\nu\in(\sigma_k)} |\zeta_N(\xi_{\nu k})|^{4/\lambda} \ll T^{(2-2\sigma_k+\varepsilon)/(2/r\lambda)+(3(1-\sigma_k)(2-\sigma_k))((r\lambda-2)/r\lambda)},$$

where

$$(2-2\sigma_k+\varepsilon)\frac{2}{r\lambda} + \frac{3(1-\sigma_k)}{2-\sigma_k}\frac{r\lambda-2}{r\lambda} < 1. \tag{24}$$

Substituting this and (22) into (19), we obtain

$$\sum_{\nu\in(\sigma_k)} M_{\nu k}^{2/\lambda}(r_3) = O(T^{2-2\sigma_k+\varepsilon}) + O(T^{(2-2\sigma_k+\varepsilon)(2/r\lambda)+(3(1-\sigma_k)/(2-\sigma_k))((r\lambda-2)/r\lambda)}).$$

It is not hard to show that

$$2-2\sigma_k+\varepsilon < (2-2\sigma_k+\varepsilon)\frac{2}{r\lambda} + \frac{3(1-\sigma_k)}{2-\sigma_k}\frac{r\lambda-2}{r\lambda}.$$

Therefore,

$$\sum_{\nu\in(\sigma_k)} M_{\nu k}^{2/\lambda}(r_3) \ll T^{(2-2\sigma_k+\varepsilon)(2/r\lambda)+(3(1-\sigma_k)/(2-\sigma_k))((r\lambda-2)/r\lambda)}. \tag{25}$$

Let us now estimate the second inner sum in (17). By definition,

$$M_{\nu k}(r_1) = \max_{s\in\Gamma_{1\nu}^{(k)}} |f(s)| = \max_{s\in\Gamma_{1\nu}^{(k)}} |\zeta^{2\lambda}(s) - \zeta_N^{\lambda}(s)|$$

and the condition $\operatorname{Re} s > 1$ holds for any point s on the circle $\Gamma_{1\nu}^{(k)}$. If we denote by $\eta_{\nu k}$ a point on the circle $\Gamma_{1\nu}^{(k)}$ where $|f(s)|$ attains its maximum, the condition $\operatorname{Re}\eta_{\nu k} = \sigma_\eta > 1$ also holds for it, and so at this point $\zeta^{2\lambda}(s)$ expands into the absolutely convergent series

$$\zeta^{2\lambda}(\eta_{\nu k}) = \sum_{n=1}^{\infty} \frac{\delta_\lambda(n)}{n^{\eta_{\nu k}}} \quad\text{and}\quad \zeta_N^2(\eta_{\nu k}) = \sum_{n\leqslant N^2} \frac{\beta_\lambda(n)}{n^{\eta_{\nu k}}},$$

where $\beta_\lambda(n) \leqslant \delta_\lambda(n) \leqslant \tau_4(n)$ and $\beta_\lambda(n) = \delta_\lambda(n)$ for $n = 1,\dots,N$. Then, by Hölder's inequality for $p = (2-\lambda\beta)/(1-\beta)$ (here

$$(2-\lambda\beta)/(2-2\beta) > 1 \quad\text{and}\quad (2-\lambda\beta)/(1-\beta) > 2 > 1,$$

and so Hölder's inequality is applicable) and by (1), we have

$$\sum_{\nu\in(\sigma_k)} (M_{\nu k}(r_1))^{2(1-\beta)/(2-\lambda\beta)}\cdot 1$$

$$\leqslant \left\{\sum_{\nu\in(\sigma_k)} \left[(M_{\nu k}(r_1))^{2(1-\beta)/(2-\lambda\beta)}\right]^{(2-\lambda\beta)/(1-\beta)}\right\}^{(1-\beta)/(2-\lambda\beta)} \left\{\sum_{\nu\in(\sigma_k)} 1\right\}^{(1+\beta-\lambda\beta)/(2-\lambda\beta)}$$

$$= \left(\sum_{\nu\in(\sigma_k)} \left|\sum_{n>N} \frac{b_\lambda(n)}{n^{\eta_{\nu k}}}\right|^2\right)^{(1-\beta)/(2-\lambda\beta)} T^{(3(1-\sigma_k)/(2-\sigma_k))((1+\beta-\lambda\beta)/(2-\lambda\beta))}, \tag{26}$$

where

$$\frac{1-\beta}{2-\lambda\beta}+\frac{1+\beta-\lambda\beta}{2-\lambda\beta}=1, \qquad \frac{1-\beta}{2-\lambda\beta}+\frac{3(1-\sigma_k)}{2-\sigma_k}\frac{1+\beta-\lambda\beta}{2-\lambda\beta}<1,$$

$$M_{\nu k}(r)=|\zeta^{2\lambda}(\eta_{\nu k})-\zeta_N^2(\eta_{\nu k})|=\left|\sum_{n>N}\frac{b_\lambda(n)}{n^{\eta_{\nu k}}}\right|,$$

$$b_\lambda(n)\leqslant\tau_4(n), \quad (2-\lambda\beta)/(1-\beta)>1, \quad \operatorname{Re}\eta_{\nu k}=\sigma_\eta>1.$$

To estimate the sum in the parentheses in (26), we partition the interval $(N,+\infty)$ into the intervals $(N,2N]$, $(2N,4N]$, $\dots$, $(2^{i-1}N,2^iN]$, $\dots$ and apply Lemma 4 to this sum in each interval:

$$\sum_{\nu\in(\sigma_k)}\left|\sum_{N<n}\frac{b_\lambda(n)}{n^{\eta_{\nu k}}}\right|^2=\sum_{\nu\in(\sigma_k)}\left|\sum_{n=N+1}^{2N}\frac{b_\lambda(n)}{n^{\eta_{\nu k}}}\right|^2+\dots+\sum_{\nu\in(\sigma_k)}\left|\sum_{n=2N+1}^{2^2N}\frac{b_\lambda(n)}{n^{\eta_{\nu k}}}\right|^2+\dots$$

$$=\sum_{i=1}^{\infty}\sum_{\nu\in(\sigma_k)}\left|\sum_{n=2^{i-1}N+1}^{2^iN}\frac{b_\lambda(n)}{n^{\eta_{\nu k}}}\right|. \tag{27}$$

Setting $b_\lambda(n)=0$ for $n\leqslant 2^{i-1}N$, we have

$$\sum_{\nu\in(\sigma_k)}\left|\sum_{n=2^{i-1}N}^{2^iN}\frac{b_\lambda(n)}{n^{\eta_{\nu k}}}\right|^2=\sum_{\nu\in(\sigma_k)}\left|\sum_{n=1}^{n=2^{i-1}N}\frac{b_\lambda(n)}{n^{\eta_{\nu k}}}\right|^2$$

$$\ll(T+2^iN)(\delta^{-1}+\ln 2^iT)\sum_{n=2^{i-1}N+1}^{2^iN}\frac{|b_\lambda(n)|^2}{n^{2\sigma_\eta}}\left(1+\log\frac{\log 2^{i+1}N}{\log 2n}\right). \tag{28}$$

Since $2^iN+2\leqslant 2n\leqslant 2^{i+1}N$, we have

$$1+\log\frac{\log 2^{i+1}N}{\log 2n}\leqslant\log\frac{e\log 2^{i+1}N}{\log(2^iN+2)}\ll 1;$$

$$\sum_{n=2^{i-1}N+1}\frac{|b_\lambda(n)|^2}{n^{2\sigma_\eta}}\ll\sum_{n=2^{i-2}N+1}^{2^iN}\frac{1}{n^{2\sigma_\eta-2\varepsilon}}\ll\int_{2^{i-1}N+1}^{2^iN}\frac{dx}{x^{2\sigma_\eta-2\varepsilon}}\ll\frac{1}{(2^iN)^{2\sigma_\eta-1-\varepsilon}}.$$

Substituting all of this into (28), for this sum we obtain

$$\ll\frac{(T+2^iN)(\delta^{-1}+(2^iN)^\varepsilon)}{(2^iN)^{2\sigma_\eta-1-\varepsilon}}\ll\frac{T}{(2^iT)^{2\sigma_\eta-1-\varepsilon}}+\frac{1}{(2^iN)^{2\sigma_\eta-2-\varepsilon}}.$$

We set $N=T^{1/r}$. Then from (27) we obtain

$$\sum_{\nu\in(\sigma_k)}\left|\sum_{N<n}\frac{b_\lambda(n)}{n^{\eta_{\nu k}}}\right|^2\ll\sum_{i=1}^{\infty}\frac{T}{(2^iT^{1/2})^{2\sigma_\eta-1-\varepsilon}}$$

$$=\frac{T}{T^{(2\sigma_\eta-1-\varepsilon)/r}}\sum_{i=1}^{\infty}\frac{1}{(2^{2\sigma_\eta-1-\varepsilon})^i}\ll T^{1-(2\sigma_\eta-1-\varepsilon)/r}.$$

Substituting this into (26), we find that

$$\sum_{\nu\in(\sigma_k)}(M_{\nu k}(r_1))^{2(1-\beta)/(2-\lambda\beta)}$$

$$\ll T^{(1-(2\sigma_\eta-1-\varepsilon)/r)((1-\beta)/(2-\lambda\beta))+(3(1-\sigma_k)/(2-\sigma_k))((1+\beta-\lambda\beta)/(2-\lambda\beta))}. \tag{29}$$

Since $2\sigma_\eta - 1 - \varepsilon > 1$, we have $1 - (2\sigma_\eta - 1 - \varepsilon)/r < 1$. Therefore, since

$$\frac{1-\beta}{2-\lambda\beta} + \frac{1+\beta-\lambda\beta}{2-\lambda\beta} = 1,$$

we have

$$\left(1 - \frac{2\sigma_\eta - 1 - \varepsilon}{r}\right)\left(\frac{1-\beta}{2-\lambda\beta}\right) + \frac{3(1-\sigma_k)}{2-\sigma_k}\,\frac{1+\beta-\lambda\beta}{2-\lambda\beta} < 1. \tag{30}$$

The expression on the left side of (30) decreases as σ_η increases. Therefore, in what follows it suffices to consider the case where $1 - (2\sigma_\eta - 1 - \varepsilon)/r > 0$. Hence $\sigma_\eta < (r-1)/2 + \varepsilon$. Substituting (25) and (29) into (17), we obtain

$$\sum_A \int_{I_\nu} |f(s)|\,dt \ll \ln T \sum_{k=1}^{l} \{T^{(2-2\sigma_k+\varepsilon)(2/r\lambda)+(3(1-\sigma_k)/(2-\sigma_k))((r\lambda-2)/r\lambda)}\}^{\lambda\beta/2}$$

$$\times \{T^{(1-(2\sigma_\eta-1-\varepsilon)/r)((1-\beta)/(2-\lambda\beta))+(3(1-\sigma_k)/(2-\sigma_k))((1+\beta-\lambda\beta)/(2-\lambda\beta))}\}^{(2-\lambda\beta)/2}. \tag{31}$$

Since $0 < \lambda\beta/2 < 1$, $0 < (2-\lambda\beta)/2 < 1$, and $\lambda\beta/2 + (2-\lambda\beta)/2 = 1$, on the basis of (24) and (30) we find that the exponent of T is less than 1; (31) is an estimate of the second sum on the right side of (5). We estimate the third sum in (5) by Hölder's inequality, by setting $p = 2/\lambda$, where $2/\lambda > 1$:

$$\sum_B \int_{T_2^{(j)}}^{T_1^{(j+1)}} |\zeta(s)|^{2\lambda}\,dt = \int_T^{2T} \chi(t)\,|\zeta(s)|^{2\lambda}\,dt$$

$$\leqslant \left[\int_T^{2T} \left(|\zeta(s)|^{2\lambda}\right)^{4/2\lambda} dt\right]^{2\lambda/4} \left[\int_T^{2T} \chi(t)\,dt\right]^{(4-2\lambda)/2}$$

$$= \left(\int_T^{2T} |\zeta(s)|^4\,dt\right)^{\lambda/2} \left(\int_T^{2T} \chi(t)\,dt\right)^{(2-\lambda)/2}.$$

Hence by Lemma 5 and (1) we have

$$\sum_B \int_{T_2^{(j)}}^{T_1^{(j+1)}} |\zeta(s)|^{2\lambda}\,dt = O(T^{\lambda/2+(3(1-\sigma)/(2-\sigma))((2-\lambda)/2)+\varepsilon}), \tag{32}$$

where

$$\frac{\lambda}{2} + \frac{3(1-\sigma)}{2-\sigma}\,\frac{2-\lambda}{2} < 1.$$

This is an estimate of the third sum in (5). In (11) we set $N = T^{1/r}$. Then for the first sum in (5) we obtain

$$\sum_A \int_{T_1^{(j)}}^{T_2^{(j)}} |\zeta_N(s)|^2\,dt = C(\sigma,\lambda)T + O(T^{1-(2\sigma-1-\varepsilon)/r})$$

$$+ O\{T^{3(1-\sigma)/2(2-\sigma)+\varepsilon}(T^{1/2} + T^{1/2-(2\sigma-1-\varepsilon)/r} + T^{(2-2\sigma+\varepsilon)/r})\}.$$

Since $r \geqslant 2$ and $1/2 < \sigma < 1$, we have $1/2 - (2\sigma-1-\varepsilon)/r < 1/2$ and $2 - 2\sigma + \varepsilon/r < 1 - \sigma + \varepsilon/2 < 1/2$. Therefore,

$$\sum_A \int_{T_1^{(j)}}^{T_2^{(j)}} |\zeta_N(s)|^2\,dt = C(\sigma,\lambda)T$$

$$+ O(T^{1-(2\sigma-1-\varepsilon)/r}) + O(T^{1/2+3(1-\sigma)/2(2-\sigma)+\varepsilon}), \tag{33}$$

where $1/2 + 3(1-\sigma)/2(2-\sigma) < 1$.

Introducing the notation $3(1-\sigma)/(2-\sigma) = Q$ and $3(1-\sigma_k)/(2-\sigma_k) = Q_k$ and substituting (33), (32), and (31) into (5), we obtain

$$\int_T^{2T} |\zeta(s)|^{2\lambda}\, dt = C(\sigma,\lambda)T + O(T^{1-(2\sigma-1)/r+\varepsilon}) + O(T^{1/2+Q/2+\varepsilon}) + O(T^{\lambda/2+(2-\lambda)Q/2+\varepsilon}) + O\left(T^{\varepsilon}\sum_{k=1}^{l} T^{\varphi k}\right), \tag{34}$$

where

$$\varphi_k = \varphi(\sigma_k, \sigma_\eta, \lambda) = \left[(2-2\sigma_k+\varepsilon)\frac{2}{r\lambda} + \frac{r\lambda-2}{r\lambda}Q_k\right]\frac{\lambda\beta}{2} + \left[\left(1-\frac{2\sigma_\eta-1-\varepsilon}{r}\right)\frac{1-\beta}{2-\lambda\beta} + \frac{1+\beta-\lambda\beta}{2-\lambda\beta}Q_k\right]\frac{2-\lambda\beta}{2}; \tag{35}$$

$$0<\lambda<2, \quad r = r(\lambda) = [2/\lambda]+1, \quad 1/2 \leqslant \sigma_k < \sigma, \quad 1 < \sigma_\eta \leqslant (r+1)/2, \quad \beta = (1-\sigma+a)/(1-\sigma_k+a). \tag{36}$$

Let us find a function $\varphi_0 = \varphi_0(\lambda)$ of a simpler form such that $\varphi_k < \varphi_0(\lambda) < 1$.

The expression (35) can be represented in the form

$$\varphi_k = \frac{1}{r}\left[\beta\left(\frac{3}{2} - 3\sigma_k - Q_k + \sigma_\eta + \frac{\varepsilon}{2}\right) - \sigma_\eta + \frac{1}{2} + \frac{\varepsilon}{2}\right] + \frac{1}{2}(Q_k+1) + \frac{\beta}{2}(Q_k-1).$$

We have

$$\begin{aligned}\varphi_k &< \frac{1}{r}\left[\beta\left(\frac{3}{2} - 2\sigma_k - Q_k + \sigma_\eta\right) - \sigma_\eta + \frac{1}{2}\right] + \frac{\varepsilon}{2r}(1+\beta) + 1\\ &< 1 + \frac{\varepsilon}{2r}(1+\beta) + \left(\frac{1}{2} - \sigma_\eta\right)(1-\beta) + \frac{\beta}{r}(2-2\sigma_k-Q_k)\\ &< 1 - \left(\frac{1-\beta}{2r} - \varepsilon\right), \qquad 2-2\sigma_k - Q_k < 0.\end{aligned}$$

For the function $\varphi_0(\lambda)$ we take

$$\varphi_0 = 1 - \kappa_0, \tag{37}$$

where

$$\kappa_0 = \kappa_0(\sigma,\lambda) = (1-\beta)/2r - \varepsilon, \qquad 0 < \beta(\sigma) < 1, \quad r = [2/\lambda]+1.$$

To obtain an asymptotic formula for the integral

$$\int_1^T |\zeta(s)|^{2\lambda}\, dt,$$

we partition $[1, T]$ into intervals of the form $[T/2^i, T/2^{i-1}]$. Then by (34) we have

$$\int_{T/2^i}^{T/2^{i-1}} |\zeta(s)|^{2\lambda}\, dt = C(\sigma,\lambda)\frac{T}{2^i} + O(T^{1-(2\sigma-1)/r+\varepsilon}) + O(T^{1/2+Q/2+\varepsilon}) + O(T^{\lambda/2+(2-\lambda)Q/2+\varepsilon}) + O(T^{\varphi_0+\varepsilon}).$$

Summing over $i = 1, \ldots, [\ln T]$, we obtain

$$\int_1^T |\zeta(s)|^{2\lambda}\, dt = C(\sigma,\lambda)T + O(T^{1-(2\sigma-1)/r+\varepsilon}) + O(T^{1/2+Q/2+\varepsilon}) + O(T^{\lambda/2+(2-\lambda)Q/2+\varepsilon}) + O(T^{\varphi_0+\varepsilon}). \tag{38}$$

Since the exponents in all the remainders in (38) are stricly less than one, our theorem is proved.

Bibliography

1. I. M. Vinogradov, *Foundations of the theory of numbers*, 7th ed., "Nauka", Moscow, 1965; English transl. of 5th ed., Dover, New York, 1954; of 6th ed., Pergamon Press, 1955.

2. ______, *The method of trigonometrical sums in the theory of numbers*, 4th ed., "Nauka", Moscow, 1980; English transl. of 3rd ed. Statist. Publ. Soc., Calcutta, 1978.

3. ______, *The method of trigonometrical sums in the theory of numbers*, Trudy Mat. Inst. Steklov. **23** (1947); English transl., Interscience, 1954.

4. A. A. Karatsuba, *Principles of analytic number theory*, "Nauka", Moscow, 1975; Spanish transl., "Mir", Moscow, 1979.

5. Hugh L. Montgomery, *Topics in multiplicative number theory*, Lecture Notes in Math., Vol. 227, Springer-Verlag, 1971.

6. Karl Prachar, *Primzahlverteilung*, Springer-Verlag, 1957.

7. E. C. Titchmarsh, *The theory of functions*, 2nd ed., Oxford Univ. Press, 1939.

8. ______, *The theory of the Riemann zeta-function*, 2nd rev. ed., Clarenden Press, Oxford, 1951.

9. A. E. Ingham, *Mean-value theorems in the theory of the Riemann zeta-function*, Proc. London Math. Soc. (2) **27** (1927/28), 273–300.

10. D. R. Heath-Brown, *The fourth power moment of the Riemann zeta function*, Proc. London Math. Soc. (3) **38** (1979), 385–422.

Translated by R. LENET

ABCDEFGHIJ–AMS–8987654